Environmental Flow Assessment

Environmental Flow Assessment

Methods and Applications

John G. Williams
Consultant, Petrolia, California

Peter B. Moyle
University of California, Davis

J. Angus Webb
University of Melbourne

G. Mathias Kondolf
University of California, Berkeley, & Université de Lyon

ADVANCING RIVER RESTORATION AND MANAGEMENT

WILEY Blackwell

This edition first published 2019
© 2019 John Wiley & Sons Ltd

The right of John G. Williams, Peter B. Moyle, J. Angus Webb and G. Mathias Kondolf to be identified as the authors of this work has been asserted in accordance with law.

Registered Office(s)
John Wiley & Sons, Inc., 111 River Street, Hoboken, NJ 07030, USA
John Wiley & Sons Ltd, The Atrium, Southern Gate, Chichester, West Sussex, PO19 8SQ, UK

Editorial Office
9600 Garsington Road, Oxford, OX4 2DQ, UK

For details of our global editorial offices, customer services, and more information about Wiley products visit us at www.wiley.com.

Wiley also publishes its books in a variety of electronic formats and by print-on-demand. Some content that appears in standard print versions of this book may not be available in other formats.

Library of Congress Cataloging-in-Publication Data

Names: Williams, John G., 1941– author. | Moyle, Peter B., author. | Webb, J. Angus, 1971– author. |
    Kondolf, G. Mathias, author.
Title: Environmental flow assessment : methods and applications/ John G. Williams, Peter B. Moyle,
    J. Angus Webb, G. Mathias Kondolf.
Description: Hoboken, NJ : John Wiley & Sons, Inc., 2019. | Includes bibliographical references and index. |
Identifiers: LCCN 2018049251 (print) | LCCN 2018056504(ebook) | ISBN 9781119217398(Adobe PDF) |
    ISBN 9781119217381 (ePub) | ISBN 9781119217367 (hardcover)
Subjects: LCSH: Stream measurements. | Stream ecology. |Environmental impact analysis.
Classification: LCC GB1201.7 (ebook) | LCC GB1201.7 .W552019 (print) | DDC 551.48/3–dc23
LC record available at https://lccn.loc.gov/2018049251

Cover Design: Wiley
Cover Images: © Anton Petrus / Getty Images

Set in 9.5/12pt Minion by SPi Global, Pondicherry, India

Printed and bound by CPI Group (UK) Ltd, Croydon, CR0 4YY

10   9   8   7   6   5   4   3   2   1

Contents

About the authors, ix

Series foreword, xi

Preface, xiii

Acknowledgements, xv

1 An introduction to environmental flows, 1
  Summary, 1
  1.1 What are environmental flows?, 1
  1.2 Why EFA is so hard; scientific issues, 2
    1.2.1 Stream ecosystems are dynamic and open, 2
    1.2.2 Fish evolve, 3
    1.2.3 Streams adjust, 4
    1.2.4 Climate changes, 4
    1.2.5 Populations vary, 5
    1.2.6 Habitat selection is conditional, 5
    1.2.7 Spatial and temporal scales matter, 5
  1.3 Why EFA is so hard: social issues, 6
    1.3.1 Social objectives evolve, 6
    1.3.2 Science and dispute resolution, 7
    1.3.3 Water is valuable, 7
    1.3.4 Managers or clients often want
       the Impossible, 7
  1.4 Why EFA is so hard: problems with the literature, 8
  1.5 Why EFA is so hard: limitations of models
     and objective methods, 8
    1.5.1 Models and environmental flow
       assessment, 8
    1.5.2 Objective and subjective methods, 9
  1.6 Conclusions, 9

2 A brief history of environmental flow
  assessments, 11
  Summary, 11
  2.1 Introduction, 11
  2.2 The legal basis for environmental flows, 12
  2.3 The scope of environmental flow assessments, 13
  2.4 Methods for quantifying environmental flows, 14
  2.5 Conclusions, 20
     Note, 20

3 A primer on flow in rivers and streams, 21
  Summary, 21
  3.1 Introduction, 21
  3.2 Precipitation and runoff, 22

  3.3 Flow regimes, 22
    3.3.1 Describing or depicting flow regimes, 22
    3.3.2 Variation in flow regimes across climates
       and regions, 25
    3.3.3 Anthropogenic changes in flow
       regimes, 28
    3.3.4 Hydrologic classifications, 29
  3.4 Spatial patterns and variability within streams, 30
    3.4.1 Spatial complexity of flow within stream
       channels, 30
    3.4.2 The variety of channel forms, 31
    3.4.3 Lateral connectivity with floodplain
       and off-channel water bodies, 33
    3.4.4 Bed topography and hyporheic
       exchange, 36
  3.5 Managing environmental flows, 37
  3.6 Conclusions, 38

4 Life in and around streams, 39
  Summary, 39
  4.1 Introduction, 39
  4.2 Structure of stream ecosystems, 40
    4.2.1 Across-channel gradients, 40
    4.2.2 Upstream–downstream gradient, 41
  4.3 Adaptations of stream organisms, 43
    4.3.1 Morphological adaptations, 43
    4.3.2 Physiological adaptations, 44
    4.3.3 Behavioral adaptations, 45
  4.4 Adapting to extreme flows, 46
  4.5 Synthesis, 47
  4.6 Environmental flows and fish assemblages, 47
  4.7 Conclusions, 49

5 Tools for environmental flow assessment, 51
  Summary, 51
  5.1 Introduction, 51
  5.2 Descriptive tools, 52
    5.2.1 Graphical tools and images, 52
    5.2.2 Stream classifications, 53
    5.2.3 Habitat Classifications, 54
    5.2.4 Species classifications, 55
    5.2.5 Methods classifications, 55
  5.3 Literature reviews, 55
  5.4 Experiments, 56
    5.4.1 Flow experiments, 56
    5.4.2 Laboratory experiments, 56
    5.4.3 Thought experiments, 56

5.5   Long-term monitoring, 58
5.6   Professional opinion, 59
5.7   Causal criteria, 60
5.8   Statistics, 60
        5.8.1   Sampling, 61
        5.8.2   Sampling methods, 61
        5.8.3   Hypothesis testing, 61
        5.8.4   Model selection and averaging, 62
        5.8.5   Resampling algorithms, 62
5.9   Modeling, 63
        5.9.1   Abundance–environment relations, 64
        5.9.2   Habitat association models, 65
        5.9.3   Drift-foraging models, 65
        5.9.4   Capability models, 66
        5.9.5   Bayesian networks, 66
        5.9.6   Hierarchical Bayesian models, 69
        5.9.7   Dynamic occupancy models, 70
        5.9.8   State-dependent life-history models
                  and dynamic energy budget models, 71
        5.9.9   Hydraulic models, 71
        5.9.10   Hydrological models, 72
        5.9.11   Temperature models, 72
        5.9.12   Sediment transport models, 72
        5.9.13   Other uses of models in EFA, 73
5.10   Hydraulic habitat indices, 73
5.11   Hydrological indices, 75
5.12   Conclusions, 75

6 Environmental flow methods, 77

        Summary, 77
6.1   Introduction, 77
        6.1.1   Hydrologic, habitat rating, habitat
                  simulation, and holistic
                  methods, 78
        6.1.2   Top-down and bottom-up approaches, 78
        6.1.3   Sample-based methods and whole-system
                  methods, 78
        6.1.4   Standard-setting and incremental
                  approaches, 79
        6.1.5   Micro-, meso-, and river-scale methods, 79
        6.1.6   Opinion-based and model-based
                  methods 79
6.2   Hydrological methods, 80
        6.2.1   The tennant method and its
                  relatives, 80
        6.2.2   Indicators of hydraulic alteration (IHA), 81
6.3   Hydraulic rating methods, 82
6.4   Habitat simulation methods, 83
        6.4.1   Habitat association models, 84
        6.4.2   Bioenergetic or drift-foraging models, 88
6.5   Frameworks for EFA, 92
        6.5.1   Instream flow incremental methodology
                  (IFIM), 92
        6.5.2   Downstream response to imposed flow
                  transformation (DRIFT), 95
        6.5.3   Ecological limits of hydraulic alteration
                  (ELOHA), 97
        6.5.4   Adaptive management, 102
        6.5.5   Evidence-based EFA, 104
6.6   Conclusions, 107

7 Good modeling practice for EFA, 109

        Summary, 109
7.1   Introduction, 109
7.2   Modeling practice, 110
        7.2.1   What are the purposes of the modeling?, 110
        7.2.2   How should you think about the natural
                  system being assessed?, 111
        7.2.3   What data are or will be available, and how
                  good are they?, 111
        7.2.4   How will the available budget be distributed
                  over modeling efforts, or between modeling
                  and data collection, or between the
                  assessment and subsequent monitoring?, 112
        7.2.5   How will the uncertainty in the results
                  of the modeling be estimated
                  and communicated?, 112
        7.2.6   How will the model and model development
                  be documented?, 113
        7.2.7   How will the models be tested?, 113
        7.2.8   How good is good enough to be useful?, 113
        7.2.9   Who will use the results of the modeling,
                  and how will they be used?, 113
        7.2.10   Do you really need a model?, 113
7.3   Behavioral issues in modeling for EFA, 114
7.4   Data-dependent activities in developing estimation
        models, 115
7.5   Sampling, 118
        7.5.1   General considerations, 118
        7.5.2   Spatial scale issues in sampling, 119
        7.5.3   Cleaning data sets, 119
7.6   On testing models, 120
        7.6.1   The purpose of testing models, 120
        7.6.2   Why testing models can be hard, 120
        7.6.3   The problem with validation, 120
        7.6.4   The limited utility of significance tests, 121
        7.6.5   Tests should depend on the
                  nature of the method being applied, 122
        7.6.6   Models should be tested multiple ways, 122
        7.6.7   The importance of plausibility, 123
        7.6.8   The importance of testing models with
                  independent data, 123
        7.6.9   The quality of the data limits the quality
                  of the tests, 123
        7.6.10   The importance of replication, 123
        7.6.11   Models should be tested against other
                  models, 123
7.7   Experimental tests, 126
        7.7.1   Flow experiments, 126
        7.7.2   Behavioral carrying-capacity tests, 128
        7.7.3   Virtual ecosystem experiments, 128
7.8   Testing models with knowledge, 129
7.9   Testing hydraulic models, 129
7.10   Testing EFMs based on professional judgement, 130
7.11   Testing species distribution models, 131
        7.11.1   Goodness of fit, 132
        7.11.2   Prevalence, 132
        7.11.3   Imperfect detection, 133
        7.11.4   Spatial scale and other complications, 133
7.12   Conclusions, 141
        Note, 142

8 Dams and channel morphology, 143

Summary, 143

8.1 Introduction, 143

8.2 Diagnosing the problem and setting objectives, 145

8.3 Managing sediment load, 146

    8.3.1 Existing dams, 146

    8.3.2 Proposed dams, 147

    8.3.3 Obsolete dams, 150

8.4 Specifying morphogenic flows, 152

    8.4.1 Three common approaches to specifying morphogenic flows, 152

    8.4.2 Clear objectives needed, 153

    8.4.3 Magnitude, 153

    8.4.4 Duration, 155

    8.4.5 The hydrograph, 155

    8.4.6 Seasonality, 156

    8.4.7 Recurrence, 158

8.5 Flows for managing vegetation in channels, 159

8.6 Constraints, 159

    8.6.1 Minimizing cost of foregone power production and other uses of water, 159

    8.6.2 Preserving spawning gravels, 160

    8.6.3 Preventing flooding and bank erosion, 161

8.7 Conclusions, 161

9 Improving the use of existing evidence and expert opinion in environmental flow assessments, 163

Summary, 163

9.1 Introduction, 163

9.2 Overview of proposed method, 164

9.3 Basic principles and background to steps, 165

    9.3.1 Literature as a basis of an evidence-based conceptual model, 165

    9.3.2 Translate the conceptual model into the structure of a Bayesian belief network, 166

    9.3.3 Quantify causal relationships in the BBN using formal expert elicitation, 166

    9.3.4 Update causal relationships using empirical data, 166

9.4 Case study: golden perch (*Macquaria ambigua*) in the regulated Goulburn River, southeastern Australia, 168

    9.4.1 Evidence-based conceptual model of golden perch responses to flow variation, 168

    9.4.2 Bayesian belief network structure of the golden perch model, 168

    9.4.3 Expert-based quantification of effects of flow and non-flow drivers on golden perch, 169

    9.4.4 Inclusion of monitoring data to update the golden perch BBN, 171

9.5 Discussion, 172

    9.5.1 Improved use of knowledge from the literature, 172

    9.5.2 Improving the basis of Bayesian networks for environmental flows, 173

    9.5.3 Hierarchical Bayesian methods as best practice, 174

    9.5.4 Piggy-backing on existing knowledge, 175

    9.5.5 Resourcing improved practice, 175

    9.5.6 Accessibility of methods, 176

9.6 Summary, 176

10 Summary conclusions and recommendations, 177

10.1 Conclusions and recommendations, 177

    10.1.1 Confront uncertainty and manage adaptively, 177

    10.1.2 Methods for EFA, 178

    10.1.3 Recommendations on monitoring, 180

    10.1.4 Recommendations for assessments, 181

10.2 A checklist for EFA, 182

Literature cited 185

Index 215

About the authors

John G. Williams is a retired consultant with a PhD in physical geography who has published on botany, climatology, hydrology and salmon biology as well as on environmental flow assessment. He has served as an elected director of a water management district in California, and as special master for important litigation regarding environmental flows. He can be reached at jgwill@frontiernet.net.

Peter B. Moyle is Distinguished Professor Emeritus at the University of California, Davis. He has been working on flows and fish issues since the 1970s. He is particularly proud of his role in designing a flow regime to benefit fish, plants and birds for Putah Creek, near the UCD campus. (See https://watershed. ucdavis.edu/cws-wfcb-fish-conservation-group.)

J. Angus Webb is an Associate Professor at the University of Melbourne, Australia. He leads the Ecohydraulics laboratory group in the Water, Environment and Agriculture Program within the Melbourne School of Engineering, and is heavily involved in the monitoring, evaluation and adaptive management of environmental flows being delivered under Australia's Murray–Darling Basin Plan. He was awarded the 2012 Early Career Achievement Award from the Australian Society for Limnology, and the 2013 Award for Building Knowledge in Waterway Management from the Australian River Basin Management Society. (See www.ie.unimelb. edu.au/research/water/)

G. Mathias Kondolf is a fluvial geomorphologist and environmental planner at the University of California Berkeley and a fellow at the Collegium, Institute for Advanced Study at the University of Lyon, France. He works on sustainable river management and restoration, including managing sediment in regulated rivers. (See https://riverlab. berkeley.edu.)

Series foreword

Advancing river restoration and management

The field of river restoration and management has evolved enormously in recent decades, driven largely by increased recognition of ecological values, river functions and ecosystem services. Many conventional river-management techniques, emphasizing strong structural controls, have proven difficult to maintain over time, resulting in sometimes spectacular failures, and often a degraded river environment. More sustainable results are likely from a holistic framework, which requires viewing the "problem" at a larger catchment scale and involves the application of tools from diverse fields. Success often hinges on understanding the sometimes complex interactions among physical, ecological and social processes.

Thus, effective river restoration and management require nurturing the interdisciplinary conversation, testing and refining of our scientific theories, reducing uncertainties, designing future scenarios for evaluating the best options, and better understanding the divide between nature and culture that conditions human actions. It also implies that scientists should communicate better with managers and practitioners, so that new insights from research can guide management, and so that results from implemented projects can, in turn, inform research directions.

This series provides a forum for "integrative sciences" to improve rivers. It highlights innovative approaches, from the underlying science, concepts, methodologies, new technologies and new practices, to help managers and scientists alike improve our understanding of river processes, and to inform our efforts to steward and restore our fluvial resources better for a more harmonious coexistence of humans with their fluvial environment.

G. Mathias Kondolf,
University of California, Berkeley

Hervé Piégay
University of Lyon, CNRS

Preface

In a 2010 review, Arthington et al. remarked that: "There is now wide recognition that a dynamic, variable water regime is required to maintain the native biodiversity and ecological processes characteristic of every river and wetland ecosystem. Yet it remains a challenge to translate this 'natural flow regime' paradigm into quantitative environmental flow prescriptions for individual reaches from source to sea" (citations omitted). This book is about methods and approaches for meeting this challenge.

Environmental flow assessment is largely about flow, as the name suggests, but not just about flow. Other biotic and abiotic factors influence flowing water ecosystems, and environmental flow assessment (EFA) needs to take them into account. And, EFA is a social process, probably more than a scientific process. We treat EFA mostly as a kind of applied ecology, but we do not ignore the complications arising from human nature.

People working on EFA have diverse backgrounds, so we expect the same of readers of this book. Some will see themselves primarily as managers, rather than as scientists or engineers, and many will be familiar mainly with one region or even one stream system. Therefore, we have included material that will seem elementary to some readers, mostly to emphasize the variety of stream ecosystems that are the subject of assessments. Similarly, although we expect that many readers will already know a lot about EFA, we have tried to avoid assuming that they do. And, we do not try to be comprehensive. For example, we say little about riparian systems, and almost nothing about estuaries, although dealing with them is an important part of the overall problem. Rather, we try to elaborate an approach or point of view that can be applied generally.

We take a more critical attitude about methods for EFA than other books on the same subject, such as Locke et al. (2008) or Arthington (2012). We make recommendations, but we explain the shortcomings of the methods we recommend, as well as of those we don't. Part of our motivation in writing this book is concern about careless use of models in EFA, and we deal with that at length. Reluctance to criticize others' work is generally an admirable trait, but not in science, where it is part of the job, provided it is not mean-spirited.

It is an unhappy truth that many scientific papers have been published that should not have been, and many published research findings are false (Ioannidis 2005). There are various reasons for this, and a major one is flawed statistical analyses, especially overreliance on and misuse of statistical significance tests. Ioannidis wrote about the biomedical literature, but the same applies in environmental sciences. For example, Bolker et al. (2009) found problems with 311 of 537 applications of generalized linear mixed models in articles on ecology and evolution, and our impression is that papers on EFAs tend to exhibit a lower level of statistical understanding, and to receive poorer reviewing on statistical matters, than papers in related fields. We discuss and illustrate statistical problems with methods for EFA and related studies, but at a conceptual level, without getting into the technical details.

Geographically, the western USA, and especially California, is overrepresented in the book, as are salmonids. This seems parochial, and it is, but the western USA is highly diverse geographically, salmonids have diverse life-histories, and most of the literature on EFA deals with salmonids. Since three of us have lived and worked in California for decades, we are more familiar with EFA as it is actually done in California than elsewhere, so our California bias results largely from following the advice to "write what you know." However, we are broadly familiar with EFA elsewhere, and recommend an approach developed in Australia.

On language, we follow more recent (and more appropriate) usage and refer to "environmental flows" instead of "instream flows," but we do not intend any change in meaning with this terminology. We have tried to write in plain language, and to avoid overly technical or overblown academic writing such as the following, which we did not make up: "Temporary streams naturally experience flow intermittence and hydrologic discontinuity that act to shape fish community structure," or worse: "Thus, theoretically, although habitat suitability curves underpinning area-weighted suitability indices apparently invite the intervention of modeling approaches, the more complex and less-definite relations between physical habitat and ecological response may reduce this potential, with correspondence at best, treated probabilistically."

Why would anyone who has something to say use such language? We expect that some readers will disagree with some of what we write, but we have tried to write it clearly.

With one exception, separate authorship is not listed for the various chapters, although readers with any sense of language will notice immediately that the writing styles varies. Each chapter has a main author, but each of us has read, commented on, and approved the others. The exception, Chapter 8, Dams and Channel Morphology, was written by fluvial geomorphologist Mathias Kondolf and collaborators from his research group in Lyon, France: Remi Loire, Hervé Piégay, and Jean-Réné Malavoi, who are thus listed as co-authors for the chapter.

Overall, our somewhat lofty goal is to give users (and students) of environmental flow methods a better understanding of the tools they are using, and especially where they may fall short. Methods for EFA are constantly evolving, especially analytical tools. Practitioners would be well served to be more critical of existing well-used methods, and to investigate alternatives coming on line. The more EFAs reflect reality, the more likely they will provide useful information, to the benefit of both flowing-water ecosystems and human populations that derive so much benefit from them.

Acknowledgements

The ideas presented in Chapter 9 stem largely from development work undertaken in two Australian Research Council Linkage Projects (LP100200170, LP130100174) and eWater Cooperative Research Centre projects. We acknowledge the contributions of the many staff and students involved. We thank Genevieve Smith, in particular, for allowing the use of her Master of Environment research project as the case study presented therein. The material in Chapters 1, 2, 4, and 6 expands on worked funded by the California Energy Commission, Public interest Energy Research Program, through the Center for Watershed Sciences, University of California, Davis. Preparation of Chapter 8 was partly supported by the Collegium, Lyon Institute for Advanced Studies, University of Lyon, and the EURIAS Fellowship Programme and the European Commission (Marie-Sklodowska-Curie Actions – COFUND Programme – FP7).

CHAPTER 1

An introduction to environmental flows

Summary

Environmental flows are flows in a river required to sustain aquatic ecosystems and other beneficial uses of free-flowing rivers. Environmental flow assessment is a general term for studies that can inform management of flows. Such assessments are surprisingly difficult to do right, constrained by the natural variability of the environment through which rivers flow and the diverse needs of organisms that live there. They are also made difficult by social constraints that pit human demands for water against those of the environment, and by aspects of human behavior.

1.1 What are environmental flows?

The 2007 Brisbane Declaration of the 10th International River Symposium and Environmental Flows Conference states that: "Environmental flows describes the quantity, timing, and quality of water flows required to sustain freshwater and estuarine ecosystems and the human livelihoods and well-being that depend upon those ecosystems." We will use this definition, taking "freshwater ecosystems" to include riparian areas. "Instream flows" is an older term that means much the same thing, but we prefer "environmental flows" because it implies a broader view of what should be assessed; instream flow assessments historically have been concerned mainly with the physical environment of only a few species, especially salmonids. We take environmental flow assessment (EFA) to be the process of trying to translate the Brisbane definition into usefully precise estimates of environmental water needs and the effects of modified flows on ecosystems and human well-being, to inform decisions such as:

- Whether to reserve some portion of the flow in a stream for environmental uses, and if so, how much, and on what kind of schedule;
- How effects of an existing project on streams or estuaries can be mitigated (or not) by releases of environmental flows or restrictions on water withdrawals;
- Whether and how to modify existing water projects to improve environmental conditions;
- Whether and how to build a new water project.

Environmental flow assessment is hard to do well. This book is about the scientific and social difficulties with EFA and how to address them as best one can. In this chapter, we first explain why EFA is so difficult, and address problems with the EFA literature.

Environmental Flow Assessment: Methods and Applications, First Edition. John G. Williams, Peter B. Moyle,
J. Angus Webb and G. Mathias Kondolf.
© 2019 John Wiley & Sons Ltd. Published 2019 by John Wiley & Sons Ltd.

1.2 Why EFA is so hard; scientific issues

1.2.1 Stream ecosystems are dynamic and open

Twenty-some years ago, three of the authors of this book participated in a small workshop on environmental flow assessment at the University of California at Davis, which concluded that "...currently no scientifically defensible method exists for defining the instream flows needed to protect particular species of fish or aquatic ecosystems" (Castleberry et al. 1996). Despite major progress with analytical and statistical methods over the last 20 years, especially those described in Chapter 9, we still believe that at best an EFA should be regarded as a first cut, to be implemented within the context of adaptive management. Why is this problem so hard? Scientists have a truly wonderful understanding of the nature of energy and matter, the evolution of the universe, the atomic structure and properties of molecules, the structure and activities of cells, the origin of species and the evolutionary relationships among organisms, and much more. Why, then, is it so hard to assess the consequences of taking some of the water out of a stream, or changing the timing or temperature with which water flows down the stream?

The reasons have been known for some time: ecosystems are open, dynamic systems that are "...in a constant state of flux, usually without long-term stability, and affected by a series of human and other, often stochastic, factors, many originating outside of the ecosystem itself" (Mangel et al. 1996, p. 356). For such reasons, Healey (1998) argues that questions such as "How much can a river's hydrology be altered without endangering its ecological integrity?" are trans-scientific, sensu Weinberg (1972); trans-scientific questions: "... can be stated in the language of science but not answered by the traditional means of science." These ideas have been restated recently by Harris and Heathwaite (2012) and by Boyd (2012, p. 307): "Predicting the dynamics of real ecosystems – or even of components of these ecosystems – will remain beyond the reach of even the best ecosystem models for the foreseeable future."

A long-term study on the South Fork Eel River in Northern California (Box 1.1) illustrates these points. Although the highly predictable seasonality of flow is a major factor structuring the food web in that river, year-to-year variation in the timing and magnitude of high-flow events results in substantial variation in the structure of the food web and its response to mobilization of the bed by high flows; for practical purposes, predictions of the response can only be probabilistic, not deterministic.

As another example, consider the valuable and well-managed sockeye salmon fishery in Bristol Bay, Alaska, for which long-term catch records are available for three major fishing districts, corresponding to areas of spawning and rearing habitat. The catch is a good proxy for the number of spawning fish, known since about 1950 (Hilborn et al. 2003). Although there has been little human disturbance in the spawning and rearing areas except for climate change, the relative contributions to the catch from the different districts has varied widely over time, as described by Hilborn et al. (2003, p. 6567):

> The stability and sustainability of Bristol Bay sockeye salmon have been greatly influenced by different populations performing well at different times during the last century. Indeed, no one associated with the fishery in the 1950s and 1960s could have imagined that Egegik would produce over 20 million fish in 1 year, nor could they imagine that the Nushagak would produce more than the Kvichak, as it has in the last 4 years. It appears that the resilience of Bristol Bay sockeye is due in large part to the maintenance of all of the diverse life history strategies and geographic locations that comprise the stock. At different times, different geographic regions and different life history strategies have been the major producers. If managers in earlier times had decided to focus management on the most productive runs at the time and had neglected the less productive runs, the biocomplexity that later proved important could have been lost.

Hilborn et al. (2003) were thinking of fisheries management, but the same point would apply to

Box 1.1 Variable Effects of High Flows on a River Ecosystem

Eighteen years of field observations and five summer field experiments in a coastal California river suggest that hydrologic regimes influence algal blooms and the impacts of fish on algae, cyanobacteria, invertebrates, and small vertebrates. In this Mediterranean climate, rainy winters precede the biologically active summer low-flow season. *Cladophora glomerata*, the filamentous green alga that dominates primary producer biomass during summer, reaches peak biomass during late spring or early summer. *Cladophora* blooms are larger if floods during the preceding winter attained or exceeded "bankfull discharge" (sufficient to mobilize much of the river bed, estimated at $120 \, m^3 \, s^{-1}$). In 9 out of 12 summers preceded by large bed-scouring floods, the average peak height of attached *Cladophora* turfs equaled or exceeded 50 cm. In five out of six years when flows remained below bankfull, *Cladophora* biomass peaked at lower levels. Flood effects on algae were partially mediated through impacts on consumers in food webs. In three experiments [with caged fish] that followed scouring winter floods, juvenile steelhead (*Oncorhynchus mykiss*) and …[coastal roach, *Hesperoleucus venustus*] suppressed certain insects and fish fry, affecting

persistence or accrual of algae depending on the predator-specific vulnerabilities of primary consumers [that were] capable of suppressing algae during a given year. During two post-flood years, these grazers were more vulnerable to small predators (odonates and fish fry, which… [steelhead stocked in the cages always suppressed] … [As a result, the abundant grazers] had adverse effects on algae in those years. During one post-flood year, all enclosed grazers capable of suppressing algae were consumed by steelhead, which therefore had positive effects on algae. During drought years, when no bed-scouring winter flows occurred, large armored caddisflies (*Dicosmoecus gilvipes*) were more abundant during the subsequent summer. In drought-year experiments, stocked fish had little or no influence on algal standing crops, which increased only when *Dicosmoecus* were removed from enclosures. Flood scour, by suppressing invulnerable grazers, set the stage for fish-mediated effects on algae in this river food web. Whether these effects were positive or negative depended on the predator-specific vulnerabilities of primary consumers that dominated during a given summer. (Power et al. 2008, p. 263 edited for clarity)

managing the freshwater habitat in these regions; there have been major geographical shifts in productivity in this undisturbed habitat, and no one knows why.

1.2.2 Fish evolve

We are used to thinking of evolution as a slow process, but this is not always the case. Stearns and Hendry (2004) wrote that: "A major shift in evolutionary biology in the last quarter century is due to the insight that evolution can be very rapid when populations containing ample genetic variation encounter strong selection (citations omitted)." It is now clear that significant evolution can occur within

the time spans commonly considered in EFA, and fish populations may respond to changes in the environment in unexpected ways. For example, in several California rivers, releases of cold water from the lower levels of reservoirs have created have good habitat for large trout. The steelhead populations in these rivers apparently have evolved toward a resident life-history in response (Williams 2006). Where hatcheries "mitigate" for habitat lost above dams, salmonids evolve greater fitness for reproduction in hatcheries, and lower fitness for reproducing in rivers (Myers et al. 2004; Araki et al. 2007; Christie et al. 2014); significant domestication can occur in a single generation (Christie et al. 2016). If hatchery fish mix with naturally spawning fish in the river below the dam, the population of naturally spawning fish below the dam that can be supported by a given flow regime will be reduced as fitness declines.

1.2.3   Streams adjust

Alluvial or partially alluvial streams create their own channels. Anything that substantially changes flow or sediment transport in a stream, such as a new dam, will provoke geomorphic adjustments in channel size and form that will change the physical habitat, compromising assessments based on the pre-project habitat.

1.2.4   Climate changes

Long-term climate records and paleoclimatic data from tree rings and other sources show that climates have always varied over decades and centuries, and now greenhouse gas emissions are driving rapid change. One predictable change, already evident in flow data, is more winter runoff and less snowmelt runoff in mountain streams. Precipitation may increase or decrease, depending upon the region, and may become more variable. Thus, the amount and temporal distribution of

water available to be allocated between instream and consumptive uses will change, as will the temperature of the water. Methodologically, climate change confounds analytical methods that assume that the statistical properties of flow data will be stationary, i.e. not change over time (Milly et al. 2008). Predicting climate change at any particular location is even more difficult than predicting global change (Deser et al. 2012), so uncertainty about climate will add substantially to the uncertainties already faced in EFAs.

Even without major human influences, climates and flow regimes vary substantially over time, especially in arid and semi-arid regions, as shown by a plot of the 30-year running average discharge in the Arroyo Seco River in California. (Figure 1.1). Thus, the particular period of record that is available for analysis can make a major difference (Williams 2017). Probably the most famous example of this is the Colorado River Compact of 1922, which allocated the water from the Colorado River among the various states of the USA in the basin. The allocation was based on unusually high flows in the early twentieth century, and so seriously over-allocated water from the river, as noted by the National Research Council (2007, pp. 99, 103):

Figure 1.1   Thirty-year running average discharge in the Arroyo Seco River in central California. There has been no significant development in the basin. Data from the USGS gage 11152000. Source: John Williams.

From the vantage point of the early 21st century, there is now a greater appreciation that the roughly 100 years of flow data within the Lees Ferry gage record represents a relatively small window of time of a system that is known to fluctuate considerably on scales of decades and centuries. (p. 99). … Long-term Colorado River mean flows calculated over these periods of hundreds of years are significantly lower than both the mean of the Lees Ferry gage record upon which the Colorado River Compact was based and the full 20th century gage record (citation).

1.2.5 Populations vary

Populations of fish and other aquatic organisms can be highly variable in time and space (e.g. Dauwalter et al. 2009), even in stable stream environments (e.g. Elliott 1994). This makes it hard to determine population trends or whether changes in flows have done any good or harm (Korman and Higgins 1997; Williams et al. 1999). This is particularly true for anadromous fish, populations of which may be strongly affected by ocean conditions that vary from year to year (e.g. Lindley et al. 2009). Within short sections of streams, abundance can vary strongly over periods of days (e.g. Bélanger and Rodríguez 2002), so assessments of habitat quality based on fish density can be unstable.

1.2.6 Habitat selection is conditional

Environmental flow assessments are often based on the assumptions that providing more of the kind of habitat where fish are found will increase the population of fish. The assumption may be sound, provided that it is tempered by biological understanding, by appropriate choice of spatial scale in the assessment, and by the recognition that habitat selection is conditional; in other words, fish can only select habitat that is available to them, and habitat selection at fine spatial scales can be affected by many factors, including habitat at coarser spatial scales, population density, competition, season, water temperature, cloud cover, and even discharge (Chapter 7). It is also necessary to consider how much of a particular kind of habitat a population of a given size needs, and to recognize that other factors altogether may determine abundance. Habitats affect populations through their effects on births, deaths, growth, and migration.

1.2.7 Spatial and temporal scales matter

The response times of the resources of concern complicate EFAs. Biotic communities may take decades to respond detectably to management actions, or the response may change over time. For example, the population of Sacramento River spring Chinook salmon initially increased after the construction of Shasta Dam (Eicher 1976), but later collapsed (Williams 2006), probably because of interbreeding with fall Chinook salmon. This problem is particularly acute for fish that use spatially dispersed and distinct habitats over the course of their life cycles, when only some of the habitats are affected by the actions.

Even if the inquiry concerns physical habitat, response times may still present problems. Events such as scouring floods that seem to destroy habitat in the short term may create other habitat, such as deep pools, in the long term. Anything that substantially changes sediment transport in a stream, such as a new dam that blocks sediment transport or modifies flows, will provoke geomorphic adjustments in channel size and form that will change the physical habitat.

Spatial scales also matter, for example in assessments of habitat selection (Cooper et al. 1998; Welsh and Perry 1998; Tullos et al. 2016). Factors that seem to drive habitat selection at a fine spatial scale may explain relatively little at a coarser spatial scale (Fausch et al. 2002; Durance et al. 2006; Bouchard and Boisclair 2008). As an additional complication, organisms can select habitat at multiple scales. In a classic observational study, Bachman (1984, p. 9) wrote that:

The mean home-range size of 53 wild brown trout was $15.6\,m^2$ (SE, 1.7) as determined from minimum-convex polygons encompassing 95% of the scan sighting of each fish each year. … Typically, foraging sites were in front of a submerged rock, or on top of but on the downward-sloping rear surface of a rock

… From there the fish had an unobstructed view of oncoming drift. While a wild brown trout was in such a site, its tail beat was minimal … indicating that little effort was required to maintain a stationary position even though the current only millimeters overhead was as high as 60–70 cm s^{-1}. Most brown trout could be found in one of several such sites day after day, and it was not uncommon to find a fish using many of the same sites for three consecutive years.

Thus, the trout selected habitat on a scale of centimeters with respect to the rock, on a scale of meters with respect to incoming drift, and a scale of tens of meters with respect to home range; further study might have shown selection of home ranges on a scale of hundreds or thousands of meters.

1.3   Why EFA is so hard: social issues

1.3.1   Social objectives evolve

Like ecosystems, societies are not stable equilibrium systems; social attitudes and objectives also evolve, as do environmental laws and regulations, and the evolution is rapid relative to the duration of major water-development projects. We are old enough to remember the resurgence of environmental concern in the 1960s that laid the basis for much of current environmental law in the USA, such as the Clean Water Act, the Endangered Species Act, and the National Environmental Policy Act. Environmental concerns also affected judicial decisions. For example, in 1971, in *Marks v. Whitney* (6 Cal.3d 251), a decision about tidelands in Tomales Bay, the California Supreme Court broadened the uses that are protected by the Public Trust to include providing environments for birds and marine life, and scientific study. This decision did not come from abstract legal reasoning, but rather from the political mood of the time. In pertinent part, the decision states that:

Public trust easements are traditionally defined in terms of navigation, commerce and fisheries. They have been held to include the right to fish, hunt, bathe, swim, to use for boating and general recreation purposes the navigable waters of the state, and to use the bottom of the navigable waters for anchoring, standing, or other purposes (citations omitted). The public has the same rights in and to tidelands. … The public uses to which tidelands are subject are sufficiently flexible to encompass changing public needs. In administering the trust the state is not burdened with an outmoded classification favoring one mode of utilization over another (*citations omitted*). There is a growing public recognition that one of the most important public uses of tidelands – a use encompassed within the tidelands trust – is the preservation of those lands in their natural state, so that they may serve as units for scientific study, as open space, and as environments which produce food and habitat for birds and marine life, and which favorably affect the scenery and climate of the area. …

This broadening of trust uses was extended to navigable lakes and streams and their tributaries in 1983 in *National Audubon Society v. Superior Court* (33 Cal.3d 419), concerning environmental flows in Rush Creek, a tributary to Mono Lake. The Audubon decision and the environmental attitudes it reflected also gave new life to existing legislation affecting environmental flows, such as Fish and Game Code sec. 5937, discussed in Chapter 2. Changing social attitudes also change the practical effect of environmental laws. Monticello Dam on Putah Creek in California releases water for re-diversion 10 km downstream. These releases support a trout fishery, which, together with recreational uses of the reservoir, was long thought to meet any environmental obligations arising from the project, including Fish and Game Code sec. 5937. Over time, however, native fishes that were formerly regarded as "trash fish" came to be valued, and litigation resulted in revised environmental flow releases to protect them (Moyle et al. 1998).

Similar changes have developed elsewhere, although the nature and pace of the change has varied among nations and regions. South Africa, for example, experienced sudden advances in the relevant law and methods for EFA in the euphoric period after Nelson Mandela ushered in a peaceful end to apartheid. Together with scientists in Australia, where semi-arid conditions made methods such as the physical habitat simulation system (PHABSIM) clearly unsuitable, South African scientists developed holistic methods (Arthington et al. 1992a). These were applied in Australia when the need for multi-state planning in the Murray–Darling basin, underscored by the Millennium Drought, brought about major changes in water law that called for shifting allocations of water from consumptive uses to the environment (Skinner and Langford 2013).

1.3.2   Science and dispute resolution

Environmental flow assessments almost always occur within the context of disputes over water, and the resolution of these disputes will involve trade-offs and balancing, and often negotiation. For this reason, the main publication on the use of the Instream Flow Incremental Methodology (Bovee et al. 1998) deals extensively with negotiation and dispute resolution. We do not deal with these aspects of the flow-setting process in this review, since we are not experts in them, although we recognize that effective negotiation and dispute resolution are critical aspects of protecting environmental flows. However, it is also important to keep in mind that science and dispute-resolution are separate endeavors that have different rules for settling questions.

Distinctions among human activities often break down in the details, but generally, science settles questions by testing hypotheses or models with data. Procedures for doing this may be generally agreed upon, but they are always subject to criticism, alternatives can always be put forward, and conclusions are always subject to change in light of new evidence. In legal or political disputes, on the other hand, questions can also be settled by the parties agreeing to an answer, and in legal disputes this answer may be final, at least for the parties involved, regardless of new evidence that may emerge. For example, the parties in a dispute over water may agree that the results of a study of part of the stream in question will be taken as representative of the whole. This will not wash in science. Science and dispute-resolution

both have major roles in EFA, but it is important to keep them separate.

In the regulatory world, disputes are supposed to be resolved, which requires that decisions be made in reasonable time. This produces a tension between science and dispute-resolution. Adaptive management, discussed in Chapter 6, can be viewed as a way to reduce this tension, but it will not do away with it.

1.3.3   Water is valuable

Because water is valuable, disputes over the allocation between environmental and other uses are often intense. Mark Twain allegedly said that "Whiskey is for drinking; water is for fighting over," and, even if the quote is not authentic, the comment rings true. If consultants or agency staff on one side of a dispute see their job simply as furthering the interests of their client or employer, then consultants and staff on the other side have little choice but to do the same, resulting in "combat biology." Something similar results from the tendency of people in a dispute, as social animals, to see their side as in the right, and to accept the opinions of others on their side as correct, with opinions of those on the other side as suspect at best. It is hard to conduct a dispassionate assessment in these circumstances.

1.3.4   Managers or clients often want the impossible

In disputes, properly describing uncertainty can be problematic. Scientists working on EFA normally work for someone else, usually a manager in an agency or consulting firm, or sometimes a specific client, and often in the context of disputes over water. Often, the manager or client will want more definitive results than the state of the science allows. Experience shows that there are scientists who will provide such results, or even the particular results desired, and this presents yet another difficulty for those wanting to do honest work. We wish we had a solution for this problem, but we do not.

A related "real world" problem for EFA is that specialists in the field may build a career around one method or another, and become personally invested in it. They are then resistant to criticisms of the method that cannot be accommodated by minor changes in it. As Upton Sinclair famously wrote, "It is difficult to get a man to understand something, when his salary depends on his not understanding it."

Lest this recitation of difficulties seem too gloomy, we reiterate Healey's (1998) point that people do know quite a lot about fish and riverine ecosystems. We do have a lot of background knowledge and analytical tools with which to think about environmental flow assessment. The rub, however, is that we cannot do a good job of EFA without clear thinking, and clear thinking is as hard to do as it is essential. Therefore, we should do the best we can, be clear about what we did and did not do and why, and try to work in an adaptive framework that will allow changes in management as new information and understanding become available.

1.4   Why EFA is so hard: problems with the literature

There are several literatures on environmental flow assessments or on matters highly relevant to them. It is common to distinguish peer-reviewed journals from agency or consulting reports, but there are also important distinctions among peer-reviewed journals. Roughly, there is a more academically oriented literature, largely in ecological or hydrologic journals, and a more applied fisheries literature, with surprisingly little overlap between them. There are also many relevant papers in journals on geomorphology, engineering, and statistics, and a large literature on habitat selection in wildlife journals. Unfortunately, there are now also "pay to publish" journals that will print almost anything. Even among legitimate journals, these distinctions matter, because the quality of the reviewing tends to vary. Generally, the reviewing for the academically more prestigious journals is more rigorous, but the reviewers for these journals may not be as familiar with the details of a particular topic as reviewers for the relevant specialty journals. The distinction that really matters is whether journal

articles or agency reports are based on good logic, methods, and evidence.

Peer review is an important part of scientific quality control, but it is far from perfect and many deeply flawed articles are published. Ioannidis (2005) described this problem for biomedical research in an influential article entitled "Why most published research findings are false," and the problem has received considerable attention since. For example, the Open Science Collaboration (2015) recently reported that replication of the work reported in 100 psychology articles from leading journals showed that most reported findings were not substantiated. There are various reasons for this unfortunate state of affairs, including conscious or unconscious bias by the investigators, and misuse of statistical methods (the latter is a common one). We are not statisticians, but we often see obvious statistical problems with papers dealing with EFA. The upshot is that even the scientific literature needs to be approached carefully and critically, and those of us who are not experts at statistics should cultivate good relations with people who are. Even apart from statistical issues, we should read the literature with the question "Why should I believe that?" always in mind. Skepticism is particularly justified in reading the EFA literature, as the history of EFA shows.

1.5   Why EFA is so hard: limitations of models and objective methods

1.5.1   Models and environmental flow assessment

Models are essential tools for environmental flow assessment, but are often misused (Chapter 7). The proper use of models is to help people think, even for well-studied physical systems. Consider weather forecasting. The National Weather Service forecasters in our area base their forecasts on the results of three and sometimes four different models, using their knowledge of how well each model handles particular kinds of weather, and the plausibility of each model result. Proper use of models requires a good understanding of the model, the data at hand, the system being modeled, and the questions being addressed. A remark by the economist Thomas Piketty seems applicable to ecosystems: "Models can contribute to clarifying logical relationships between particular assumptions and conclusions but only by oversimplifying the real world to an extreme point. Models can play a useful role but only if one does not overestimate the meaning of this kind of abstract operation" (Piketty 2015, p. 70). Inevitably, models embody simplifications of the world, based on the aspects of the world that we (or someone) believe are important for the problem at hand. That is, we model the way that we think the world works, but we should remember that the world has no obligation to work that way. The invaluable thing that models do is to show us the logical consequences of our thinking, or, for estimation models, to show us how well the data support our thinking.

1.5.2 Objective and subjective methods

A few decades ago, it was common for scientists to promote "objective" methods for analyzing problems, generally by applying some numerical model. This conceit has largely been given up, in the face of persuasive arguments that modeling always involves subjective choices. For example, Brenden et al. (2008) used regression trees analysis to develop a classification system for stream segments in Michigan, partly out of concern that a classification based on expert opinion could be hard to defend. However, they deliberately selected a similarity threshold of 0.6, largely because it generated a system "that had good agreement with a previously completed expert-opinion delineation of stream segments" (p. 1622). In modeling for EFAs, there are always subjective choices about what to include in the model and how

to do so (Kondolf et al. 2000). Subjectivity will enter into EFAs, whether we want it to or not; the question is whether the subjectivity will be recognized and taken into account.

Just as science should inform EFAs, EFAs should inform science. That is, studies conducted for EFA should be so conducted as to add to the general body of knowledge, and there should be feedback regarding questions and uncertainties that loom large in assessments and may be amenable to traditional scientific inquiry. Thus, the reasoning and assumptions underlying: environmental flow models (EFMs) should be stated explicitly, as should the reasoning underlying environmental flow decisions. In particular, it should be possible to tell what kinds of evidence or new understanding would justify a change in the assessment or the decision.

1.6 Conclusions

For at least two reasons, environmental flow assessment is not just science: the main question it asks may be trans-scientific, and usually the question is asked in the context of dispute-resolution. These qualities are not unique to EFA, but rather apply to ecosystem management generally, which has been called a "wicked problem" accordingly (DeFries and Nagendra (2017). Science can and should inform environmental flow assessment, and EFMs should be consistent with scientific practice. However, the limits to what science can contribute should be recognized. Ecosystems are enormously complicated, and it is not realistic to expect that standard methods can be devised by which EFA can be successfully accomplished without good data, careful thought and informed judgment.

CHAPTER 2

A brief history of environmental flow assessments

Summary

Attempts to quantify flow needs for fish and the environment have long followed two approaches: embodying existing knowledge into some method, and developing new understanding of the species or ecosystem in question. Methods have tended to become increasingly elaborate with the development and continued improvement of computers. However, while holistic methods have garnered considerable interest in recent years, their complexity and cost has resulted in renewed interest in simple but conservative hydrological methods that can be applied widely and quickly.

2.1   Introduction

The brief history given here emphasizes the early period of environmental flow assessment (EFA), in the mid-twentieth century, which is not well described in the recent literature. It also emphasizes California, since we are most familiar with developments here, and California was in the forefront of developing methods for EFA, but we try to cover the main trends generally. The review of environmental flow methods in Chapter 6 elaborates some of the points touched on here.

Broadly, there have long been two responses to the challenge of EFA. One is using existing knowledge to specify what flows should be, based either on professional opinion or on some model. As an example, Menchen (1978, p. 4) used assumptions about the area of spawning habitat required to produce a given number of Chinook salmon, and simple statistical analyses between spring flows and adult returns, to specify flow targets for the Tuolumne River, a tributary to the San Joaquin River in California:

> Spawning gravel area requirements for runs of 32 000 to 52 000 spawners are fairly straightforward. An area of 1 000 000 square feet will accommodate 32 000 spawners on the Tuolumne. Fifty thousand spawners require 1 562 000 square feet. … outflows during March through June of around 4 000 cfs are required to produce 32 000 adults and of around 7 500 cfs are required for runs of 50 000 adults.

The other approach is developing better scientific understanding of aquatic ecosystems. As an early example, the state of California began an interagency Delta Fish and Wildlife Protection Study to clarify

Environmental Flow Assessment: Methods and Applications, First Edition. John G. Williams, Peter B. Moyle, J. Angus Webb and G. Mathias Kondolf.

the environmental consequences of a major state water project. According to the abstract of the first report of this work (Kelley 1965):

> The Delta Fish and Wildlife Protection Study was organized in 1961 to investigate the effects of future water development on fish and wildlife resources dependent upon the Sacramento–San Joaquin River estuary, and to recommend measures to protect and enhance these resources. The investigations described in this bulletin were designed to answer a number of specific questions relevant to water development plans and also to start us toward an understanding of the estuary's ecology. The bulletin describes the results of about 2 years of collecting and 1 year of analysis on zooplankton, zoobenthos, and fishes of the middle or bay portion of this estuary and on zooplankton and zoobenthos of the upper portion that is known as the Delta.

Unfortunately, the State of California has restricted this ecological approach mainly to the Delta, but it has been pursued elsewhere, for example the studies in the state of Michigan, discussed in Chapter 6. The rationale for this approach was nicely stated by Anderson et al. (2006, p. 317): "To many, the research agenda we are proposing will appear similar to that of much of 'basic' aquatic ecology. This is no accident; we contend that successfully providing for [environmental flow needs] in streams and rivers requires understanding how these systems work." It seems to us that both kinds of response are needed, and, as argued in later chapters, can be reconciled by adaptive management.

2.2 The legal basis for environmental flows

Environmental flow assessment presupposes a way to act on the results, so a logical precursor to EFA is some way to protect environmental flows. The evolution of such protection has varied among and within states and nations, depending largely on the relation between the supply and demand for water, and also on the state of economic development and on legal systems. Again, California was in the forefront, at least on paper.

Since 1915, California law has required in Fish and Game Code section 5937 that "The owner of any dam shall allow sufficient water at all times to pass through a fishway, or in the absence of a fishway, allow sufficient water to pass over, around, or through the dam to keep in good condition any fish that may be planted or exist below the dam." In practice, this law was often ignored, because the politics of the day favored development of water resources over environmental concerns (Börk et al. 2012). Thus, San Clemente Dam, built on the Carmel River in the mid-1920s, had a fishway to pass migrating steelhead, but the owner was not required to release water into the stream when the reservoir level was below the fishway entrance, as happened every summer. In mid-century, a need to release water for fish would have affected a proposed federal dam on the San Joaquin River, and in 1951 the California Attorney General obligingly issued an opinion that the section applied only to "water in excess of what is needed for domestic and irrigation purposes," effectively gutting the statute until the politics changed.

The politics and applicable laws changed with the environmental movement of the 1960s and 1970s. In 1974, the Attorney General in California issued a new opinion regarding section 5937, effectively reversing the 1951 opinion (Börk et al. 2012). In 1983, a California Supreme Court decision, *National Audubon Society v. Superior Court* (33 Cal.3d 419), effectively allowed section 5937 to be applied retroactively, although it required a balance between environmental and consumptive uses of water. The California Environmental Quality Act (CEQA) and the National Environmental Policy Act (NEPA) were enacted in 1970, the Clean Water Act in 1972, and the Endangered Species Act in 1973. These and other statutes required a broad assessment of the environmental effects of water projects, and provided the mean to do something about them.

The timing of changes in legal protections for environmental flows has varied among countries and regions. In the United Kingdom, a Water Resources Act providing for minimum flows was enacted in 1963, and strengthened in 1991 (Petts 1996). In South Africa, progressive water laws were enacted in the late 1990s in the heady period after the fall of apartheid. International agreements have become important recently, for example, the European Union Water Framework Directive (Hering et al. 2010), and the Great Lakes Compact (discussed in Chapter 6). The national government in Australia relied on its international conservation obligations for authority to impose basin-wide management of the Murray–Darling river system with the Water Act of 2007, which accordingly provided for environmental flow targets as well (Skinner and Langford 2013).

2.3 The scope of environmental flow assessments

Although some early assessments such as the Delta study looked at ecosystems, early flow assessments and flow standards mostly concerned salmonids; environmental flows were reserved for non-game fishes in only 3 of 45 cases in California reviewed by Hazel (1976). However, environmental laws such as NEPA and CEQA required a broader scope. Although some environmental flow assessments of the 1980s dealt only with one or a few species, assessments for controversial projects needed to comply with the new laws. The lead author of this book was personally involved with two. One involved studies during the early 1980s to support an application for a new dam on the Carmel River, in coastal California. The river supports a species of Pacific salmon, steelhead, that is a popular sport fish. However, the studies dealt not just with steelhead through the freshwater part of their life cycle, but also with aquatic invertebrates, riparian vegetation, migrating birds, fluvial geomorphology, and recreation. In the 1989 trial of *Environmental*

Defense Fund et al. v. East Bay Municipal Utility District, which considered the environmental effects of proposed diversions from the American River in California, evidence was presented regarding Chinook salmon and other fishes, riparian vegetation, fluvial geomorphology, water temperature, recreation, and the public health consequences of diverting water at a downstream location. These studies were not bundled into a named methodology, but they were nevertheless comprehensive.

Although a broad range of environmental effects were considered in these cases, the quality of the consideration was often lacking. The record from the 1989 trial of *EDF et al. v. EBMUD* documented the state of the art at the time. This was a major case, and the parties went 'all out' to present evidence supporting their positions, and to critique evidence that contradicted them. In the decision, the judge lamented the "fundamental inadequacy" of the evidence on environmental flows that had been presented. Importantly, the decision imposed interim environmental flows that were intended to be precautionary, and ordered the parties to cooperate in a program of studies to clarify what the flows should be (Somach 1990; Littleworth and Garner 1995; Williams 1995). Thus, the decision embodied a form of adaptive management, which had been proposed initially for management of salmon fisheries by Walters and Hilborn (1976) and elaborated for other environmental issues by Walters (1986), Walters and Holling (1990), Ludwig et al. (1993), Lee (1994) and others. Essentially, adaptive management recognizes that estimates of the consequences of management actions are uncertain, so that management actions are necessarily experiments, and makes learning an objective of management. Castleberry et al. (1996), with specific reference to *EDF et al. v. EBMUD*, called for the adaptive management of environmental flows, and such calls have become conventional. Unfortunately adaptive management has proved to be difficult to implement, but an exemplary implementation in Australia is described in Chapter 6.

2.4   Methods for quantifying environmental flows

When flow releases were required in the USA during the mid-twentieth century, they were based on professional opinion (Giger 1973). For example, when Los Padres Dam was built on the Carmel River in the late 1940s, the California State Water Resources Control Board (SWRCB) required a minimum release of $0.14 \, m^3 \, s^{-1}$ (5 cfs), based on the professional opinion of Leo Shapovalov of the Department of Fish and Game. Shapovalov was an accomplished biologist who co-authored a landmark study of coho salmon and steelhead (Shapovalov and Taft 1954). Nevertheless, his recommended flow release was approximately the release rate required by the purpose of the dam, which was to store water from winter rains and release it through the summer dry season for re-diversion at the San Clemente Dam, farther downstream. Thus, the requirement really was only that the water be released at a steady rate, and likely reflected Shapavalov's understanding of the political balance of power of the time.

Many new dams were being built in the western USA after World War II, and, as a consequence, work on more structured approaches to EFA began in this region in the late 1940s (Tharme 2003). The state of the relevant science and methods to apply this knowledge to EFA was well described in a review of environmental flow needs of salmonids by Richard Giger for the Oregon Wildlife Commission (Giger 1973). Most of the biological issues of current concern had been identified and studied to some degree, at least for salmonids (see Northcote 1969), but Giger thought that not enough was yet known, and that none of the existing approaches to EFA were adequate, even when assessments were concerned with only a few species (p. 103):

> Based on this literature review, it seems clear that an ideal method for establishing rearing flow recommendations should take into account a broader range of environmental needs of fish than is considered in current methodologies. Such a method should

attempt to mesh as many requirements as practical in such a way that an optimal balance is achieved. Of particular significance might be the elements of shelter and food supply. Present knowledge of the stream ecology of young salmonids, however, does not appear adequate for extensive development of new methodology having the degree of factual basis needed to overcome present criticisms.

Reading between the lines, Giger thought that an approach developed by biologists for the California Department of Fish and Game, and described by Kelley et al. (1960), was the best of the existing methods, although the department did not pursue it (p. 87):

> Their plan was to quantify the amounts of food producing, shelter, and spawning area present in streams at different flows, and from this information recommend flows that would maintain or enhance fish populations. … The techniques presented meshed important biological factors such as food production and cover with streamflow in a logical fashion.

Many other methods were also being developed and applied. Generally, the search was for "objective" methods that provided biologically sensible recommendations, rather than "subjective," opinion-based methods. For example, Smith (1973) described suitability criteria for depth and velocity for various salmon species. Wesche (1976) worked on a method for assessing the relationship between flow and cover for rainbow and brown trout in Wyoming streams, which he thought limited populations. Tenant (1976) proposed a method using percentages of the mean annual flow as criteria, for example that 30% of the mean average flow would provide "excellent" habitat. This was less arbitrary than it may seem, at least for the upper Midwest, since the criteria were based on a good deal of field observation, and he considered other resources besides fish: the title of his paper is "Instream flow regimens for fish, wildlife, recreation and related environmental resources." The Tenant method and other hydrologically based methods have been widely applied. Other methods looked for kinks in curves of stage plotted over discharge, or at geomorphic features such as riffle-pool ratio, features providing cover, etc. The idea was to identify flows below which habitat conditions would deteriorate quickly, and so identify minimum flows that should be protected.

Still other proposed methods were based on microhabitats, defined in terms of depth, velocity, and substrate; Giger (1973, p. 103) did not think much of them:

> It is difficult to visualize how microhabitat data, as presently constituted, would be adequate of itself for use in determining flow levels. Such data might, however, be one valuable element of an advanced methodology for recommending flows, particularly if certain species are to be emphasized in the recommendations. Microhabitat information presently consists of limited physical data not easily related to food and shelter requirements. Neither can the data be used in a spatial context. Overall, there seem to be a number of problems facing the use of such data in flow recommendations, in terms of both adequacy and implementation.

At about this time, computers were coming into common use, allowing for methods that previously were too computationally demanding to be practicable. Hydraulic models useful for predicting the water surface elevation (stage) of a stream at a given discharge also became available, although assumptions in the models were more appropriate for very high flows than for ordinary flows. Building computer models for EFA was a logical next step, and Waters (1976) described such a model in the Proceedings of the Symposium Specialty Conference on Instream Flow Needs (Box 2.1).

According to Waters, his model was based on the Kelley et al. (1960) approach, although he needed to turn it into a microhabitat model for application on a computer. Nevertheless, it did take microhabitat for "food production" into account. Implementation of the model used transects located in "representative reaches" rather than by proper statistical sampling, but the model did calculate and display confidence intervals around estimated quantities, a practice that subsequent workers evidently thought unnecessary. The "most useful" products of the program were plots of "weighting factors" showing the relative value of microhabitat features at a particular discharge.

Also in 1976, the US federal government established an interagency Cooperative Instream Flow Service Group funded mostly by the Environmental Protection Agency, although it was under the sponsorship of the U.S. Fish and Wildlife Service. This group soon developed a computer model

Box 2.1   The 1976 Conference on Instream Flow Needs

A word is in order about the 1976 Conference on Instream Flow needs. There was a great rush of applications to build hydropower facilities in the mid-1970s because of the energy crisis of the time, and a corresponding demand for methods for EFA, that now had to meet the standards of the 1970 NEPA and various new state laws such as the CEQA. In response, the American Fisheries Society and the Power Division of the American Society of Civil Engineers sponsored this major conference, and the various papers in the two volumes of the proceedings which document the thinking of the time. For example, on the range of factors that should be considered, a representative of the U.S. Forest Service noted that "Of course, the instream flow needs include fisheries, recreation, esthetics, stream-side vegetation, and so on" Resler (1976, p. 626). The Instream Flow Council, with permission from the American Fisheries Society, has made an electronic copy of the conference proceedings available on their web site (http://www.instreamflowcouncil.org/resources/ifc-publications/afs-publications). Anyone with a serious interest in the history of environmental flows would do well to browse through these volumes.

that combined habitat suitability curves, similar to Water's plots of weighting factors, with one of several hydraulic models, to estimate a statistic called weighted usable area (WUA). By 1979, the model was called PHABSIM, for physical habitat simulation system. PHABSIM typically weights areas using suitability curves for depth, velocity, and substrate, although curves for other variables such as cover or velocity gradient are sometimes used as well. This became the most widely used method in North America and several other countries (Tharme 2003), and is described and critiqued in Chapter 6. Along with PHABSIM, the Cooperative Instream Flow Group also developed a more comprehensive framework for EFA, called the Instream Flow Incremental Flow Methodology, or IFIM, and the two are often confused. As discussed in Chapter 6, this framework has attractive features, but in actual applications only PHABSIM was commonly used.

PHABSIM rapidly gained acceptance. An elegant laboratory study by Kurt Fausch (1984) showed that the positions selected by juvenile salmonids could be predicted from bioenergetics considerations, and probably this seemed to give the approach intellectual weight, although it actually undercut it, as described in Chapter 5. Reiser et al. (1989) presented the results of a survey asking agencies in the USA and Canada about the methods that they used at the time: PHABSIM was at the top of the list. Nevertheless, there were cogent early criticisms of PHABSIM that have never been properly answered (Mathur et al. 1985; Shirvell 1986; Scott and Shirvell 1987), and a dearth of good evidence for a strong relationship between WUA and the abundance or biomass of fish, as discussed in Chapter 7. As explained by R. J. Behnke of Colorado State University (Behnke 1986, p. 8):

> The great advantage of [PHABSIM] over other methodologies is its ability to quantitatively display changes in WUA (assumed to represent the habitat quality of target species) with changes in flow, which can be plotted on an actual or proposed hydrograph. This allows negotiators to discuss trade-offs and mitigation for proposed projects in a quantitative manner. As such, [PHABSIM] was quickly embraced by federal agencies as a long-sought savior to their problem of

quantification of gains or losses to the biological system from flow changes. For many, the hard question of what does WUA relate to, was ignored or not even considered. When the question was asked and tested, the results were a disillusionment to many and a confirmation to those who were aware of the limitations of prediction discussed above.

However, PHABSIM also had academic defenders, for example James Gore, who with John Nestler published an influential defense of the method (Gore and Nestler 1988), albeit mixed with mild criticism, which justified the lack of correspondence between WUA and biomass, and invoked (erroneously, in our view)[1] an argument by a theoretical ecologist about population models (Levins 1966) to excuse PHABSIM's lack of realism.

As another selling point, PHABSIM seemed to be more rigorous and scientific than other methods of the time, and it was done on a computer! People did not have computers in their homes at that time, and computers had a mystique about them that is now hard to imagine. In negotiations over by-pass flows, the engineers for the hydropower companies would have reams of computer printout showing the energy costs of increasing the by-pass flows, and, with PHABSIM, the environmental agencies could also have reams of computer output showing the associated habitat benefits.

Fairly quickly, however, developments in ecology and fluvial geomorphology further challenged the scientific underpinnings of PHABSIM. PHABSIM and other early methods were based partly on the idea that streams could be regarded, geomorphically and ecologically, as equilibrium systems. "Formerly, the dominant paradigm was that of an ecosystem that was stable, closed, and internally regulated and that behaved in a deterministic manner" (Mangel et al. 1996, p. 356). Thus, provided that the stream were "fully seeded" with some species, then density-dependent mortality would quickly bring the population into equilibrium with the physical habitat provided by the channel, so that one could assume that biological data collected in one or a few seasons also reflected longer term conditions. Similarly, if the channel were in "dynamic equilibrium," measurements from a sample of transects would continue to provide a good description of the reach, even the channel changed. Riffles might become pools, and pools riffles, as meanders progressed, but over a reach the pattern would remain the same.

Under the new paradigm, however, the ecosystem was regarded as open, "usually without long-term stability, and affected by a series of human and other, often stochastic, factors, many originating outside of the ecosystem itself" (Mangel et al. 1996, p. 356). For streams, for example, more attention was given to the "intermediate disturbance hypothesis" (e.g. Ward and Stanford 1983), according to which biodiversity peaks at some intermediate level of habitat disturbance. Evidence for the shortage of coarse sediment in many regulated streams, and growing understanding of the importance of large wood to stream ecosystems, similarly drew attention to the importance of physical disturbances for maintaining stream ecosystems, and the term "disturbance regime" entered the EFA lexicon (e.g. Resh et al. 1988). Implicitly, this line of thought undercut the idea of streams as equilibrium systems, at least over short and medium time scales.

Nevertheless, after initial resistance, water diverters and their consultants apparently realized that the many options available in PHABSIM allowed it to be used as a vehicle for negotiations; arguments about more or less flow in the river could be phrased in terms of options that would change the shape of the curve of WUA over discharge. Thus, PHABSIM was generally accepted by participants in environmental flow controversies, reducing the incentive to develop better alternatives. Although the popularity of the method waned temporarily in the late 1990s in response to a new wave of criticism (e.g. Castleberry et al. 1996; Williams 1996), PHABSIM and its variants remain one of the most commonly used method in the USA, New Zealand, and several European countries.

Besides PHABSIM, the U.S. Fish and Wildlife Service also developed "habitat suitability index models" for various species of fish and wildlife. These were simulation models. In an index model for brown trout, these were described as (Raleigh et al. 1986, p. iii):

The Habitat Suitability Index (HSI) models presented in this publication aid in identifying important habitat variables. Facts, ideas, and concepts obtained from the research literature and expert reviews are synthesized and presented in a format that can be used for impact assessment. The models are hypotheses of species–habitat relationships, and model users should recognize that the degree of veracity of the HSI model, SI graphs, and assumptions may vary according to geographical area and the extent of the data base for individual variables. After clear study objectives have been set, the HSI model building techniques presented in U.S. Fish and Wildlife Service (1981) and the general guidelines for modifying HSI models and estimating model variables presented in Terrell et al. (1982) may be useful for simplifying and applying the models to specific impact assessment problems. Simplified models should be tested with independent data sets, if possible. Statistically-derived models that are an alternative to using Suitability Indices to calculate an HSI are referenced in the text.

The index model described by Raleigh et al. (1986) included 18 variables, including maximum temperature, minimum dissolved oxygen, percent shade, pH, and peak flows. Although this approach seems more comprehensive and "ecological" than PHABSIM, index models did not catch on for fishes, probably because unlike PHABSIM, they did not "quantitatively display changes in [habitat] with changes in flow." Index models continue to be used for wildlife and, curiously, for oysters (e.g. Sonia et al. 2013), but often have not fared well when tested against data (e.g. Hubert and Rahel 1989; Farmer et al. 2002).

In another approach, there were various attempts to fit regression models to data on fish abundance and stream habitat variables, reviewed by Fausch et al. (1988); these can be called "capability models" (Korman et al. 1994). In a sense, these were statistical versions of the habitat suitability index models. If such models included flow or variables such as stream width that change with flow, then the models could be used for EFA. However, the models lacked generality, in that models fit to data from one stream commonly failed to make decent predictions of abundance in other streams, so this approach to EFA was largely abandoned. Probably this was a mistake; Fausch et al. (1988) pointed out many technical problems such as overfitting that could have been remedied at the time, and more suitable and powerful statistical modeling approaches, discussed in later chapters, are now available.

Meanwhile, technological developments allowed for elaboration of existing methods. Computer power increased and costs decreased, making 2-D hydraulic models practicable for habitat modeling (e.g. Leclerc et al. 1995). "Smart" instruments vastly improved data collection, and computer graphics allowed for impressive displays of modeling results. However, although computers made novel and powerful statistical methods available, they were adopted only slowly for EFA, and then often improperly, as discussed in later chapters. Ahmadi-Nedushan et al. (2006) reviewed frequentist statistical methods for evaluating habitat suitability partly "to propose new approaches using existing statistical methods." Increased computer power also made Bayesian methods more practicable (Clark 2005), but these also have been taken up only slowly, and the same was true of recursive methods such as the bootstrap (Efron and Tibshirani 1991, 1993; Williams 1996). Generally and unfortunately, EFA has been characterized by statistical backwardness.

Three new themes in EFA appeared in the 1990s: drift foraging models, mesohabitat models, and holistic approaches. Kurt Fausch (1984), Hughes and Dill (1990), and Hill and Grossman (1993) used the same basic approach to tie microhabitat selection to bioenergetics. This has become an active area of work, as discussed in Chapter 6, but to date has seen little use in actual flow assessments, although the potential for doing so was recognized early on (e.g. Baker and Coon 1997).

Mesohabitat models apply suitability curves at the scale of "hydromorphologic units," such as pools, riffles and runs. MesoHABSIM (Parasiewicz 2001) has been applied mainly in the eastern USA, but also in Europe (Parasiewicz et al. 2013). Analyzing mesohabitat use avoids some of the spatial and temporal scale problems of microhabitat models, and is intuitively more satisfactory for streams where fishes are not predominantly drift-feeders. In Europe, CA-SiMiR, a mesohabitat model using fuzzy logic, was developed by the Institute of Hydraulic Engineering at the University of Stuttgart in the early 1990s, and has since been applied in some rivers on other continents as well (Noack et al. 2013). CASiMiR also uses suitability curves with categorical values (e.g. low, medium, high) but with the wrinkle that the curves can overlap, so that a particular value of a habitat variable can correspond with one probability to one category, and with complementary probability to the adjacent category. Marsili-Libeli et al. (2013) describe another fuzzy logic model that was applied to two rivers in Italy.

Holistic approaches are based on the developing scientific consensus that maintaining a semblance of the natural pattern of seasonal flows is critical for sustaining aquatic ecosystems, and that various components of the annual hydrograph have definable ecological functions. This "natural flow paradigm" was articulated particularly by Poff et al. (1997), but the same basic idea was expressed earlier in terms of the need for channel maintenance flows and migration, spawning and rearing flows, etc. In the USA, the natural flow idea was not quickly developed into a full-blown "methodology," although a project by the Nature Conservancy, a major environmental group, developed the 'range of variation' approach, using a computer program to assess changes in flow regimes in terms of 33 statistics (Richter et al. 1996, 1997).

In Australia and South Africa, holistic methodologies based on the natural hydrograph were developed, largely by university scientists, in which a target flow regime would be defined in terms of such features as low flows, including periods of no flow, the first major flood of the wet season, medium-sized floods, etc., as described in the report of an international workshop (Arthington et al. 1992a). As the authors noted, the approach assumes that some water can be diverted without too much effect on the aquatic ecosystem, so that "…all other things being equal, the extant biota and functional integrity of the ecosystem can be maintained." However, they also emphasized that more need to be learned about stream ecosystems, and without using the term, advocated adaptive management. The authors were explicit about the need for a new approach, particularly for semi-arid streams:

> Several methodologies have been developed for assessing the environmental water requirements of which the best known and most sophisticated is the Instream Flow Incremental Methodology (IFIM) developed by the United States Fish and Wildlife Service (Bovee 1982). The physical habitat component of IFIM (PHABSIM in the USA and RHYHABSIM in New Zealand; see Jowett 1982) and other methods involving transect analysis are based on the evaluation of habitat availability for particular species at different levels of stream discharge. While such methods may be useful in relatively stable, homogenous rivers for assessing the requirements of economically important fish species such as trout, they are seldom useful at a community or ecosystem level. Thus, in South Africa and Australia, where there are relatively few economically important species, the measurement of habitat requirements is only one aspect of environmental water allocations (see papers in Ferrar 1989; Teoh 1989).
>
> In addition, semi-arid rivers typical of these southern hemisphere regions are generally extremely heterogeneous geomorphologically, and highly variable and unpredictable hydrologically (McMahon 1986, 1989; Davies and Day 1989). They cannot realistically be characterized by transect-based methods at a few points along the system. These and other theoretical objections to the Instream Flow Incremental Methodology have been review by Mathur et al. (1985), Scott and Shirvell (1987), Bain and Boltz (1989) and Arthington et al. (1992b).

In South Africa, progressive water laws enacted in the late 1990s provided the opportunity for putting holistic approaches into effect. In response, South African scientists developed the Building Block Methodology (King et al. 2008). This is a structured, bottom-up assessment process centered on a workshop with participation by scientists and engineers with a broad range of expertise, and the development of several possible flow scenarios to allow for broad participation in the process leading to a final decision. With further development, this became the DRIFT methodology (downstream response to imposed streamflow transformation; King et al. 2003, 2004), reviewed in Chapter 6.

Water is highly valuable in arid and semi-arid Australia, and most rivers were badly depleted by diversions early on. Accordingly, EFA in Australia has been associated with river rehabilitation and reducing diversions for direct human use, putting a premium on scientifically credible methods.

Various methods have been used in Australia. The Murray Flow Assessment Tool (MFAT) is a simulation model developed by the CSIRO that uses suitability curves based on expert opinion for fishes, vegetation, and water birds, and also models algal growth (Young et al. 2003). The approach is rather like PHASIM, except that the suitability curves are defined in terms of hydrological variables instead of hydraulic variables. Unfortunately, like PHABSIM, MFAT has options for combining response curves that can have substantial effects on the results, and overall it seems not to perform well. Describing a test of the tool, Lester et al. (2011a) noted that "Overall MFAT scores were poorly correlated with the fish assemblages sampled during the Sustainable Rivers Audit in the upper and lower Mitta Mitta and Yarrawonga to Edwards River catchments. This suggests that either habitat suitability is poorly defined within the model, or that habitat is not the primary driver of the measured fish assemblages." CSIRO has developed another simulation model for the Murray–Darling Basin (Yang 2010), but more recent work in the basin has turned to other approaches, such as a classification tree grouping of hydrological and biological data described by Lester et al. (2011b) and management focused on "umbrella environmental assets" described by Swirepik et al. (2015), that applies a holistic approach to selected environmentally important areas.

In the Australian state of Victoria, flow targets are generally set initially using the FLOWS methodology, described by DEPI (2013). FLOWS is based on natural flow paradigm, and considers components of the flow regime such as: cease to flow; low (base) flow; freshes and pulses; high flows; bankfull flows; overbank flows. Importantly, a hypothesis-based monitoring program was developed for selected rivers from the conceptual models relied in the assessments. As an example of hypotheses to be tested, Cottingham et al. (2005) offer that a flow pulse (e.g. equivalent to bankfull discharge) with a duration of three to four days will: mobilize and flush fine sediments from the bed substrate; scour filamentous algae and biofilm from the bed; and increase habitat diversity and availability and, ultimately, increase macroinvertebrate and fish diversity and abundance. Finally, Bayesian methods were contemplated for analyzing the monitoring data, so that flow targets could be updated. An "evidence-based" elaboration of FLOWS is described in Chapters 6 and 9.

ELOHA, short for ecological limits of hydraulic alteration (Poff et al. 2010), has been the most widely applied holistic approach to EFA. Arthington et al. (2012) describe a major test of ELOHA in Queensland, discussed in Chapter 6, and Kendy et al. (2012) describe nine case studies in the USA. According to Kendy et al. (2012, p. 5): "Major components of ELOHA include a hydrologic foundation of streamflow data, classification of natural river types, flow–ecology relationships associated with each river type, and river condition goals" that should be applied adaptively. The method is intended for regional applications, based on the idea that flow-ecology relationships can be developed for classes of rivers, so that expensive stream-by-stream assessments could be avoided (Poff et al. 2010).

In practice, applications of ELOHA seem a mixed bag; two examples, one seemingly more satisfactory than the other, are described in Chapter 6. However, it does seem clear that applying ELOHA is neither simple nor inexpensive (Richter et al. 2012), and in consequence there has been renewed interest in simple but conservative hydrological methods that can be applied widely and quickly for initial planning of water projects (Hoekstra et al. 2011; Richter et al. 2012; Caissie et al. 2015).

2.5 Conclusions

Various methods for EFA have been and still are used: professional opinion, "rules of thumb" based on hydrological data or stage–discharge relationships, computer-based micro- and mesohabitat simulation methods, statistical methods, and "holistic" methods based on preserving salient aspects of annual hydrographs. Early applications of EFA mostly involved professional opinion, although in some cases major study programs were undertaken to inform decisions regarding the development and operation of water projects. Computer-based microhabitat methods were developed in the 1970s and one of them, PHABSIM (or variants of it), quickly became popular in the USA, parts of Europe, and New Zealand, although simpler, hydrologically based methods were often used for smaller or uncontroversial smaller projects. In the twenty-first century, holistic methods, developed in Australia and South Africa, have been used increasingly, as have Bayesian methods, but simple hydrologic methods remain useful for planning studies, or where stress on streams ecosystems is low. Despite many advances in methods, predictions of the environmental effects of flow alteration are still uncertain, so adaptive management remains the best practice.

Note

[1] Gore and Nestler (1988, p. 99) noted that "Levins (1966) made the profound statement, but one often ignored by modelers, that 'population models cannot simultaneously maximize generality, realism, and precision.'" They went on to claim that "IFIM as a concept and the PHABSIM as a tool maximize generality and precision at the expense of reality." This seems off-point. An example of a population model that sacrifices realism to generality and precision, Levins cited the Volterra predator–prey equations, which seem rather different from PHABSIM. Levins thought about models differently than users of PHABSIM: he wrote that (p. 430): "The validation of a model is not that it is 'true' but that it generates good testable hypotheses relevant to important problems."

CHAPTER 3

A primer on flow in rivers and streams

Summary

Flow in streams is a result of complex processes and patterns, which result in the rate and temporal patterns of flow through a watercourse, and the spatial patterns of depth and velocity within a channel. The flow transported by the stream is a function of many characteristics of its catchment: precipitation, infiltration, subsurface flow, surface runoff, and groundwater interactions. All these are influenced in turn by climate, underlying bedrock, geomorphology, soils, vegetation cover, and land-use patterns, especially the level of impervious cover and how much of it is directly connected to receiving channels by pipes or channels.

3.1 Introduction

This chapter presents a basic and brief overview of temporal and spatial patterns of flow in streams, to provide background for thinking about environmental flows. We do not attempt a comprehensive review of all possible flow patterns, but rather illustrate the range of flow regimes in natural and regulated rivers. When evaluating flows, four general aspects need to be kept in mind.

- First, natural flow regimes vary widely from region to region and from river to river within a region. Human alterations to natural streamflows likewise range widely, from increased flashiness due to urbanization of a catchment, to complete reversal of seasonal patterns downstream of large dams. Thus, although regional experience with environmental flows can be enormously useful, it does not obviate the need to assess conditions in each stream individually.
- Second, because hydrology is usually taught and modeled from an engineering perspective, complexities in flow patterns that are biologically important may be ignored or averaged to solve problems commonly encountered in engineering, such as how large to size a culvert or how extensive overbank flooding will be.
- Third, streams carry not just water, but also carry sediment, wood, live organisms, and nutrients. All of these need to be considered for setting environmental flows.
- Finally, management of environmental flows can be constrained by artificial factors such as the capacity of reservoir outlets or the rate at which flow can be modified as water moves down a channel.

Environmental Flow Assessment: Methods and Applications, First Edition. John G. Williams, Peter B. Moyle, J. Angus Webb and G. Mathias Kondolf.
© 2019 John Wiley & Sons Ltd. Published 2019 by John Wiley & Sons Ltd.

3.2 Precipitation and runoff

Flow in streams results from precipitation. In many environments, most of the precipitation falling on a catchment will infiltrate into the soil, and most of this, roughly two-thirds, will be transpired by plants or evaporate. However, this generalization does not apply to areas of exposed clay or unfractured bedrock, where infiltration may be nil, or to basins with large areas of impermeable surfaces such as roofs and pavement (discussed below). The precipitation that runs off in direct response to a rainstorm is called stormflow, while the more stable flow in streams, sustained by subsurface drainage and the intersection of the water table with the stream channel, is called baseflow. Storm runoff in response to rain in most natural environments is dominated by subsurface flow and direct runoff from saturated zones adjacent to stream channels (Figure 3.1).

Precipitation patterns and their resulting runoff patterns vary widely, depending on climate and geology. In colder climates, snow falls in cold-weather months, accumulates, and then melts during warm spring/summer weather or in the rains of relatively warm storms. Thus, snowmelt runoff does not track precipitation closely, but is delayed. For steams draining glaciers, the delay is much greater. By contrast, runoff from rainfall typically tracks precipitation closely, unless the underlying bedrock is porous basalt or limestone. Rain can fall year-round or only seasonally, which produce perennial or seasonal flow regimes. With prolonged periods of drought between rains, as naturally occur in Mediterranean and semi-arid climates, hillslope sources can become exhausted and water tables drop, resulting in very low base-flows or dry channels. These patterns in flow regimes are important for environmental flow assessment (EFA), because native aquatic and riparian organisms have evolved to exploit or at least tolerate them.

3.3 Flow regimes

3.3.1 Describing or depicting flow regimes

Flow regimes can be described statistically or graphically. Both ways are useful, but graphical descriptions are often more informative. Different graphical descriptions emphasize different aspects of flow regimes, and so are useful for different purposes.

Hydrographs, plots of discharge against time, are a natural and common way to depict flow regimes that emphasize the seasonal pattern of flow. If flow responds directly to rainfall, it can be useful to plot hydrographs together with plots of rainfall against time (hyetographs – plotted upside down from the top of the figure), to show the relationship (Figure 3.2). Other common graphics are box plots, flow–duration curves, and flood frequency curves. Box plots are useful for showing interannual variation in the flow of a stream, or comparing the flow in multiple streams, for which hydrographs are not well suited. Flow-duration curves are useful graphical summaries of the temporal distribution of flows at a point on a stream, generally at a stream gage. The plots have percentage of time on one axis, flow values on the other, and a curve showing the percentage of time that the flow in a stream is equal to or greater than specific values on the flow axis. Commonly, the discharge values are on a logarithmic scale. Somewhat similar to flow-duration curves, flood frequency curves show the percentage of years in which a given discharge is equaled or exceeded (Figure 3.2).

Flow-duration curves and flood frequency plots summarize past flows, but are used primarily to predict future conditions. This use is based on the assumption that the factors influencing flows, the climate and conditions in the catchment, are stable, so that future flows can be estimated from past flows. Unfortunately, climate change undercuts this assumption (Milly et al. 2008), as can changes in the catchment, as discussed below. The resulting uncertainty about future flows increases the difficulty of EFA.

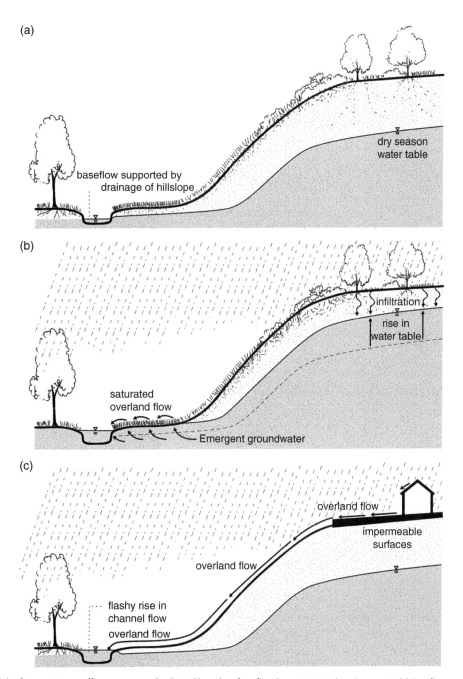

Figure 3.1 Stormwater runoff processes are dominated by subsurface flow in most natural environments. (a) Baseflow condition, in which slow drainage from the hillslope maintains flow in the stream. (b) During rainstorms, infiltrating rainwater causes the water table to rise under the hillslope, increasing the hydraulic gradient of the water table and driving groundwater into the stream. Once the floodplain area adjacent to the stream channel is saturated, it can no longer infiltrate water, so all the rain falling there runs off as overland flow. (c) If the land surface is paved over, infiltration is reduced or eliminated, and rain runs directly to the stream as overland flow. This urban runoff flows rapidly to the stream as turbulent open-channel flow, and thus reaches the stream in a fraction of the time it would take subsurface stormflow to reach it. Moreover, the urban runoff is of poor quality, having picked up contaminants from streets, parking lots, and roofs along the way (Figure prepared by Jen Natali).

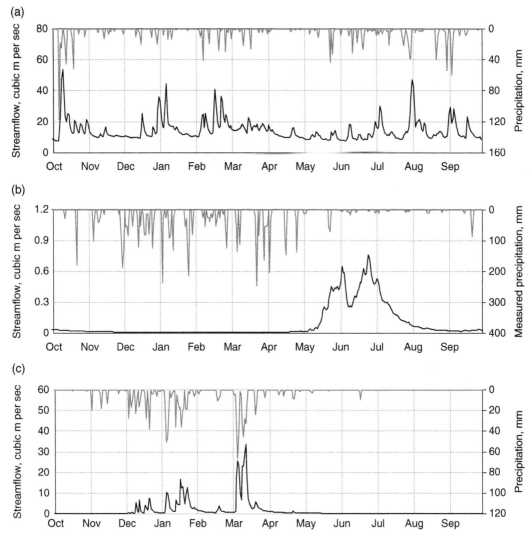

Figure 3.2 Contrasting hyetogaphs (rainfall) and hydrographs (streamflow) for: (a) humid-Atlantic climate Big Coldwater Creek near Milton, Florida (US Geological Survey gage 02370500, for water year 1971), which receives some precipitation virtually every month of the year, (b) Michigan Creek near Cameron Pass, Colorado, in the Rocky Mountains (US Geological Survey gage 06614800, for water year 2009) which displays a classic montane snowmelt hydrograph, with runoff (in warm summer months) delayed from the precipitation (from winter snows), and (c) Big Sulphur Creek near Cloverdale, California (US Geological Survey gage 11463200, for water year 2016), a Mediterranean-climate stream in the wine country north of San Francisco, is dominated by runoff in response to winter rains followed by low baseflows during the summer drought. Flow data from the US Geological Survey; precipitation data from NOAA Climate Data Center sites at Pensacola Regional Airport, FL (site USW00013899), Rustic, CO (Rustic 9 WSW, site USC00057296), and Healdsburg, CA (Healdsburg 3.7 WNW, site US1CASN0074). Precipitation is plotted from the top of the charts downward, flow from the bottom of the chart upward. (Plot courtesy of Matt Deitch, University of Florida, used by permission). Box-and-whisker plots (left) and flow duration curves (right) of mean daily flows for: (d) Big Coldwater Creek near Milton, Florida (drainage area 614 km², period of record, water years 1939-2018), (e) Michigan Creek near Cameron Pass, Colorado (drainage area 3.9 km², for water years 1973–2018), and (f) Big Sulphur Creek near Cloverdale, California (drainage area 221 km², for water years 1957–2018). Data from the US Geological Survey; plot courtesy of Jennifer Natali, University of California Berkley.

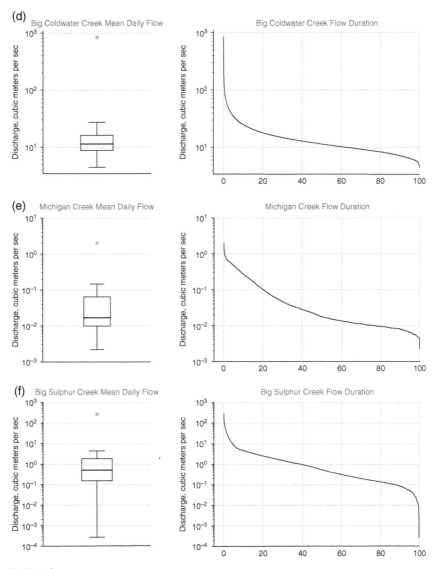

Figure 3.2 (*continued*)

3.3.2 Variation in flow regimes across climates and regions

The precipitation and runoff processes prevailing in a given region produce characteristic hydrographs, as illustrated in Figure 3.2, which contrasts hyetographs and hydrographs of humid-Atlantic climate, montane snowmelt, and Mediterranean-climate streams. The basin of Big Coldwater Creek (northwest Florida) receives some precipitation virtually every month of the year, and its hydrograph fluctuations reflect the inputs from individual storms. Michigan Creek (Colorado Rocky Mountains) displays a classic snowmelt hydrograph, with runoff (in warm summer months) delayed from the precipitation (from winter snows). Big Sulphur Creek (in the wine country north of San Francisco,

California) is dominated by runoff in response to winter rains followed by low baseflows during the summer drought, characteristic of its Mediterranean climate. In humid-Atlantic climates, such as the mid-Atlantic seaboard of the USA or the western coastal areas of northern Europe, there is rain (or soon-melted snow) year-round, which maintains strong base flows, and means that streams can experience their highest flows of the year in almost any season, derived from different types of storms. The ratio of maximum to minimum monthly flows is typically less than ten. In contrast, in Mediterranean-climate regions (e.g. the Mediterranean basin, much of California and parts of Chile, western Australia), the ratio of maximum to minimum monthly flows is commonly 20–200.

Although Mediterranean-climate regions are considered dry climates, they often receive as much precipitation as humid regions (Deitch et al. 2017). For example, Healdsburg on the Russian River in Sonoma County, California, has more annual rainfall (1094 mm, 1960–2010) than Lafayette in Tippecanoe County, Indiana (941 mm, 1960–2010). However, Indiana receives enough rainfall year-round to farm without irrigation, while Sonoma County experiences a dry summer and relies on rivers, reservoirs and groundwater to support irrigated agriculture (Deitch and Dolman 2017; Deitch et al. 2009). The high variability in seasonal discharge results from rainfall runoff, in response to nearly all annual precipitation falling in winter and spring, followed by a predictable drought period extending from summer through the fall (Gasith and Resh 1999), causing many streams to naturally dry up or contract to isolated pools. Interannual variation in flow is much greater in Mediterranean regions as well: the coefficient of variation for Healdsburg is 33.5 % compared to only 14.1 % for Lafayette. Both prolonged droughts and extreme wet years are common, driven by the El Niño Southern Oscillation (ENSO) as documented for Australia (Bond et al. 2008), California (Dettinger et al. 2011), and Chile (Waylen and Caviedes 1990).

The strong seasonality of Mediterranean-climate precipitation patterns is moderated in rivers whose drainage basins include either mountains that are sufficiently high to develop snowpack, or porous bedrock such as basalt. In the Ebro River basin of Spain, tributaries from the Pyrenees to the north have a significant snowmelt component, while tributaries draining the lower-elevation mountains in the southern part of the basin are entirely rainfall-runoff driven (Batalla et al. 2004). A similar contrast is evident in the Sacramento–San Joaquin River basin in California, where tributaries draining the Sierra Nevada (in the eastern part of the basin) have important snowmelt hydrograph components while tributaries draining the lower-elevation Coast Ranges (in the western part of the basin) are entirely rainfall-runoff fed (Figure 3.3). Even within fully rainfall-dominated Mediterranean-regions of coastal California, there are pronounced latitudinal gradients, such as Sonoma County north of San Francisco, which receives 1000 mm of annual precipitation on average, contrasted with Santa Barbara County in the south with less than 500 mm of mean annual precipitation. Flow in Battle Creek, which drains basaltic terrain on the west side of volcanic Mt. Lassen in the northern Sacramento, is moderated both by snowmelt and by geology. The maximum recorded flow in a 55-year record is less than nine times the minimum flow.

A given river can have contributions from different runoff mechanisms along its length, as illustrated in Figure 3.3 for the Merced River, California. The upper reaches of the Merced above the Pohono Bridge gage exceed 1180 m in elevation and snowmelt is the dominant runoff mechanism, while with distance downstream, runoff is dominated more and more by lower-elevation, rainfall-dominated tributaries, as reflected in the hydrograph for the gage at Snelling, at 99 m elevation (drainage area 2750 km²). In contrast, across the Central Valley from Snelling, the hydrograph of Orestima Creek is entirely rainfall-dominated, with peak flows in the winter, directly following rainfall in the basin (Figure 3.3).

The lower floodplain reaches of many temperate and especially tropical rivers overflow their channels frequently and for periods of weeks or months. Within the Sacramento and San Joaquin Rivers and

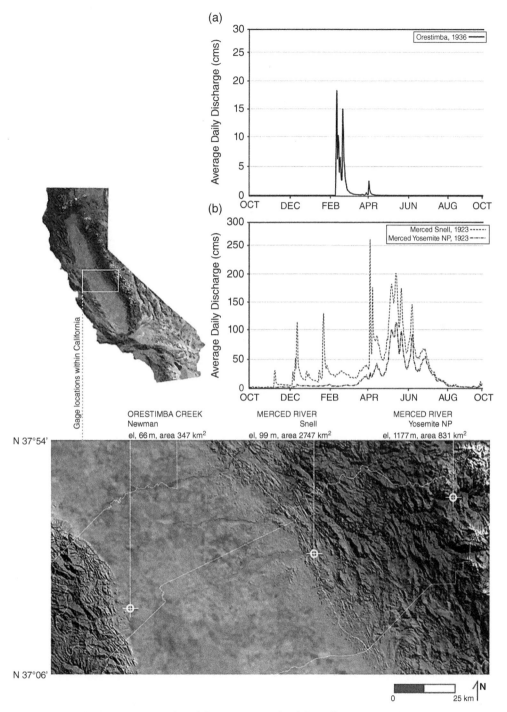

Figure 3.3 Hydrographs showing gradients from fully Mediterranean (rainfall-runoff dominated) to fully montane (snowmelt-influenced) streams in the Sacramento–San Joaquin River system of California, with (a) rainfall-dominated hydrograph for Orestimba Creek at Newman (elevation 66 m), which drains the Coast Ranges on the west side of the San Joaquin Valley; and (b) hydrographs for the Merced River at Pohono Bridge (1177 m elevation), with a snowmelt hydrograph, and at Snelling (99 m elevation), which shows the addition of rainfall-runoff components (from lower-elevation tributaries) overlain upon the snowmelt hydrograph. (c) Map shows gage locations and their relation to topography, reflecting the greater snowmelt component for the high-elevation gage and orographic effects more generally. Source: Kondolf et al. (2013), used by permission; data from U.S. Geological Survey website.

their tributaries as a whole, upper reaches might flood valley floors only rarely, but the downstream reaches of these rivers are flanked by floodplains that were historically inundated most years, for weeks or months during wet years, resulting in an "inland sea" (Kelley 1989). As discussed further below, these periods of inundation are ecologically important, e.g. for growth of juvenile fish. Similarly, the floodplain of the Apalachicola River, Florida, was naturally inundated for weeks or months, and this inundation supported high biological diversity and productivity, and through contributions of particulate organic matter and dissolved nutrients, the floodplain inundation supported the estuarine ecosystem of Apalachicola Bay, renowned for its oysters (Chanton and Lewis 2002). Extended periods of overbank flooding have been well documented on large tropical rivers, which experience a "predictable and prolonged flood pulse," to which organisms have adapted… "Because many vertebrates living in the main channel depend on the floodplain for food supply, spawning, and shelter, they have developed strategies to utilize periodically available habitats" (Junk et al. 1989).

Despite these prolonged and ecologically important periods of overbank flow along many large rivers, a conceptual model that overbank flooding occurs only every 1.5 or 2 years is widely accepted, especially among non-geomorphologists. This idea developed from research in fluvial geomorphology undertaken on temperate-climate rivers, with many seminal papers on the relations between flow and channel form (from the 1950s and 1960s) based on humid-climate and snowmelt rivers with relatively stable channels (e.g. Leopold and Maddock 1953; Wolman and Miller 1960). The flow that just fills the channel is termed the "bankfull" discharge, and while the flow needed to fill this channel has a wide range of return periods, from far less than one year to a decade or more (Williams 1978), the notion that bankfull discharge should equate to the Q1.5 is widely used by practitioners, and the Q1.5 is often conflated with the morphologic bankfull flow, even when that latter does not correspond to the former in reality. Clearly the conceptual model that

rivers stay within their banks except for relatively brief excursions overbank every 1.5–2 years does not encompass the reality of prolonged overflow onto the floodplain in downstream reaches of large rivers. The conceptual model of the Q1.5 as the dominant channel-forming discharge also does not hold for rivers in drier climates, whose form tends to be more influenced by less frequent, larger events (Hecht 1994; Wolman and Gerson 1978).

3.3.3   Anthropogenic changes in flow regimes

Climate change is likely to affect different regions differently, so generalizations are difficult, but as one effect, more precipitation will fall as rain, and less as snow, as has already been documented (Trenberth 1999). Meltwater from glaciers will also be affected. Even in areas where precipitation remains the same, warmer temperatures result in more evapotranspiration, so less water reaches streams. Unfortunately, it is more difficult to predict regional effects of climate change than to predict globally averaged effects (Deser et al. 2012), and the extent of future climate change will depend on future efforts to reduce carbon dioxide emissions.

Dams and diversions obviously affect hydrographs, but climate change and changes in land cover or groundwater pumping can do so as well. By lowering water tables, groundwater extraction commonly reduces baseflows. With increasing human disturbance resulting in devegetation, soil compaction, and land coverage with impervious surfaces (pavement, rooftops), less rain infiltrates and more runs off immediately across the land surface, as overland flow. As a result, streams become flashier, with higher peak stormflows, and reduced baseflows due to reduced recharge of the groundwater that supports baseflow. These effects are well documented in many urbanized catchments, such as Mercer Creek in western Washington state (Konrad and Booth 2002). As the catchment of Mercer Creek became urbanized, less rainfall infiltrated and more ran off directly into the creek's tributaries and, increasingly, into storm drains that were replacing and supplementing natural creek channels. The result was an increase in annual peak

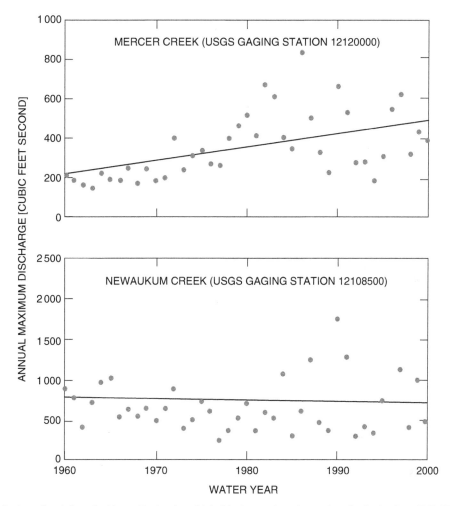

Figure 3.4 Annual peak flows for Mercer Creek, whose 31-km² basin experienced extensive urbanization from 1960–2000, and for nearby Newaukum Creek, whose 70-km² basin drains mostly rural lands. Data from the U.S. Geological Survey in original (imperial) units, ft³s⁻¹ (one ft³s⁻¹ is equivalent to 0.0283 m³s⁻¹). Source: From Konrad (2003).

flows over time, an increase that did not occur in the peak flows of Newaukum Creek, a comparably sized basin nearby draining a predominantly rural landscape over this time period (Figure 3.4). The change in runoff processes is reflected in the storm hydrographs of the two streams, in which the greater unit storm runoff in the urbanized basin is clearly visible, as is the higher unit baseflow in the rural basin (Figure 3.5).

3.3.4 Hydrologic classifications

Because hydrologic characteristics tend to vary by region, in response to climate, topography, and underlying geology (Poff et al. 2006), classifying streams by their flow regimes can be useful, and such classifications have been developed for at least 15 countries, states or other regions with the past decade alone (e.g. Snelder et al. 2009 for France, Monk et al. 2011 for Canada). As an example, Lane

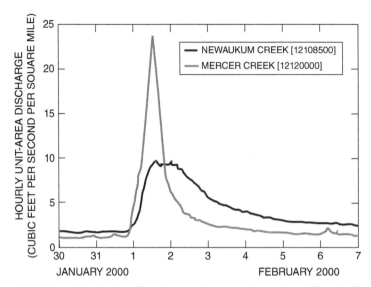

Figure 3.5   Hydrographs for Mercer (urbanized) and Newaukum creeks, expressed as unit runoff in original (imperial) units, ft³ s⁻¹ mi⁻² (equivalent to 0.0109 m³ s⁻¹ km⁻²). Not only are peak flows higher in the urbanized basin, but unit baseflows are lower as well. Source: Data from U.S. Geological Survey, from Konrad (2003).

et al. (2017) used a cluster analysis to identify eight flow classes in California: snowmelt, high-volume snowmelt and rain, low-volume snowmelt and rain, winter storms (rain), groundwater, perennial groundwater and rain, and flashy ephemeral rain. Kennard et al. (2010b) used Bayesian mixture modeling to develop a classification system for Australian streams. As discussed in Chapter 6, the ecological limits of hydrologic alteration (ELOHA) framework for EFA is intended to develop initial flow prescriptions for classes of streams, rather than individual streams, and so requires such a classification system. Hydrologic classifications have the great virtue of illustrating how different natural hydrographs can be within a given region and across regions, and when coupled with biological information they can provide an initial basis for specifying environmental flows within a context of adaptive management. However, as discussed in Chapter 5, they are human constructions, not descriptions of natural entities, and should be applied with caution.

3.4   Spatial patterns and variability within streams

The spatial patterns of depth, velocity, and substrate are important factors for the habitat of many stream fishes. Habitat simulation methods for EFA try to account for these patterns of flow within channels, which result primarily from the geometry of the channel. Flows to or from shallow groundwater are also important, especially in gravel or sand-bedded streams, but are harder to simulate and are less frequently taken into account.

3.4.1   Spatial complexity of flow within stream channels

Several common methods for EFA involve simulations of the spatial variation of depth and velocity within streams, as discussed in Chapter 6. These methods typically use flow models from engineering that do not capture the full spatial complexity of flow in stream channels, which is commonly not appreciated.

The complexity of patterns in natural channels is illustrated by the relatively tranquil Solfatara Creek, a 5-m-wide gravel-bed stream that drains 62 km² in Yellowstone National Park, Wyoming, USA. In a 20-m-long reach (coarse sand to medium gravel substrate, average channel slope 0.001), Whiting and Dietrich (1991) measured depth and velocity across 11 channel transects spaced 2 m apart, in such detail that measuring each transect took about a day. The measurements were made under steady-flow conditions at a relatively low flow (about one-third of the bankfull stage).

Despite its apparent tranquility, the velocity field was quite complex, displaying large vertical and horizontal variations within given transects as well as between closely spaced transects (Figure 3.6) (Whiting and Dietrich 1991). The large variation in channel form and velocity distributions from one transect to the next illustrates the spatial sampling problems inherent in any transect-based method for evaluating instream flows: The results will vary substantially depending on the precise location of transects. The measured velocity fields show that vertical velocity profiles often deviate substantially from the logarithmic profile commonly assumed (Figure 3.7), as has been noted elsewhere (e.g. Dingman 1989; Beebe 1996), with the highest velocities sometimes near the bed (e.g. transects 1 and 2). Although measurements of velocity at 0.6 depth (i.e. 40 % of the vertical distance above the bed) are widely used as a measure of average column velocity, based on the assumed logarithmic profile, the actual profiles measured by Whiting and Deitrich indicate that these may give a poor approximation of the true column velocity. The velocity at 0.6 depth overestimated the vertically averaged velocity in most cases (the median difference was about 6 %) but underestimated it by almost 60 % at some verticals in transect 10, where the flow deepened after passing over a mid-channel bar (Kondolf et al. 2000). In steep streams with large roughness elements, velocity profiles would be even more complex.

Details of the flow can vary in important ways even where the general patterns are similar. This is illustrated in transects 1–4, which have approximately the same shape and general lateral distributions of velocity, with higher velocity in the deeper part of the channel. Yet the velocity gradients in transects 1 and 2 were quite different from those in transects 3 and 4. In transects 1 and 2, the vertical gradient was almost nonexistent near the outside of the bend but became very steep under the high-velocity core, near the bottom. Such steep gradients do not occur in transects 3 and 4 (Figure 3.6). Such velocity gradients matter to fish (Bachman 1984; Heggenes 1994, 1996; Jenkins 1969), so such differences are important.

3.4.2 The variety of channel forms

The classic meandering river channel, with deep pools at the outside of bends and shallow riffles midpoint between the bends, occurs widely in nature but is certainly not a universal pattern in natural rivers. However, such meandering channels have become widely accepted as the model of what rivers "should" be like. There are probably many reasons for this, but part of the explanation may lie in the aesthetic preferences for curved channels (and curved forms generally) (see Kondolf 2006 for a summary). Hogarth (1753) proposed that the sinuous line was an ideal form of beauty, and such forms were found in water features constructed on estates of that era and more recently in twentieth-century river-restoration projects (Figure 3.8) (Podolak and Kondolf 2015). At a recent conference of freshwater scientists, the 308 attendees at a plenary session were given a sheet of paper with a straightened channel depicted and asked to draw a restored stream. Over 80 % drew single-thread meandering channels, most showing two repetitive meander patterns of one or one and a half wavelengths, with a mean sinuosity of 1.23; only 8 % drew multi-thread channels (Wilson et al. 2019). In many respects, the single-thread meandering channel is likely to have the simplest hydraulics of natural rivers, aside from straight channels, which are relatively infrequent in nature.

Bedrock-controlled rivers are typically steep, but in virtually all cases will have complex hydraulics by virtue of the irregularities of the bedrock outcrops

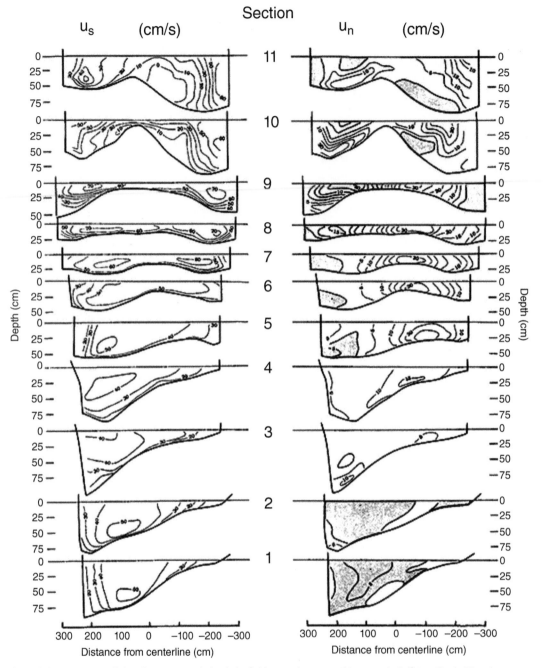

Figure 3.6 Downstream (u_s) and cross-stream (u_n) velocity fields at sections spaced 2 m apart in Solfatara Creek, Wyoming, reprinted from Whiting and Dietrich (1991). Isovels (lines of equal velocity) are at 10-cm s^{-1} intervals; shaded areas indicate flow toward the left bank. Downstream isovels range from 0 to 70 cm s^{-1}, cross-stream isovels up to 20 cm s^{-1} to the left and up to 30 cm s^{-1} to the right. The high-velocity core near the bottom in sections 1 and 2 (50 cm s^{-1} downstream) moves up and splits going over the bar in sections 7–10, with downstream velocity peaking at more than 70 cm s^{-1} in sections 8 and 9. Velocity is highest near the right bank in section 11 (60 cm s^{-1}), with a secondary maximum (>50 cm s^{-1}) forming to the left of the bar. Water close to the right side of the bar in section 11 is eddying upstream (<0 cm s^{-1}). Section numbers increase in the downstream direction. See text for site description. Source: Whiting and Dietrich (1991), used by permission; further discussion in Kondolf et al. (2000).

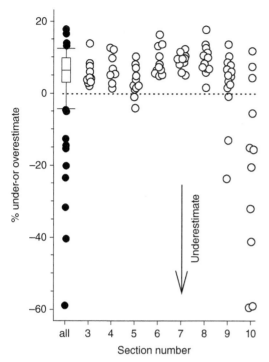

Figure 3.7 Percentage differences between the estimates of water velocity that were obtained by measuring velocity at 0.6 depth and those that were obtained by averaging the detailed velocity measurements made by Whiting and Dietrich (1991) for eight of the sections shown in Figure 3.6. Positive differences indicate that the measured velocity at 0.6 depth is greater than the average of the detailed measurements. Each circle represents one measurement vertical; the box plot summarizes the differences for all sections. Source: Kondolf et al. (2000), used by permission.

(Figure 3.9). One message from Figure 3.9 relevant to environmental flows is the breadth of the range of natural channel patterns and forms. While simplifications can be useful in understanding certain aspects of river behavior and resulting hydraulic conditions in which fish and other organisms must make their way, overly simplified models of river geomorphology can also lead us to miss important aspects of a specific, real river – aspects that may not accord with the simplified model imposed. Given the tremendous complexity of flow patterns documented in a simple meandering stream such as Sofatara Creek (which would plot toward the left-hand end of the third row of Figure 3.9), we could only expect greater complexity and unpredictability in other, more complex channels shown in Figure 3.9.

3.4.3 Lateral connectivity with floodplain and off-channel water bodies

In most mid-latitude rivers, flows stay within the stream banks except during floods, which are commonly regarded as abnormal and harmful. From an environmental perspective, however, floods and overbank flows are normal and vital for natural ecosystems (Opperman et al. 2017). As noted above, even in the mid-latitudes, regular floods can be important; before the construction of extensive levees, the annual flood of the Sacramento River from snowmelt runoff historically inundated vast areas of the valley bottom, providing valuable habitat for juvenile salmon and other animals (Williams 2006). Before construction of the Aswan High Dam, the annual flood of the Nile River was renowned for renewing fertility of the floodplain by depositing fresh sediment. As noted above, overbank flows are especially important in downstream reaches of large river systems and in tropical rivers, where floodplains can be inundated for months, and the river ecology cannot be considered in isolation from that of the floodplain (Junk 1997; Junk et al. 1989; Opperman et al. 2017). The diversity of habitats in a river system is much greater when the floodplain and off-channel water bodies are included, in addition to the main channel itself.

and bed controls. Even fully alluvial channels, i.e. streams flowing through erodible alluvium (and thus often termed 'self-formed' channels) have a wide variety of forms owing to variations in the principal governing factors of flow, sediment load and sediment caliber, and channel slope. Church (2006) developed a classification of alluvial channel features and forms based on sediment transport processes, such as distinguishing between threshold channels in which sediment transport is mediated by a surface layer of gravel, versus labile (typically sand-bed) channels in which sediments are frequently and easily mobilized

Line of Beauty

Plate 1:49

Plate 1:50

Plate IIa

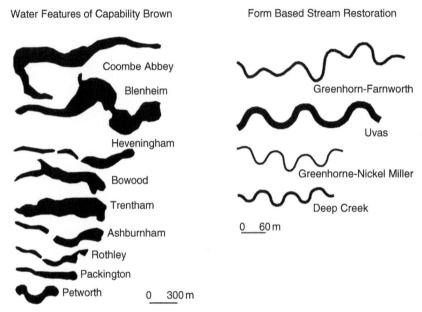

Water Features of Capability Brown

Coombe Abbey

Blenheim

Heveningham

Bowood

Trentham

Ashburnham

Rothley

Packington

Petworth

0   300 m

Form Based Stream Restoration

Greenhorn-Farnworth

Uvas

Greenhorne-Nickel Miller

Deep Creek

0   60 m

Figure 3.8 Three sets of three sets of "S" curves compared based on sinuosity and symmetry: Hogarth's (1753) "line of beauty", compared to two sets of river designs: Capability Brown's eighteenth-century river beautification projects on English estates, and twentieth-century stream-restoration designs in North America. Source: Podolak and Kondolf (2015), copyright © Landscape Research Group Ltd, reprinted by permission of Taylor & Francis Ltd, http://www.tandfonline.com on behalf of Landscape Research Group Ltd. For discussion of examples shown, see Podolak and Kondolf (2015).

Hydrologic and ecological connectivity are often considered in terms of three dimensions: longitudinal (i.e. up- and down-river), lateral (normal to the flow direction, e.g. channel–floodplain), and vertical (surface and shallow groundwater) (Figure 3.10) (Kondolf et al. 2006). The lateral connectivity is especially dependent on the flow regime, to have adequate flows to inundate the floodplain and to connect secondary channels. These periods of floodplain inundation are critically important in providing nutrients in the riverine foodweb and providing foraging for juvenile fish, as well documented on floodplains of many tropical rivers (Junk et al. 1989), as well as in the Yolo Bypass of the lower Sacramento River system

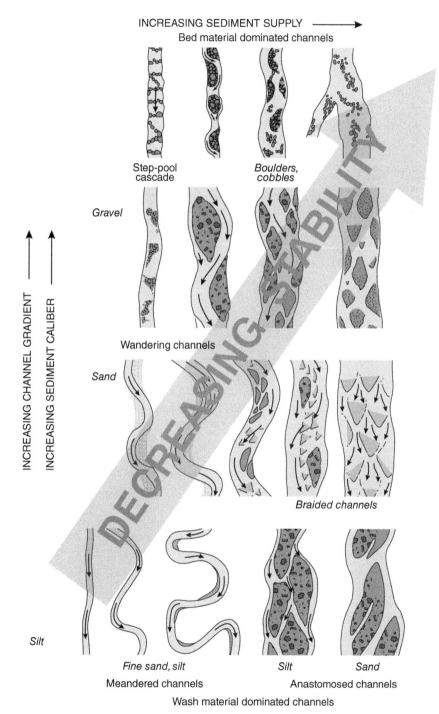

Figure 3.9 Alluvial channel form and its principal governing factors. Shading is indicative of sediment character. Source: Figure 6 in Church (2015). Reprinted by permission of Springer-Nature.

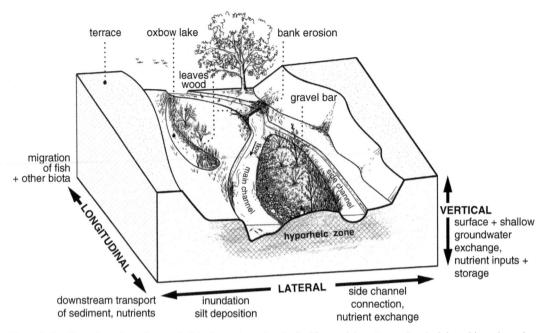

Figure 3.10 Three dimensions of connectivity in river systems: longitudinal (up- and downstream direction), lateral (e.g. channel–floodplain), and vertical (exchanges between surface and shallow groundwater). Figure prepared by Jen Natali.

(Feyrer et al. 2006) and the floodplain of the lower Cosumnes River (Jeffres et al. 2008; Opperman et al. 2017). A much-shared photograph from the latter environment shows the greater growth rates of juvenile fish on the inundated floodplain than in the main channel (Figure 3.11). Similarly, floodplain water bodies are hotspots of biological diversity in river systems, and high flows are essential to periodically connect these off-channel water bodies to the main channel (Petts and Amoros 1996; Ward et al. 1999).

Overbank flows are particularly at risk from the development of watersheds, especially from dams and levees. As the example of the Nile shows, reducing flows during the natural flood threatens critical ecological functions and services. Levees can have the same effect. Of course, there are counterbalancing benefits. Generally, the benefits from development are easier to quantify than the ecological functions and services that are put at risk, and quantifying the latter is part of the challenge of EFA.

3.4.4 Bed topography and hyporheic exchange

The importance of interchanges between surface and shallow subsurface waters has been increasingly documented within the past two decades. In gravel and sand-bed rivers, shallow subsurface waters within the interstices of coarse sediments are termed the hyporheic zone, and the organisms that inhabit this zone are termed the hyporheos. Exchanges between the hyporheic zone and surface waters can assume tremendous ecological importance, for example, to maintain favorable conditions for incubation of salmonid embryos and to maintain invertebrate communities in intermittent streams (Boulton et al. 1998). Surface-hyporheic interchanges (Kondolf et al. 2006) are encouraged by irregularities in the gravel bed surface, with the classic riffle-pool morphology inducing infiltration along the upstream end of the riffle (or "tail of the pool") (Cardenas et al. 2004). These exchanges can be critically important

Figure 3.11 Photograph showing comparative growth rates of juvenile Chinook salmon (*Oncorhynchus tshawytscha*) reared in an enclosure on the mainstem Cosumnes River (left) versus those reared in an enclosure on a vegetated part of the floodplain (right), both after 54 days in their respective habitats. Source: Jeffres et al. (2008), used by permission.

to maintaining flow of fresh, oxygenated water past incubating salmonid embryos, and also to moderating temperatures in streams subject to high daytime temperatures (Figure 3.12).

3.5 Managing environmental flows

Managing environmental flows is normally more complicated than turning the valves on the outlet structures of dams. The flows released from the dam will interact with changes in the channel form downstream (i.e. expansions, contractions, shallowing or steepening of channel gradient), seepage into and from shallow groundwater, tributary inputs (water and also sediment, nutrients, organisms, wood, contaminants, etc.), diversions into backwaters and overflow onto the floodplain, infrastructure (e.g. dams, training structures), sediment inputs such as landslides, etc. Moreover, even if such downstream influences are well documented today, some

are subject to change with time, such as tributary inputs. As a result, with distance downstream, the flows released from a dam will change in magnitude and quality, and the pattern of these changes may evolve over time. This simple fact adds a layer of complexity to any program of flow releases and should be accounted for to avoid problems such as the erroneous assumption by Sabo et al. (2017) that river stage quantified at Stung Treng on the Mekong River would hold at the confluence of the Tonle Sap River several hundred kilometers downstream (Williams 2018).

While the changes that flow releases undergo as they travel downstream in a river are important, there are also many constraints on the flows that can be released from a dam. As discussed further in Chapter 8, potential releases can be limited by factors such as the capacity of the dam's outlet structures, requirements to maintain reservoir levels for recreation or for flood control, cost of lost hydroelectric power generation, and even the requirement

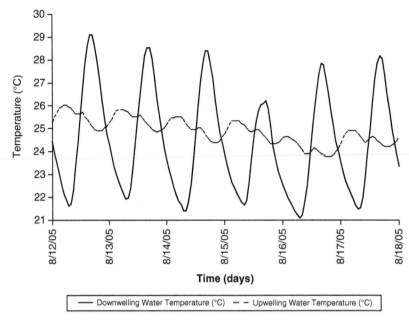

Figure 3.12   Diurnal temperature fluctuations in surface waters downwelling into a gravel riffle (solid line) and in hyporheic water emerging from the riffle gravel downstream (dashed line) for a six-day period in August 2005 near river mile 6.9 on Deer Creek, California. Source: Tompkins (2006), used by permission.

to avoid inundating endangered or threatened species that have encroached into the channel area that would naturally have been inundated by floods (Schmidt et al. 1999).

3.6   Conclusions

Flow in streams results from precipitation and run-off, both of which are temporally and spatially variable. Hydrographs characteristic of different regions reflect these runoff processes and their timing, and native species are adapted to these patterns. Over-bank flows are part of the natural flow regime of most rivers, especially tropical rivers. Thus, various hydrograph components are associated with specific geomorphic and ecological processes. Regional hydrologic classifications have been developed to try to provide a basis for assessing natural baseline flows against which impairment from dams and diversions can be measured. Within stream channels, the flow field can be complex, even in relatively simple channels with tranquil flow. Overall, flow is a much more complex phenomenon than is generally recognized. As a result hydraulic models of streams need to be applied with caution, because they commonly assume (or select) simple channel geometries, despite the wide range of natural channel forms.

CHAPTER 4

Life in and around streams

Summary

1 Rivers and streams have both upstream–downstream and lateral gradients of habitats, which affect the distribution of stream organisms.
2 Flowing waters are challenging places to live, with highly variable flows, strong unidirectional currents, and constantly changing conditions. Yet streams also support diverse organisms from bacteria and aquatic plants to fishes, herons, and otters. The diverse organisms have diverse, interactive adaptations, often distinguished by morphological features evolved for living in current, such as flattened bodies.
3 A good way to understand how fish and other animals maintain populations in difficult environments like streams is to look at how they allocate energy available to them, using a bioenergetic approach.
4 Stream organisms respond to rapid environmental change through behavioral adaptations, including microhabitat selection, dispersal, and reproduction but there are limits to how much they can adapt to human-mediated flow regimes.
5 Setting flow standards is complicated by the fact most stream organisms live in complex, locally adapted, species assemblages. Most stream fishes show a high degree of adaptability to changing conditions and to other species, but also have limits. As a result, establishing environmental flows requires visionary goals, based on solid research and monitoring.

4.1 Introduction

Rivers and streams (lotic systems) vary widely across the globe, from the Arctic to the tropics, and the mountains to the plains (Chapter 3). The plants and animals (biota) adapted these systems vary widely as well. Much of that variability has to do with the flow regime, whether natural or altered (Dodds 2002). Environmental flow assessment (EFA) is mainly about responses of the biota to changes in the flow regimes. Therefore, especially for readers who are not biologists, this chapter briefly reviews some basics of fish biology and stream ecology. Moyle and Cech (2003) and Dodds (2002, 2009) provide more detail.

Studies on the ecology of stream organisms tend to focus on "natural" streams, especially wadeable streams in the mid-latitudes. This is a biased sample.

Environmental Flow Assessment: Methods and Applications, First Edition. John G. Williams, Peter B. Moyle, J. Angus Webb and G. Mathias Kondolf.
© 2019 John Wiley & Sons Ltd. Published 2019 by John Wiley & Sons Ltd.

Another bias in our studies is that few streams and rivers in the world are without human influence. Their biota reflects that influence. Nevertheless, the variability and diversity of stream organisms means that providing environmental flows that actually favor the local or desired biota, usually fishes, requires understanding how the affected stream ecosystem works; this is not an easy task.

With these caveats, we briefly describe how river and streams are structured and adaptations of stream organisms in relation to flow. Next we discuss how assemblages of organisms in rivers and stream are organized and how that organization is affected by changes in flow regime.

4.2   Structure of stream ecosystems

Most stream ecosystems are characterized by upstream–downstream gradients of organisms that track channel gradients of flow, substrate, and water quality. For most streams, the upstream–downstream gradient can be broadly divided into a headwater erosional zone, an intermediate zone, and a depositional zone, each with a more or less distinctive biota. This zonation is reflected in the Stream Continuum conceptual model (Vannote et al. 1980), which is widely used as the basis for ecological studies, although based on temperate streams. Streams, however, are not a simple continuum of characteristics and biota, especially when the effects of tributaries, topography, and local climate are taken into account. This has led to development of many alternative models (Arthington 2012, chapter 6).

4.2.1   Across-channel gradients
Stream channels are complex, but from a habitat perspective they can be simplified into an across-channel gradient consisting of a permanent channel zone, a floodplain/riparian zone, and a hyporheic zone, where water flows (usually) through alluvial deposits underneath the channel and the floodplain. The **permanent channel zone**, with its complexity of side channels, backwaters, and in-channel structural

features is the principal habitat for fish and other aquatic vertebrates, as well as for diverse assemblages of invertebrates, including mollusks, crustaceans and aquatic insects. The **floodplain zone** is characterized by water-loving terrestrial vegetation, most prominently large trees that often form complex, highly structured plant assemblages in **riparian** (streamside) areas (Opperman et al. 2017). The riparian plants in turn support large populations of birds, mammals and other vertebrates, as well as abundant invertebrates, often adults of species with aquatic larvae. The riparian forest occupies the edges of the geomorphically active floodplain. This floodplain is a wide mosaic of dynamic habitat created by the power of flood-flows, which scour some areas, while depositing sediment in others.

When flooded, floodplains are often major feeding and spawning areas for fishes, as well as habitat for diverse waterfowl, especially during migration. Floodplains may also have many permanent features that support specialized biota, such as oxbow lakes. Floodplains associated with tropical rivers support vast forests that can be flooded for months at a time, creating extraordinary productivity that supports a high biomass of vertebrates, from fish to hippos. This productivity, long recognized for tropical rivers, is increasingly being recognized for other river systems as well. For example, the Yolo Bypass, a farmed floodplain of the Sacramento River, California, is highly favorable habitat for juvenile salmon and migratory birds when flooded during winter months, when temperatures are cool and the floodplain is not being farmed. The salmon achieve nearly maximum growth rates while residing in fields that grow rice in summer (Katz et al. 2017).

Below the floodplain and channel is the **hyporheic zone**, which supports invertebrates specialized for living a life in underground nooks and crannies. In some cases, water flows though the hyporheic zone even when surface flows have stopped. For fishes that bury their eggs, such as salmonids, the hyporheic zone is an important habitat, because survival of eggs and larvae depends on a steady flow of well-oxygenated water through the substrate.

Small fishes and invertebrates such as crayfish often burrow into the top of the hyporheic zone for cover.

It is important to recognize that these three zones, while seemingly distinct, are closely linked (Moyle and Cech 2003; Opperman et al. 2017). For example, hyporheic flow can carry nutrients from decaying salmon carcasses to the roots of riparian trees, and also carry nitrogen fixed by bacteria associated with the roots of alders to the surface stream. Adults of aquatic insects, including those emerging from deep in the hyporheic zone, are important food for riparian birds, while terrestrial insects are often important food for fish and amphibians. Fish move on to the inundated floodplain to feed on terrestrial invertebrates, as well as blooms of zooplankton present in nutrient-rich backwaters and ponds. Life in the three zones is also closely linked to the general zones in the upstream–downstream gradient.

4.2.2 Upstream–downstream gradient

We emphasize that the general along-channel gradient scheme described here, while useful conceptually, has many exceptions, such as streams that go dry in their middle reaches, streams that start in low-gradient swamps and forests, or streams that start as outflows of lakes or springs. Spring-fed streams in particular have little erosional power and so have stable substrates, resulting in a much lower diversity of organisms, but higher biomass, than streams driven by precipitation (Lusardi et al. 2016). Steady releases of cold water from deep reservoirs often make streams below dams resemble spring-fed streams; this can be a problem for designing environmental flows intended to support a high diversity of organisms.

Erosional Zone. This zone is occupied by small, high-gradient, rocky streams, usually with cold water, at least in the cooler, mountainous parts of the globe. Lower in the zone, the channel is often characterized by a regular sequence of riffles, glides, and pools, diverse habitat for the biota. As the name indicates, high flows scour the bed in this region, constantly creating "fresh" habitat. The riparian zone is typically narrow but the erosive nature of the flows

causes many trees to fall in to the stream, creating structure and cover that fish and invertebrates favor. In the northern hemisphere the characteristic fishes of this zone are trout (Salmonidae) and sculpins (Cottidae), at least in more mountainous areas. Trout are streamlined, muscular swimmers (which is why we like to catch and eat them); they typically feed on aquatic insects drifting in the water column (drift) but terrestrial insects from ants to beetles are also eaten, as are small fish like sculpins. Sculpins are small, flat, bottom-dwellers that live under rocks and logs and prey on aquatic invertebrates. They eat small trout on occasion and are eaten by bigger trout. In many trout streams, one or more species of minnow (Cyprinidae) and suckers (Catostomidae) may also be present. Studies show a high degree of ecological segregation among the species, which is reflected in their very different morphologies (Figure 4.1). In streams where headwaters are in hills, plains, and swamps rather than mountains, the erosional zone may hardly be present, replaced by muddy-bottomed streams that contain species tolerant of warm water such as green sunfish and fathead minnows. Nevertheless, most headwater streams during periods of high flow deliver nutrients, sediment, and other ecosystem building-blocks to downstream reaches (see Chapter 3).

Intermediate Zone. This is the zone that characterizes long segments of smaller rivers and tributary streams, where gradients are moderate, water is warm and often turbid, and pools and runs are the dominant physical habitat, at least in summer. Streams in this zone are also characterized by regular pool–riffle sequences where the gradient picks up. Riparian zones can vary from dense forest to open grasslands, although today most of the riparian areas are farmed or host urban areas. Flooding can be extensive, but gradients are high enough to make it very likely that water will move fairly quickly downstream, especially if hastened by levees. In eastern North America, for watercourses that are not too badly modified, the zone is home to a diversity of small fishes, mainly minnows, darters (Percidae), catfishes (Ictaluridae) and similar fishes, that

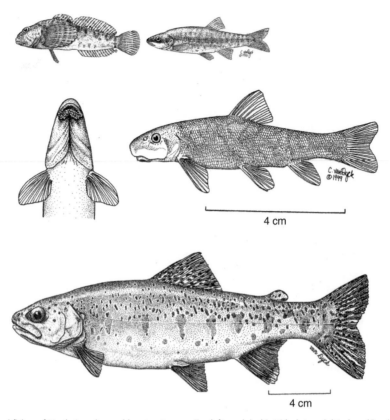

Figure 4.1 Typical fishes of North American cold-water streams. Top left, sculpin (Cottidae); top right, dace (Cyprinidae); middle, sucker (Catostomidae) showing subterminal mouth; and bottom, trout (Salmonidae). Source: Drawings by C. M. van Dyck from Moyle (2002).

vary from river to river. Often larger fishes, such as suckers (Catostomidae), make migrations up these waters for spawning and rearing, during high spring flows. In many rivers, soft-bottom-loving invertebrates such as crayfish and mussels (Unionidae) are or were abundant. Food webs are complex, and many fish and invertebrates show feeding specializations. As might be expected, the zone and its fauna merge gradually into the erosional zone above and the depositional zone below.

Depositional Zone. As the name implies, sediment carried by rivers is deposited on the valley floor as the stream spreads out and loses energy. The characteristic stream in the depositional zone is a large river, in natural conditions meandering

back and forth across a broad, forested floodplain. The water in such rivers tends to be deep, warm and turbid, flowing over sandy to muddy bottoms, often laced with beds of aquatic plants. Backwaters and oxbow lakes are among the most productive habitats. Usually, most of a region's fish fauna can be found in this zone, ranging from small insectivores and plankton feeders, to large, deep-bodied, bottom-feeding forms, such as common carp and suckers (Catostomidae), to predators such as eels (Anguillidae), bass (*Micropterus*), and catfishes. Many of the fishes are specialists, such as plankton-feeding shad (Clupeidae) and paddlefish (Polyodontidae). In tropical rivers, this region supports a high diversity of fishes, including species that migrate on and off the floodplains. Invertebrates include diverse mussels and abundant species of burrowing mayflies (Ephemeridae), as well as diverse assemblages of invertebrates living on aquatic plants. The riparian zones, where they still exist, are typically wet forests that are often flooded for months at a time, supporting many species of birds and other life.

As indicated earlier, any kind of zonation scheme, like the one above, is likely to break down under close inspection, but such schemes are useful as conceptual models. In particular, they help to explain the effects of dams on stream biota. Dams can "reset" a stream, so that the erosional zone is immediately below the dam, supporting trout, sculpin, and other cold-water fishes, often creating "tailwater" fisheries for trout. The intermediate zone reappears downstream, if sufficient water is released to support it, with a diverse fish fauna. The depositional zone remains the same, in theory. A major problem with adapting this model to a regulated river is that most sediment is captured by the dams, so there is not enough sediment in play to restore most river functions; the "erosional zone" becomes sediment-starved ("hungry water"), its rocks imbedded, locked together, so they provide little habitat for aquatic invertebrates or benthic fishes. Recovery to conditions and biota more approaching the original conditions comes only when tributaries add sediment and flows to the system. The serial discontinuity hypothesis, however, notes that multiple dams on a river system cause repeated resets, resulting in the lower sections having a low abundance and diversity of aquatic species, even if they have environmental flows (Ward and Stanford 1983).

4.3 Adaptations of stream organisms

Streams and rivers are difficult places to live, with highly variable flows, strong unidirectional currents, and constantly changing conditions, reflected in the zonation patterns. Yet streams can also be very productive, capable of supporting a high biomass of diverse organisms from bacteria to aquatic insects to fish to herons, beaver, and otters. Thus many specific adaptations exist for living in a stream environment, which can be roughly categorized as morphological, physiological, and behavioral.

4.3.1 Morphological adaptations
The usual vision of stream organisms includes
- a sleek trout, featuring a body designed to swim efficiently against or in currents.
- A frog jumping from the bank into the water to hide underneath a log, or its stout tadpoles wriggling around in a backwater.
- Swallows and dragonflies skimming the water surface, capturing adult insects for food.
- Beaver gnawing down trees and building dams to create ponds.
- Tiny blackflies coating rocks in fast water as larvae and emerging to inflict annoying bites on passing people.
- A thin coating of algae on rocks that makes them too slippery to walk on.
- A bed of aquatic plants moving gently in the current.
- The flattened larvae of mayflies and stoneflies crawling or clinging to rocks in fast water, grazing on the algae or preying on each other.
- Giant trees that shade the stream and occasionally topple into it.
- Herds of hippopotamus snorting and blowing in an African river.

This litany of stream-adapted organisms reflects the diversity of adaptations to living in, on or near streams, the diversity of habitats used by aquatic organisms, and how interconnected these habitats and organisms are. The examples are a tiny fraction of the total richness of species that are affected by stream flows, both natural and modified. If we just look at fish, for example we find complexity in form and function (Figure 4.1). Fish are usually recognizable as fish, especially the "standard" teleost fishes like a trout: a streamlined body, large eyes, two sets of paired fins, gills, a terminal mouth, and a forked tail for propelling a muscular body. Most teleost body shapes are deviations from this basic design. Tails can be rounded or square. Mouths can open below the snout or above it. Pectoral fins can be large or small, low or high on the body. Bodies can be deep, round, or flattened. Scales and fin spines are optional. The deviations in body shape reflect differences in biology, especially habitat and diet, but still assert the original template. These differences in morphology, however, represent differences in habitat use and in likely responses to changes in flow.

The most specialized morphologies are those of organisms adapted for life in fast water, such as sculpins and stonefly larvae, that have extremely flattened bodies, or organisms that attach themselves rocks in fast water, such as blackfly larvae. Any regional fish book (e.g. Moyle 2002) gives a good idea of the diversity of stream-fish body shapes, while guides to aquatic invertebrates give a good idea of their adaptations (e.g. Voshell 2002).

4.3.2   Physiological adaptations

Streams are highly variable environments, which require physiological adaptations, such as metabolic rates and environmental tolerances, on the part of the biota that lives there. Most instream organisms can tolerate limited ranges of temperature, dissolved oxygen, water clarity, and other factors, so their distributions can often be determined by knowing their tolerances and measuring the appropriate environmental variables at different locations. Stream flow (broadly defined) often determines the location of physiologically suitable habitat; cool headwaters, for example, are typically saturated with oxygen while lowland streams can have warm backwaters with low or highly variable dissolved oxygen levels. Not surprisingly the biota of these contrasting areas can be quite different, even in the same watershed. However, aquatic animals are typically mobile and sometimes appear in places that do not seem suitable for them (by our standards).

A good way to understand how animals manage to maintain populations in difficult environments is to look at how they allocate the energy available to them, using a *bioenergetic approach*. Dodds (2009)

considers this approach to be part of the ecological law that "All organisms require energy for maintenance and reproduction (p. 29)," subject to the laws of thermodynamics. This approach has a long history of being applied to fishes with the basic bioenergetic equation being presented in numerous ways. Here we follow Dodds (2009):

$$E_{\text{intake}} = E_g + E_r + E_m + E_s + E_w$$

Thus, the energy taken in by the organism is allocated to growth (g), reproduction (r), metabolism (m), storage (s), and waste (w). Energy intake beyond what is needed for metabolism and waste is sometimes called net energy intake. It is obvious that for populations to persist, some members must be able to take in enough energy to grow and reproduce successfully. However, thinking in terms of the energy budget is useful because the terms on the right-hand side of the equation are interrelated, and some of them can be negative, so that, for example, the temperature tolerance of organisms depends in part on the amount of food available, and for a given amount of food, the growth rate depends on temperature.

A simple example of how this works can be seen in juvenile coho salmon. During their stream rearing-period they do not have to allocate energy to reproduction. Therefore, their main need is to balance growth with other functions and then survive in order to migrate to sea. In the constant conditions of a laboratory, juvenile coho have an "optimal" temperature range for growth of 12–14 °C in which the conversion rate of food to fish flesh appears to be most efficient.

However, Bisson et al. (1988) observed juvenile coho salmon rearing in a small stream in Washington where maximum weekly temperatures regularly exceeded 20 °C and daily maxima sometimes reach 29 °C for short periods. The coho thrived because the salmon had essentially unlimited food and there were no competitors or predators present. The explanation for this becomes clear if survival and growth of coho salmon is put in terms of an energy budget. Basically, a juvenile salmon will grow if it ingests more energy than it consumes carrying out activities such as searching for food or avoiding predators. If it ingests less energy than it uses for daily activities it will eventually die. Part of that energetic cost can be increased metabolic rate and stress caused by temperatures higher than the optimum. In the Washington stream, conditions were so good from a bioenergetic perspective that the juvenile salmon were able to survive temperatures only slightly below their absolute lethal temperature and grow at temperatures normally considered to be too high for existence. Presumably it helped that temperatures in the stream dropped to much lower levels at night, so digestive (metabolic) efficiency was high during darkness.

Such observations can matter for EFA. Thus, in the Shasta River, California, the California Department of Fish and Wildlife determined that temperatures in the river for threatened coho salmon should be maintained at less than 15 °C through the summer, based on a literature review (Stenhouse et al. 2012). This standard reduces the amount of water available for diversion for local agriculture. An unpublished study by Robert Lusardi (UC Davis), however, showed that juvenile coho kept in cages through the river temperature gradient actually had much higher growth and survival rates at higher temperatures because of the high abundance of invertebrate prey present in the spring-fed system. This suggests that compromises are available to benefit both fish and farmers. Bioenergetic models for use in environmental flow assessments are discussed in Chapter 6.

4.3.3 Behavioral adaptations

All aquatic organisms have complex behaviors (even plants, if you regard life-history adaptations as behavior) that enable them to adjust quickly to living in streams as the environment changes. Here we focus on the behavioral adaptations of fishes, because these are the best understood.

Microhabitat selection. Flowing waters are typically high-energy environments, and living in them requires fishes and other aquatic organisms to have adaptations that allow them to either avoid "going with the flow" or to take advantage of it. The places individual organisms select are often referred to as microhabitats, with a spatial scale of a meter or less. For most of their activities, individual organisms select microhabitats that maximize their net energy intake, subject to constraints from the risk of predation. There are several ways that stream organisms accomplish this, depending particularly on their morphological adaptations.

To avoid fast water, many fishes aggregate in in deep pools, slow-flowing side channels, among fallen trees, under rocks, or, if small in size, shallow edgewater. However, drift-feeding fish take advantage of the flow, and typically select microhabitats in slowly moving water adjacent to higher-velocity flow that brings more invertebrates within striking distance. Trout, for example, minimize time spent swimming in fast-flowing water and generally prefer to "hang out" in low-velocity areas, such as eddies behind rocks, that are adjacent to high-velocity areas. Higher velocity waters deliver drifting terrestrial and aquatic insects which they capture by making quick forays into the faster water. Similarly, some insects, such as hydropsychid caddisfly larvae, live in rocky crevices and spin "nets" to collect edible matter drifting by. However, fish and other organisms also take other factors into account when selecting microhabitat, as discussed in Chapter 7. Thus, simple measurements of fish microhabitat often used in EFAs do not adequately reflect the entire environment the fish need, such as a shallow riffle upstream that produces the drifting invertebrates on which the fish are feeding, or the presence of competitors or predators.

For instance, small fishes such as sculpins (Cottidae), darters (Percidae), and many minnows (Cyprinidae) tend to avoid high velocities by living under or between rocks in fast water, where they can forage on abundant insects and algae. The complexity of factors resulting in use of this microhabitat is illustrated by the study of Baltz et al. (1982) on interactions between two fish species that both preferred to live under/among the rocks in fairly high-gradient riffles – riffle sculpin (*Cottus gulosus*), and speckled dace (*Rhinichthys osculus*). The two species competed for space under rocks, but sculpin consistently displaced dace in laboratory studies. In the field, dace were found in less-desirable edge habitat and in downstream riffles where higher temperatures excluded the sculpins. Both species have morphological, physiological, and behavioral adaptations for living in fast water, such as enlarged pectoral fins, but sculpin are more sedentary and capable of preying on small dace, so can keep dace out of "prime" foraging areas. However, the dace have much wider temperature tolerance, so can occupy habitat in warm waters, which varies in position with season and year. Thus an EFA for dace in sculpin waters, based on a microhabitat study, would assume their preferred habitat is shallow edges, not under rocks in fast water.

Other factors also influence microhabitat choice. In slow-moving waters, schooling behavior of small fishes is often present, an anti-predator defense. Schools can avoid large objects in the water that serve as ambush sites for predators, or, alternatively, hang out near dense cover, such as a submerged tree, that can provide a refuge from predators. Water clarity can also determine in part how much small fishes stay in cover; predation probabilities are lowered in turbid waters. Similarly, schools of fish can break up at night as individuals forage independently or, in the case of migrating juvenile salmon, move downstream.

Dispersal. Another problem resolved by stream organisms is dispersal, especially for reproduction. Stream flows are unidirectional, downstream, so stream organisms have to find a way back to maintain populations in upstream areas, despite velocity barriers such as waterfalls and rapids. Presumably the reason that the dominant invertebrates in streams are insects is that the adult stages can fly or be blown upstream, while juveniles can disperse downstream by drifting in the current. Many, if not most, stream fishes have upstream spawning migrations (or at least movements). Trout and salmon are famous for their lengthy migrations to spawning areas and their oily flesh is high in fuel needed to move against currents. But even for these fish, movement through fast water

is normally done in bursts and most movement tracks low-velocity areas in the stream channel. Even the legendary leaps over waterfalls typically involve individuals finding reverse eddies that can help propel them over the falls.

Reproduction. The sites and conditions for reproduction in stream organisms are a narrow subset of the total habitat and require careful consideration when environmental flows are established. For many fishes, the first requirement is typically loose gravel or rocks where spawning fish can hide eggs, often by burying them. These spawning sites serve as rearing areas for embryos, larvae, and small juveniles, which are prone to being swept downstream. For fishes, larvae and juveniles often need shallow, low-velocity edge habitat that is warmer than the main stream and rich in tiny food organisms. As they grow, fishes generally move into deeper and faster water. This is driven in part by predation; as fish grow larger they become more desirable as prey to herons, kingfishers, snakes, and other predators, and they are most vulnerable to these predators in shallow water. Again, this indicates the importance of biological interactions in determining microhabitat use.

4.4   Adapting to extreme flows

The places fishes and other organisms occupy in streams also vary with season and flow. Stream pools where many fish hold during summer are scoured by high flows in winter. During high-flow events, stream organisms have to move deep into the bottom or bank substrates, move into side channels and floodplains, or otherwise find refuge from the combination of high velocities, high silt loads, moving rocks and debris, and altered water quality. Extreme-flow events are consequently often times of high mortality of fishes and other organisms, reflected in high variation in abundances (Gido and Propst 2012). Mortality is likely to be especially high in channelized streams where there are no refuges from high flows. In contrast, spring-fed systems, which lack scouring flows, often support high densities of fish and invertebrates, if low species richness, all year around (Lusardi et al. 2016).

4.5 Synthesis

The complexity of adaptations of stream organisms means providing effective environmental flows always has to take local and regional differences in streams and their faunas into account; this is especially true for the fishes, which are often the target of EFAs. The job is made easier by the fact fishes are all adapted for living in water, a viscous medium, and this constrains their morphology and physiology to a large degree: fish of different species living in different geographic regions tend to show adaptations to their environment that are similar. In addition, the medium places overall limitations on fish (and invertebrate) design.

Thus fishes living in headwaters or rapids tend to be streamlined for swimming in fast currents (trout, salmon) or small and flattened for living under rocks to avoid being swept away (sculpin, dace). Their physiology is adapted for living in cold water that is saturated with oxygen. In contrast, fishes living in the more riverine habitats at lower elevations have more diverse body shapes, reflecting habitat diversity and less need to fight against currents. They are often deep or heavy bodied (think bass, sunfish, tench, carp). Physiologically they are adapted for warm water, often with low or variable oxygen levels. In all environments, fish are in close contact with their surroundings because water has to flow constantly across their gills, which not only take up oxygen, but salts and pollutants as well. They also have integrated sensory systems because vision is often impaired in the murky water, so it pays to have smell, touch, taste, hearing, and electrical sensory systems all working keenly together. Not surprisingly, small changes in environment quickly trigger responses in fish to move away from harmful areas or to change behavior. Such behavior needs to be taken into account in EFAs.

In short, stream fishes and their prey live in a very dynamic environment in which persistence requires morphological, physiological, behavioral, and life-history adaptations to flowing water. The dynamic environment requires dynamic responses to change at all scales and time frames, so determining the flows needed in regulated streams to maintain a diversity of stream organisms is not an easy problem to resolve. The task becomes even more difficult when assemblages of stream organisms are considered for management, especially fishes.

4.6 Environmental flows and fish assemblages

While stream fishes can choose their environments to some degree, each species is the result of a long history of adaptation and movement in a changing landscape, modulated by interactions with other fishes. The presence of a native species in a particular place can be viewed as the species having passed through a series of filters (Figure 4.2). The broadest filter is geologic history. For every broad geographic region there is a collection of potential fish species that could have been present in a given local region or watershed, but were excluded by such things as the rise of mountain ranges or shifts in the course of rivers. Within regions, zoogeographic barriers, such as waterfalls or salt water, often exclude species from potential habitats. The fish species that pass this second filter are then segregated from one another in part by physiological constraints. Deep-bodied warm-water fishes cannot live and reproduce in cold-water habitats and vice versa, resulting in the shifts in fishes from upstream to downstream or from creeks to rivers. The fishes remaining in a given region then have to adapt to one another, through competition, predation, and symbiosis. The result is the observed natural "community" (assemblage) of fishes that is found in any given place, typically made up of fishes with a diversity of origins and body shapes.

But this natural assemblage of species is further changed by human actions such as habitat alteration

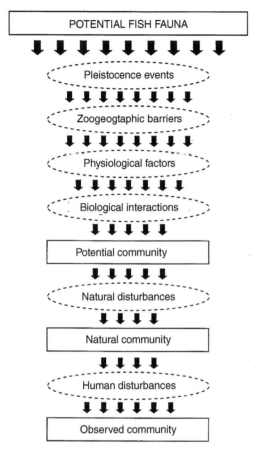

Figure 4.2 Filters (dotted boxed) that determine local fish assemblages (solid boxes) that are the focus of EFAs. Note that most contemporary fish assemblages are strongly affected by human actions, including introductions of new species, so the observed community (assemblage) is likely to be more variable than the "natural" assemblage as a result. Source: Moyle (2002), with permission from University of California Press.

and introductions of alien (non-native) fishes. Any assemblage is likely to be very dynamic as a result. Fish assemblages are rarely stable at any given location, although there may be the appearance of stability at the broad landscape level. For native fish assemblages, co-adaptation helps to promote assemblage stability, reflecting constant adaptation/evolution on the part of the fishes. Invasions of alien fishes complicate matters because native species have to adjust quickly with the tools they have (pre-existing adaptations) or be extirpated. Native species that are not extirpated in response to invasions may wind up integrated into a novel assemblage of interacting species, which can persist through time. This reflects the basic fact that most stream fishes have a long history of adapting to interactions with diverse species and to changing landscapes.

These adaptations tend to promote assemblages of fishes that are fairly predictable at the scale of most field studies (<10 years) although over the longer term the predictability diminishes. For example, Moyle and Vondracek (1985) found the fish assemblage of small California stream, Martis Creek, to be stable and highly structured after a five-year study, although two of the most abundant species were not native to the creek – brown trout (*Salmo trutta)* and rainbow trout. After 10 years of study, Strange et al. (1993) found alternative states of the assemblage existed, which depended on the effects of infrequent high-flow events. The actual structure/species composition of the assemblage at any given time was dependent on "situation-specific interactions between density-independent and density-dependent processes (p. 1)" for each species. After 30 years, Kiernan et al. (2012) found the basic fish assemblages at different sites in Martis Creek were persistent, but abundances of individual species varied widely and idiosyncratically among years, largely in response to high- and low-flow events. The alien species showed the most variation in abundance. In general, while the fishes showed some segregation by microhabitat, diet, and other variables, biotic interactions did not drive assemblage structure because of the overwhelming and differential effects of variable flows on abundances of species and life history stages.

In short, fish assemblages reflect both the geologic and recent histories of the streams in which they are found. Composition of these assemblages, an important consideration for EFAs, changes through time and space (Grossman and Moyle 1982). Changes wrought by people to streams can dramatically change the assemblage composition. Changes in flow, water quality, and channel structure can favor species usually regarded as pests, such as common carp (*Cyprinus carpio*), Japanese clams (*Corbicula fluminea)* and Eurasian watermilfoil (*Myriophyllum spicatum*), over more desirable species, usually native. Thus understanding the biology of species, assemblages, and ecosystems that you want to favor with an EFA is crucial, or unpleasant surprises will likely result.

4.7 Conclusions

Stream fishes and other organisms show remarkable abilities to adapt to changing conditions but within limits. Thus changing a flow regime usually results in changing the biota in the affected stream. The greater the change in flow regime, the greater the biotic change. This generality suggests that management flexibility should be part of any program involving EFAs, with consideration given to response to both short-term and long-term changes, including climate change. Goals that establish desirable conditions are essential. This means that understanding each affected stream and its biota through research and monitoring is necessary to maximize benefits and avoid undesirable consequences. We actually know a lot about stream ecology, and stream ecologists can make pretty good, if qualitative, predictions about the likely effects on fishes of a designed flow regime. But there is a huge need to improve these predictions through quantitative approaches, as long as the assumptions behind these approaches area acknowledged.

CHAPTER 5

Tools for environmental flow assessment

Summary

People conducting environmental flow assessments (EFAs) draw on a variety of intellectual tools, concepts and methods that have been developed mostly in science or engineering, ranging from simple graphics to classifications to complex computer models. Some of them, or their products, are parts of methods or "methodologies" for EFA. Some are more useful than others. One way or another, all of them are tools for thinking. This chapter briefly reviews various tools, methods, models, concepts, or approaches that are useful or potentially useful for EFA.

5.1 Introduction

This chapter briefly reviews various tools, methods, models, concepts or approaches that are useful or potentially useful for environmental flow assessment (EFA). Generally, they are components of methods for environmental flow assessment (EFMs). At the outset, however, we stress that the proper role of all of these is to help people think. Ultimately, successful EFA depends on clear and critical thinking; the human brain is the most important tool for EFA.

Thinking about environmental flows and conducting environmental flow assessments does not occur in a mental vacuum, but rather within a set of attitudes, assumptions, principles and scientific viewpoints that can be called a conceptual foundation (Lichatowich 1998). As examples of conceptual foundations, although they did not use the term, Frissell et al. (1997) discussed "production/exploitation" and "ecosystem/restoration" approaches to salmon management. Similarly, Bottom et al. (2005) distinguished "production thinking" and "population thinking" on the same subject (Table 5.1). Biologists, engineers, and economists are likely to bring different conceptual foundations to the task of EFA, making it harder for them to work together. Conceptual foundations matter, and it is worthwhile to make them explicit early in the assessment.

The next most important tool for EFA is the growing body of scientific knowledge of fluvial systems, the organisms that inhabit them, and the linkages among fluvial and other ecosystems. We cannot hope to do a good job of managing ecosystems or the species that depend upon them if we do not understand how they work (Anderson et al. 2006). Unfortunately, our understanding is still incomplete, as emphasized in Chapter 1.

Environmental Flow Assessment: Methods and Applications, First Edition. John G. Williams, Peter B. Moyle, J. Angus Webb and G. Mathias Kondolf.
© 2019 John Wiley & Sons Ltd. Published 2019 by John Wiley & Sons Ltd.

Table 5.1 Comparison of production thinking and population thinking in salmon management.

	Production Thinking	**Population Thinking**
Goals	Efficiency, production	Resilience, reproduction
Population Units	Arbitrarily defined	Biologically defined
Time Frame	Short	Evolutionary
Objectives	Control of survival and abundance	Conserve local populations and life-history diversity
Function of Estuary	Corridor for a single, homogenous group of salmon	Nursery area for many self-sustaining populations
Estuary Management	Control of predators, promote rapid salmon out-migration	Protect habitats of diverse life-history types

Source: After Table 2.1 in Bottom et al. (2005).

The scientific literature is the main repository of scientific knowledge. Through the process of peer review, published papers have been subject to the scrutiny of other scientists, which provides some measure of quality control. Taking a critical attitude to the literature is nevertheless important, for reasons described in a celebrated paper entitled "Why most published research findings are false" (Ioannidis 2005). Ioannidis discussed biomedical research, but most of the reasons have roots in human nature or poor statistical practice, and apply to environmental research as well. What distinguishes science is that false claims are usually corrected by subsequent work, but in the meantime practitioners and managers should take a broad and critical view of the literature, and be cautious of giving too much weight to particular studies.

Computers have played a growing role in EFA since early versions of physical habitat simulation (PHAB-SIM) were developed in the 1970s. Digital terrain data, geographic information systems, and other computer-based tools increasingly make new kinds of analyses and approaches practicable and affordable. Rapid computer processing of electronic signals has allowed the development of acoustic Doppler current profiler (ADCP) and acoustic Doppler velocimetry (ADV), which allow much more detailed measurements of velocity fields and turbulence than was possible previously, and increasing computer power has allowed drastic improvements in hydraulic or computational fluid dynamics (CFD) models,

in statistical models, in individual- or agent-based models of organisms interacting with their environments, and in dynamic energy-budget models. The internet provides access to data, literature and models that seemed unimaginable only decades ago when three of us started our scientific careers. All of these open promising new approaches for EFA. However, computer power is not a substitute for knowledge or careful thinking, and it can easily be used to no good purpose. Simply because you can do something does not make it a good idea to do so. The ease with which computers can generate "answers" underscores the importance of always asking: "Why should I believe that?"

5.2 Descriptive tools

5.2.1 Graphical tools and images

A first task in any stream assessment is getting a good description of the stream and its watershed. Old fashioned paper maps, aerial photographs and Google Satellite can be good for the purpose, and, together with existing studies of the streams in question, are the logical place to start. Analyses of older maps or photographs can provide useful information about the geomorphic context of the study and the history of any channel adjustments to existing dams and diversions. Digital terrain and other spatial data are now widely available, and allow for various GIS analyses that can be part of an EFA.

Informative graphics are important tools for communicating and for thinking (Cleveland 1985), and should be given more attention in EFAs than is common. Good software packages that make creating good graphics easy are available and should be used. The objective of each graph should be to focus attention on selected aspects of the data. Graphs that emphasize one aspect of the data tend to obscure others, so it may be appropriate to display the same data more than one way. For example, changes in commonly occurring flows can be hard to see in flow–duration curves (King and Brown 2006). Just as dense, complex prose is hard to read, so too are graphs that display too much information.

Fine-scale digital terrain data can be obtained using 3-D laser scanning (LiDAR or related methods), either remotely from airplanes, or on the ground, essentially as surveying equipment (e.g. James et al. 2007; Cavalli et al. 2008; McKean et al. 2008; Eitel et al. 2016). Lasers in the blue-green spectral region penetrate some distance into shallow, clear water, and can simultaneously generate data on the topography of the stream channel and the adjacent terrain (McKean et al. 2008; Kinzel 2009; Legleiter et al. 2016). These can be combined with georeferenced acoustic Doppler data for bed topography in deeper water, or with surveyed bed elevations. The Pacific Northwest Aquatic Monitoring Partnership has published a good survey of methods suitable for aquatic systems (Bayer and Schei 2009), but practitioners should be alert for new developments, since LiDAR and other remote sensing methods are rapidly developing and increasingly affordable. Remote-sensing methods can be used to create impressive maps and graphs that can be highly informative and will play an increasing role in EFA. Inexpensive drones now make low-altitude aerial videography and photography easy; these will also find many uses in EFA, such as producing base maps for "demonstration flow" assessments (Railsback and Kadvany 2008), or for defining sampling frames for statistically sound environmental monitoring. However, it is important to remember that these methods are only tools, and they are more likely to prove useful if you have a clear idea what questions you want them to help answer before you get the data.

5.2.2 Stream classifications

Streams differ, and people use various categories or classification systems for thinking about them. We could not think or communicate very well without such categories. Classifications of streams can be developed in various ways, based on anything from what lives there (e.g. trout streams) to the underlying substrate (e.g. bedrock or gravel-bed streams) to complicated numerical analyses using principal components and clustering algorithms (e.g. Olden and Poff 2003) or Bayesian Mixture Modeling on stream gage data (Webb et al. 2007; Kennard et al. 2010b; Sawicz et al. 2011). It is often not obvious which approach is best; the real test is how well the classification works for its intended purpose.

Classification is especially attractive for efforts to define environmental flows, or guidelines for setting environmental flows, for regional groups of streams. For example, the Ecological Limits of Hydrologic Alteration (ELOHA) approach is built on identifying types of streams for which guidance can be developed (Poff et al. 2010), and this approach has been widely applied (Kendy et al. 2012; Zorn et al. 2012). Similarly, R2 Resource Consultants (2004) classified sub-basins in the Snake River drainage to develop recommendations for environmental flows for a basin-wide stream adjudication process in Idaho. This approach makes sense; when thinking about environmental flows in a number of streams, it is natural to weigh experience in similar streams more heavily than experience in dissimilar streams.

However, it is important to keep in mind that the categories in stream classifications are human inventions, and not manifestations of some underlying natural order: "For real-world data sets, there are no 'true' classes …" (Webb et al. 2007, p. 526). Nothing about streams will generate categories as distinct as biological species, or hierarchical systems as "real" as phylogenetic trees. It is easy to forget this. For example, according to Poff et al. (2010, p. 150): "By classifying rivers according to ecologically meaningful streamflow characteristics (citations omitted), groups of similar rivers can be identified, such that within a grouping or type of river there is a range of hydrologic and ecological variation that can be considered the natural variability for that type." Since the types are defined by people, it is not clear how their variability can be "natural."

If stream classifications are developed from flow records, and ecologically relevant flow statistics are generated for the classes of stream, an initial question is how long the records need to be to yield meaningful results. An obvious answer is that this depends on the variance of the flows, and on how much uncertainty can be tolerated when the classification and statistics are used. Unfortunately, Kennard et al. (2010a) answered this question using a flawed method that underestimates the length of record needed for streams with variable flow regimes (Williams 2017). Moreover, using historical flow data to develop stream classifications that will be used to make recommendations about future flows implicitly assumes that the flow records are stationary. Given anthropogenic climate change, this is a dubious assumption (Milly et al. 2008). Although classifications can be based solely on flows (e.g. Kennard et al. 2010b), it seems questionable whether such classifications will be optimal for EFA. Flow has been called the "master variable" (Poff et al. 1997), but other attributes of streams, such as size, gradient, channel substrate, water temperature, nutrient concentrations, position with respect to migration barriers, geology and so on can also have large effects on stream ecosystems.

Problems with stream classification are of practical concern. Experience with the use (or misuse) of stream classification in stream restoration work shows that real problems can result from careless use of classification systems, especially the Rosgen (1994) system (Smith and Prestegaard 2005; Kondolf 2006; Simon et al. 2007; Elliott and Capesius 2009; Miller and Kochel 2010). Many of the difficulties involved with stream classification are described in a review by Kondolf et al. (2016) of the literature on geomorphic approaches to classifying streams. There seems

to be a human tendency to treat the types as normative, as if the attributes of streams in a given category ought to be typical for the category, or there were something somehow wrong with a stream that does not fit neatly into the classification system. There is also a tendency to assume that common attributes of a class will apply to all members of the class, which is not necessarily so: platypuses are mammals, but they do not bear live young.

Alternatively, stream classifications may be useful for directly assessing the effects on fish of changes in flow. Peterson et al. (2009) found that statistical models based on channel characteristics and discharge fit data on fish presence or abundance as well as models based on habitat characteristics. That is, they found they could by-pass the usual process of predicting changes in habitat characteristics from changes in discharge. However, the strong contrasts in the geology of their study area may account for their result; see Downes (2010) for cautions about this approach.

5.2.3 Habitat classifications

The discussion above of stream classifications also applies to classification of habitat types (riffles, pools, etc.); these constructs are probably essential and certainly useful for talking about streams, but they are to some degree arbitrary. Calling the types "physical biotopes" (Padmore 1998; Harvey and Clifford 2009) does not make them more real. The habitat classification system described by Hawkins et al. (1993) has been popular, especially in the western USA, but many others have been used as well. As with stream classifications, systems of habitat types should be assessed in terms of their utility for the task at hand, and the system of types that is optimal for any given task may vary regionally, with regional variations in geology, vegetation, flow regimes or fish fauna (Williams et al. 2004). There is often an argument for applying the classification systems widely, so that data from different areas may be more easily compared, but doing this raises the risk applying the systems where they do not work well. There may be no good resolution to this dilemma. Like channel types, habitat types have also been used to predict the presence and abundance of fish (Rosenfeld 2003). Mesohabitat simulation model (Meso-HABSIM), reviewed in Chapter 6, is a method based on abundance-environment relations developed at the spatial scale of habitat types, sometimes called mesohabitats. Thinking about habitat variables at different spatial scales is essential for understanding, and various classifications based on scale (e.g. microhabitat, mesohabitat, reach etc.) are commonly used, but again it is important to remember that these are mental constructs that we impose upon the landscape.

5.2.4 Species classifications

Unlike streams or habitats, species fit nicely into a hierarchical classification system based on evolutionary relationships, and some of these categories, such as salmonids or centrarchids, are commonly used in EFAs. However, other classification systems more similar to those for rivers and habitats are also used: sport fish, macroinvertebrates and riparian vegetation are common examples. The guild concept can be applied to species that use environmental resources in similar ways, such as macroinvertebrate shredders or grazers. Welcomme et al. (2006) developed a classification of guilds of fishes that is intended to useful for environmental assessments; it may be useful and will certainly appeal to those who like fancy names such as the eupotamonic phytophilic guild. In an extension of the guild concept, Shenton et al. (2012) defined "metaspecies," artificial species with demographic attributes thought to be representative of guilds, and used to model the likely population responses of the guilds to changes in flows. Such concepts are further examples of expert opinion determining what gets analyzed or how it gets analyzed; there is nothing wrong with this, but the reasoning should be explained.

5.2.5 Methods classifications

Environmental flow methods are also commonly classified (e.g. Tharme 2003; Annear et al. 2004), but like classifications of rivers or habitats, the classes are human inventions, and ambiguous. The Tennant Method is usually called a hydrological model (Tharme 2003) or a standard-setting model (Annear et al. 2004), but could also be called an expert opinion model, since the model criteria were based on Don Tennant's opinion of the ecological value that different percentages of the annual flow would provide. Similarly, PHABSIM is generally taken as a model of physical or hydraulic habitat (e.g. Locke et al. 2008), but is also, and is usually tested as, a species distribution model coupled with the assumption that occupancy is a good indicator of habitat quality.

5.3 Literature reviews

Literature reviews are a traditional part of science. With traditional, narrative reviews, one or a few people survey and synthesize the literature on some area of study, sometimes promoting one perspective or another (as with this book) and sometimes taking a more neutral tone. When reviews are judgmental, the assessments are subjective, and will be more or less useful, depending on the knowledge and understanding of the reviewers.

Systematic reviews are a more structured tool for getting information about the effectiveness of interventions from the scientific literature. Such reviews try to assess the support in the scientific literature for specific hypotheses, using a structured approach for extracting bits of evidence from papers, and then combing these according to rules set in advance. The approach strives to be transparent and reproducible as possible, so each step in the process is described. This kind of review is well established in some fields. For example, the Cochrane Reviews have had a substantial impact on the practice of medicine and public health, and such reviews are well suited for environmental issue as well (Sutherland et al. 2004; Pullin et al. 2009). Following the example of the Cochrane Collaboration, a non-profit organization called the Collaboration for Environmental Evidence (CEE) promotes and publishes similar reviews on the practice of environmental conservation or restoration.

Over 70 such reviews are available on the collaboration's website http://www.environmentalevidence.org.

Unfortunately, full CEE systematic reviews are too expensive and time-consuming for many environmental flow assessments, so a simplified approach known as Rapid Evidence Synthesis has emerged that seeks to maintain much of the rigor of systematic reviews, but at a fraction of the resources required (Webb 2017). One method for rapid evidence synthesis, called Eco-Evidence has been developed in Australia, and is supported by software to facilitate the reviews (https://ewater.org.au/products/ewater-toolkit/eco-tools/eco-evidence). Using Eco Evidence, Webb et al. (2013) found that a systematic review was more informative than a frequently cited narrative review (Poff and Zimmerman 2010) that used the same set of research papers, and other studies have also shown structured reviews to be a valuable tool for EFA (e.g. Webb et al. 2015b).

However, studies of ecological responses generally have weaker designs than medical studies, and given that the problems with the published literature in other fields discussed above and in Chapter 1 apply as well to the literature that informs EFA, it seems important that people scoring papers have a good understanding of statistics and experimental design.

5.4 Experiments

5.4.1 Flow experiments

Large-scale flow experiments have become an established method for EFA; in a recent review, Olden et al. (2014) listed more than a hundred, mostly involving pulse flows intended to scour channels or inundate floodplains. Some, such as flow releases into the Colorado River partly intended to restore beaches used for camping by rafters (Patten et al. 2001) have been well publicized. Other experiments have involved spawning and rearing flows, especially for salmonids; for example, Dauble et al. (2010) describe a major experiment to test migration flows for juvenile Chinook salmon in the San Joaquin River, California. Unfortunately, experience shows that doing such experiments well is not easy.

5.4.2 Laboratory experiments

The results of laboratory experiments should not be carelessly extrapolated to streams, but laboratory experiments do allow for isolating the effects of particular factors that are important in EFA, for example the effect of temperature on growth rates, in ways that are not possible in field studies. Laboratory studies can also test assumptions that are embodied in EFMs, such as the assumption that the habitat use is a good index of habitat quality, or that habitat preferences are independent of flow. As examples, in a series of experiments, Fausch (1984) showed that juvenile salmonids distributed themselves in a flume in accordance with potential energy gain and dominance rank (Figure 5.1). Although Fausch did not discuss it, Figure 5.1 also shows that density was greatest in habitat of intermediate quality, so that suitability indices developed from these data without considering rank and energy gain would have been misleading. Similarly, Holm et al. (2001) developed velocity-preference curves for juvenile Atlantic salmon in a flume at three rates of flow, and showed that that they were substantially different, in the sense that PHABSIM analyses using the curves would lead to different recommendations.

5.4.3 Thought experiments

Thought experiments, which can also be called qualitative or conceptual models, often can shed light on complicated problems, as famously demonstrated by Einstein. Some thought experiments that have been important for EFA are discussed below, and it is useful to remember that simulation models are just elaborate thought experiments (Starfield 1997; Schnute 2003). In a direct application for EFA, Armstrong and Nislow (2012) developed a conceptual model for assessing flows for Atlantic salmon and brown trout and tested it with data from a long-term study on West Brook, Massachusetts. Their result shows that a flow threshold used for flow regulation in the United Kingdom is probably too low.

IID and IFD In a fruitful thought experiment, Fretwell (1972) described the Ideal Free Distribution (IFD) and the Ideal Despotic Distribution (IDD).

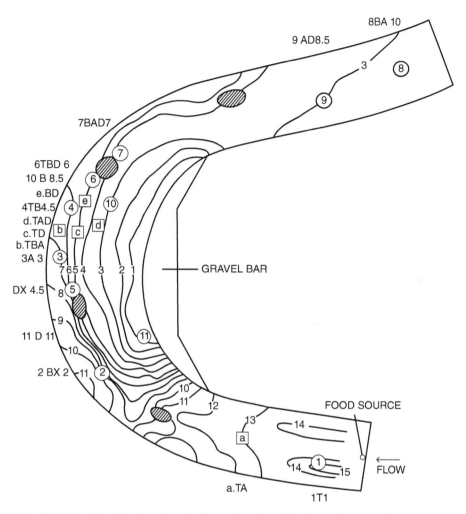

Figure 5.1 The distribution of juvenile coho salmon in a flume. There is a pool (maximum depth 15 cm) in the bend, and riffles (depth 5 cm) at either end. Contours show calculated mean potential profit (calories per hour) at focal points close to the stream bed. Hatched areas are rocks. Circles are fish positions ranked according to dominance hierarchy, and corresponding letters are fin-clip codes. Numbers preceding the fin-clips are observed dominance rank, and numbers following the codes are predicted potential profit. Squares are subordinate fish holding well above the bed that were not ranked. Source: Fausch (1984) © 2008 Canadian Science Publishing or its licensors. Reproduced with permission.

Fretwell defined the basic suitability of a patch of habitat as the expected fitness of an organism inhabiting the patch at very low population density, and assumed that the actual suitability of the patch would decline as the local population density increased. He then considered how organisms would distribute themselves if they could accurately judge the actual suitability of habitat patches, and occupy the most suitable habitat available to them. In the case of the IFD, organisms are free to occupy any patch. In this case, the population density will vary with the basic suitability of the habitat patches, but the fitness of all organisms will be approximately equal. In the IDD, on the other hand, dominant organisms exclude others from the most suitable patches, so population density does not reflect the basic suitability of habitat patches, and the fitness of members of the population will vary over patches (e.g. Figure 5.1)

Real organisms are unable to judge the suitability of available habitats with perfect accuracy, and will be more or less able to exclude others, so the ideal free and ideal despotic distributions are idealized end-members of a continuum, and real distributions of organisms will fall somewhere between them. Moreover, habitats change over time, so the distributions of habitat quality will also change. Nevertheless, this simple thought experiment has been highly influential in fish ecology (Guillermo and Healey 1999; Gibson et al. 2008; Imre et al. 2010 are examples regarding fish), and has important implications for EFMs that try to estimate habitat quality in terms of observed population distributions.

Optimal habitat ratios

Generally, the productivity of stream habitats for prey species will vary spatially, as will that for predatory species, and the distributions of productivity for the two will differ, especially if the prey species drift in the flowing water, and so are transported from one habitat patch to others. This raises the question whether there is an optimal ratio for the different kinds of habitats. A thought experiment described in Rosenfeld and Raeburn (2009) helps clarify this question (the article also describes an actual experiment involving juvenile coho salmon in an artificial stream). Juvenile coho normally occupy pools, and feed on aquatic invertebrates that mostly rear in riffles, but drift downstream (juvenile coho also eat terrestrial insects, but if the supply of these is roughly constant along the stream they can be ignored in the thought experiment). Starting from the upstream end of a riffle, the density of aquatic invertebrates in the drift will tend to increase downstream, but at declining rate, for a distance equal to the average drifting length for the invertebrates in

question. In the pool, the density of invertebrates will decrease downstream, as they are filtered from the drift by the fish. If fish abundance is limited by food, there should be some spacing of riffles and pools that is optimal for the juvenile coho.

For more complex and realistic situations, for example when there are smaller fish in riffles and larger fish in pools, or there is one species in the riffles and another in the pools, the situation gets too complex to think about without a simulation model. However, the insight remains that the ratio of habitat for invertebrates and for fish matters, and the ratio can be expected to change with flow. Better tools for dealing with this issue in EFA would be helpful, as would better data.

Percent habitat saturation (PHS)

For drift-feeding fish that defend feeding territories, a thought experiment suggests that there is some population density below which fish can find suitable territories without too much conflict, and above which they cannot. Since the size of the territories will increase with the size of the fish, that population density will depend on the size distribution of the population. Grant and Kramer (1990) proposed that PHS is a useful statistic for assessing whether populations of juvenile salmonids are limited by the area of rearing habitat, with a critical value of about 27 %, based on data from the literature. As calculated, PHS is really a length-weighted measure of density. It may need adjustment for food supply and habitat complexity, which affect territory size, and does not apply where fish are not territorial, but PHS seems like a reasonable rule of thumb for estimating whether physical habitat or some other factor is likely to limit abundance in many situations.

5.5 Long-term monitoring

Long term (≥5 years) monitoring and associated programs of study are central to adaptive management, and provide the most reliable information for testing environmental flow methods and the assumptions that underlie them; studies of Brows Beck (Elliott 1994), the South Fork Eel River (Power et al. 2008), West Brook (Davidson et al. 2010), and the Flint River (Peterson and Shea 2015) are good examples. Long-term monitoring can allow for assessing the effects of changes in flow regimes on fish populations, and in some cases for assessing the methods used to determine the changes (e.g. Kiernan et al. 2012). The scarcity of well-designed long-term monitoring programs is a serious impediment to developing better methods for EFA.

Monitoring programs traditionally measure attributes of populations, especially abundance, and attributes of habitats such as water temperature, discharge, etc. These are necessary, but attention should also be given to measuring attributes of individuals, such as growth rates or lipid content, an index of physiological condition. Detecting long-term changes in populations is difficult because of natural variability, but environmental changes are more easily detected in these organismal variables (Osenberg et al. 1994; Schmitt and Osenberg 1996; Fordie et al. 2014). The organismal variables have better statistical properties, and also provide evidence on the mechanisms by which populations may be affected.

For adaptive management, most monitoring should be linked to specific hypotheses. Basic biological monitoring is sometimes justified, especially where lack of data inhibits the development of useful hypotheses (Power et al. 2001), but there is a danger that monitoring for "status and trends" can become rote, producing data that are never thoughtfully analyzed or critically evaluated for utility. This danger can best be avoided by making data analyses that address specific questions, as well as exploratory analyses, basic parts of the monitoring program. This will also clarify also which aspects of the monitoring program are producing useful data, and which are not. Thus, monitoring should include data analysis as well as data collection.

Monitoring programs are expensive, and simulations of the monitoring, incorporating realistic amounts of uncertainty, can be an effective way to assess whether a monitoring program can provide useful answers to the questions it is designed to address, and so avoid costly mistakes (Hilborn and Mangel 1997). Sometimes monitoring can involve a census, as when all fish are automatically counted as they pass a weir, but more often monitoring involves sampling in time or space. In such cases, more information can be obtained if the monitoring program uses a probabilistic (random) design, for the reasons discussed below. Such a sampling design justifies inferences about the parts of the system that are not sampled; a design using deliberately selected samples does not.

5.6 Professional opinion

Professional opinion inevitably plays a large role in EFA, whether it is expressed in the selection and implementation of models or the analytical methods that are used, or more directly in subjective assessments of the likely response of stream ecosystems to management actions, as in the Demonstration Flow Assessment approach reviewed in Chapter 6. However, expert opinion is subject to bias and overconfidence, so there is increasing attention to the use of structured methods in getting professional opinion for assessments, often in the form of subjective probability distributions: "Because experts are human, there is simply no way to eliminate cognitive bias and overconfidence. The best one can hope to do is to work diligently to minimize its influence" (Morgan 2014, p. 7183). One way is to ask experts to estimate a most likely response, a range of plausible responses, and the probability that the true response will lie with this range; a quantitative distribution can be inferred from the answers, and used directly or as input to Bayesian methods (e.g. Webb et al. 2015a; chapter 9). However expert opinion is used, the reasoning behind the judgments should be articulated and documented, so that the judgments can be reviewed in light of new information.

5.7 Causal criteria

Much of EFA is about identifying flow–ecology relationships, but that leaves open the question whether the relationship is causal, or merely an association. This is an important but difficult question; simply trying to define causation can lead to a philosophical quagmire. Nevertheless, since EFA is part of a regulatory or management process, we can make do with a practical definition, and address the likelihood that a particular regulation or management action based on the supposed cause would have the desired result.

Showing causality even in this sense can be difficult, but is critical for supporting strong regulation and management. Epidemiologists have long struggled with similar problems, such as early attempts to show that the relationship between smoking and cancer is causal. Before the explosion of biomedical understanding in the last half century, their problems were much like the problems faced in EFA today. To help think about such problems, epidemiologists developed and debated several versions of "causal criteria" that can be used to try to sort out causal and other relationships. There are good reasons not to take these criteria too seriously (Rothman and Greenland 2005), but many epidemiologists argue that they are still useful, if only as a way to organize a weight of evidence argument for or against causality (e.g. Fedak et al. 2015). The criteria should be useful for EFA for the same purpose, and those in Table 5.2 are modified from Fedak et al. (2015) and Pickett and Wilkinson (2015). Note that, except for temporality, not all the criteria must be satisfied to make a strong argument for causality.

5.8 Statistics

Statistics, and statistical models, can be considered the science and art of getting information out of data. In a book on the bootstrap method, Efron and Tibshirani (1993) remarked that "Statistics is a subject of amazingly many uses and surprisingly

Table 5.2 "Causal criteria," or factors to consider in assessing whether a proposed flow alteration – ecological response relationship is causal.

Strength of Association	A stronger association is more likely to be causal than a weaker one (but confounding factors may make an important causal relationship seem weak).
Consistency	An association that has been shown in multiple settings is more likely to be causal.
Specificity	If an ecological response is associated with only one kind of event, the association is more likely to be causal.
Temporality	The proposed cause must precede the effect. (This criterion must be met.).
Biological gradient (dose-response)	If there is a monotonic relationship between the magnitudes of the associated factors, the association is more likely to be causal.
Plausibility	If there is a plausible argument for causality, the association is more likely to be causal.
Coherence	If several lines of evidence support causation, the association is more likely to be causal.
Consideration of alternative explanations	If there are no plausible alternative explanations, the association is more likely to be causal.
Experiment or cessation of exposure	If the proposed effect ends or lessens after the proposed cause ceases, the association is more likely to be causal.

Source: John Williams, drawing from Fedak et al. (2015) and Pickett and Wilkinson (2015).

few effective practitioners." Unfortunately, despite some exceptions, environmental flow assessment has lagged behind other fields of applied ecology in effective use of statistical methods (Ahmadi-Nedushan et al. 2006; Downes 2010; Williams 2010a), and obvious statistical problems, such as sampling without replacement in attempts to apply the bootstrap, are unfortunately common (e.g. Gard 2005; Ayllón et al. 2012). Although statistical methods are powerful, they can easily be misused (Ioannidis 2005; Bolker et al. 2009; Gelman 2015), and their utility is fundamentally constrained by the quantity and quality of information in the available data.

5.8.1 Sampling

Sampling theory is a well-developed part of statistics (e.g. Thompson 2002), and statisticians have advocated random or "probability" sampling since Neyman (1934) showed that estimates developed from stratified random sampling were more accurate than estimates from samples deliberately chosen to be "representative." Although sampling is common in EFMs, deliberate selection of "representative reaches" is the rule, and random sampling is seldom applied (Williams 2010b). Deliberate selection of representative samples is a departure from normal scientific practice, but it is unfortunately common in benthological studies (Downes 2010) as well as EFAs, and limits the quality of the information in the data. Information about the stream should be used for developing the sampling plan, but not for selecting the sample. Probability sampling does not just avoid deliberate bias; estimates from probability samples are usually more accurate than estimates from deliberate samples, and valid estimates of sampling error can only be calculated from probability samples.

Regarding the "representative reach" approach, Stevens et al. (2007, p. 12) noted that:

> One problematic issue is that a site representative of one variable is not necessarily representative of any other variable; another is that if the sites truly are representative of central tendency, then the extremes are suppressed. A major weakness of this technique is that humans fare poorly when integrating new data due to the existence of prior conceptions; this theory is supported by many experiments in cognitive psychology.

Stevens et al. (2007) also describe a method for selecting spatially balanced probability samples, called generalized random tessellation stratified (GRTS). GRTS was developed for water-quality monitoring, and seems well-suited for EFA; for example, it was used to select sites for hydraulic modeling on the Trinity River (Wright et al. 2017). Software for using GRTS is available on the web (Google "GRTS sampling"). Williams (2010b) provides general guidance for using probability sampling in EFA.

5.8.2 Sampling methods

In a few cases, such as fish passing a fish ladder, it may be possible to observe each member of a population of interest, but normally attributes of populations such as abundance, size distributions, or species compositions have to be estimated from samples. Many methods are used, such as snorkeling or other direct observation, electrofishing, various kinds of nets, etc. and which works best will depend on the situation and circumstances. Various texts and the literature provide guidance about methods, but the relative virtues of different methods are still being studied (e.g. Chamberland et al. 2014), so the methods used for an EFA should be chosen with care and attention to the recent literature. Sampling methods will entail some error, and if all the target organism are not equally susceptible to the sampling method, the samples will be biased; the errors and biases should be taken into account when the data are analyzed.

5.8.3 Hypothesis testing

Hypothesis testing in a broad sense is inherent in science and adaptive management, and as such is important for EFA, but classical hypothesis tests using p-values are generally not useful for EFA. The meaning of p-values is less clear than many suppose. According to a recent statement by the American Statistical Association, "Informally, a p-value is the probability under a specified statistical model that a statistical summary of the data (e.g. the sample mean difference between two compared groups) would be equal to or more extreme than its observed value" (Wasserstein and Lazar 2016). In typical hypothesis tests, the data of interest are compared with a null hypothesis, generally of no effect or no relationship, and if the p-value is small enough, usually less than 0.05, the result is called statistically significant and the null hypothesis is rejected.

However, almost all field studies involve more flexibility in data analysis than is consistent with the assumptions of the hypothesis test. This means that the p-values, effect sizes, and confidence intervals only apply to the data set at hand. Normally, the data will be from a sample, and the interest will be in the population from which the sample was drawn. However, when the choices in data collection and analysis from field are not specified in advance, the data do not provide a proper basis for making inferences about the population from the sample (Gelman and Loken 2013; Berry 2017). This problem is not widely appreciated, and is central to the current "replication crisis" in science (Nuzzo 2014). As noted in the American Statistical Association statement, "The widespread use of 'statistical significance' (generally interpreted as '$p < 0.05$') as a license for making a claim of a scientific finding (or implied truth) leads to considerable distortion of the scientific process." Moreover, the practical question is whether the effect is big enough to matter; showing that some model predicts whether a fish will occur in a patch of habitat better than a null model of no effect is jumping a low bar, and does not show that the model is useful for EFA, or "verified". Testing multiple hypotheses against each other, rather than against a null hypothesis, is likely to be more useful (Hilborn and Mangel 1997).

5.8.4 Model selection and averaging

As an alternative to significance tests, a better approach is to identify a set of candidate variables, and a set of statistical models, such as some kind of regression model, that include all and subsets of the variables. Then, a "model selection" procedure ranks the models based on how well they fit the data, but with a penalty term for the number of variables, since the fit will always improve as more variables are added. Burnham and Anderson (1998, 2002) popularized the Akaike Information Criterion (AIC) and a modification for small samples (AICc), but there are others such as the Bayesian Information Criterion (BIC) (Gruebber et al. 2011). Which is best in what circumstances remains a topic of debate, and none should be applied uncritically.

Model selection may be best explained with an example. Beakes et al. (2014) used the AICc to select

variables for a logistic regression model of microhabitat selection by juvenile Chinook salmon in the American River, with depth, depth squared, velocity, velocity squared, substrate size, substrate size squared, and cover as a candidate set of variables (squaring the variables with negative coefficients allows for dome-shaped responses in logistic regression). The eight top-ranking models used different sets of these variables (Table 5.3). With the AIC (or AICc, as in this case), the absolute value of the index does not matter, but smaller is better, so the models can be ranked by ΔAICc. As a rule of thumb, ΔAICc less than 4 indicates that one model is not clearly better than the other, in which case the variables from all the top-ranking models can be used, using weighted averages of the coefficients. The rationale for this "model averaging" is that the lower-ranking models contain useful information that should not be lost. Note that Model 8, which approximates the standard PHABSIM approach, ranks last. Model selection should not be applied blindly, however. It is still necessary to think carefully about the variables, and whether the selected model makes sense. It is also good practice to report the modeling results with the other sets of variables, at is done in Table 5.3.

5.8.5 Resampling algorithms

Computers now allow the use of statistical methods that use resampling algorithms such as bootstrapping (Davison and Hinkley 1997), or Monte Carlo analyses (Manly 2007). The bootstrap can be used to estimate confidence intervals for complex statistics such as PHABSIM suitability criteria (Williams 2010a; Figure 5.2). Monte Carlo procedures can be used in various kinds of simulations; Wilcock et al. (2009) describe a Monte Carlo method for estimating the uncertainty in the critical discharge for sediment transport or the cumulative transport of sediment, given assumptions about the uncertainty in parameters such as Manning's n or the critical Shields Number for the site. Random forests is a resampling method for classification (Cutler et al. 2007) that can also be used to explore model selection (Breiman 2001). Markov chain Monte Carlo

Table 5.3 AICc 95 % candidate model set and corresponding AICc score and AICc weight (W_i).

Model rank	Model parameters	AICc	ΔAICc	W_i
1	$V+V^2+D+D^2+C$	449.86	0.0	0.34
2	$D+D^2+C$	450.86	0.78	0.23
3	$V+V^2+D+D^2$	451.71	1.85	0.13
4	$D+D^2$	4.52	2.23	0.11
5	$V+V^2+D+D^2+S+S^2+C$	453.45	3.59	0.06
6	$V+V^2+C$	453.55	3.69	0.05
7	$D+D^2+S+S^2+C$	454.12	4.26	0.04
8	$V+V^2+D+D^2+S+S^2$	454.44	4.58	0.03

Source: After Table II in Beakes et al. (2014), with permission from John Wiley & Sons.
A change greater than four AICc units (ΔAICc) is evidence of model superiority. The AICc weight is a proportional measure representing the relative support estimated with AICc analysis for each competing model.
V, velocity; *D*, depth; *S*, substrate; *C*, cover.

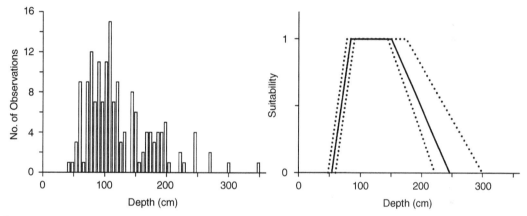

Figure 5.2 Bar graph of depth at which adult brown trout were observed (left); and a habitat suitability curve based on the observations, using data and methods in Thomas and Bovee (1993), with 95 % confidence intervals developed by bootstrapping (right). Source: John Williams.

Modeling (MCMC) is a complex resampling method used for Bayesian hierarchical modeling, which is well suited for EFA.

5.9 Modeling

Modeling is far too big a topic to cover in a brief review. The term describes a range of activities that can be categorized in various ways. For example, Hilborn and Mangel (1997), with an emphasis on ecological modeling, distinguished deterministic and stochastic models, statistical and scientific models, static and dynamic models, quantitative and qualitative models, and models used for understanding, prediction, and decision. It is also useful to distinguish simulation models, for which parameter values are taken from the literature or other sources, and estimation models, for which parameter values are estimated directly by fitting the model to the data available on the subject at hand. Here we discuss only a few points that seem particularly relevant for EFA; selecting and testing models is discussed in Chapter 7.

5.9.1 Abundance–environment relations

Abundance–environment relations (AERs) show the relationship between environmental variables such as depth or velocity and the local abundance of organisms; the habitat suitability criteria used in PHABSIM (e.g. Figure 5.2) are a good example. Most are fit to data, although some are based on expert opinion. From a statistical point of view some are more attractive than others, as discussed below. However, the inferences that can be drawn from AERs are limited, especially regarding microhabitats. In particular, it should not be assumed that the quality of habitats can be inferred from which habitats patches are occupied by the organism of interest. This seems a commonsense assumption, but reasons that it may not hold have long been known (e.g. Fretwell 1972; Van Horne 1983; Power et al. 1988), and are illustrated in Figure 5.1.

Implicitly, the approach assumes that the distribution of the animals follows the Ideal Free Distribution, discussed above; but many fish are territorial, so subordinate fish may not occupy the habitat that they would select if they could. Fish may avoid otherwise favorable habitat because of predators or competitors. Fish may be selecting habitat based on environmental features at a different spatial scale from those measured, or at several spatial scales simultaneously. In any case, the habitats that are occupied and unoccupied will depend on the abundance of fish. Moreover, habitat selection can be highly variable (Vilizzi et al. 2004), and, in addition to predators and abundance, can depend on such factors as water temperature (Hill and Grossman 1993), abundance of food (Rosenfeld et al. 2005), population density (Bult et al. 1998); habitat type (Modde and Hardy 1992); time of day (Bradford and Higgins 2001), cloud cover (Girard et al. 2003) and whether the fish is resting or feeding (Turgeon and Rodríguez 2005). For use in habitat assessment models such as PHABSIM, it is generally assumed that habitat selection will be independent of discharge, but observations, simulations and experiments contradict this (e.g. Vondracek and Longanecker 1993; Shirvell 1994; Campbell 1998; EPRI (Electric Power Research Institute) 2000; Holm et al. 2001; Heggenes 2002). Finally, it is usually assumed that managing flows to provide more habitat with features associated with greater local abundance will lead to increased populations, but the evidence for this is surprisingly slim. Populations are determined by the rates of births, deaths, and net migration, which may well be determined by other factors. Lancaster and Downes (2010a, 2010b) give a thorough analysis of the shortcomings of inferences often drawn from AERs, and a blistering response to a comment defending them (Lamouroux et al. 2010).

Resource selection functions are AERs with the property that the value taken by the function is proportional to the probability that the habitat in question will be occupied, given various assumptions described by Manly et al. (2002) and the cautions in Johnson et al. (2006). There is an enormous literature on resource selection functions or indices of habitat, mainly regarding wildlife, and for reasons that are unclear the level of statistical sophistication is generally much higher in the wildlife literature than in the freshwater fisheries literature. The same is true about testing habitat models.

Most AERs concern the central tendency in the relationships between environmental variables and the presence of the species or life stage of interest. An alternative called quantile regression (Cade and Noon 2003; Konrad et al. 2008; Knight et al. 2008) seems more useful (Figure 5.3). As noted by Lancaster and Downs (2010a, p. 389): "Given that organisms typically have wide tolerances and may not behave optimally with respect to the gradient at hand, it follows logically that AERs should be modeled as limiting responses." Essentially, the approach estimates quantiles of the distribution of the response variable over intervals of the range of the predictor variable, and fits a line to say the 95th percentile for each interval. Thus, it tries to estimate the relation between the predictor and response variables when other factors are not limiting. Quantile regression

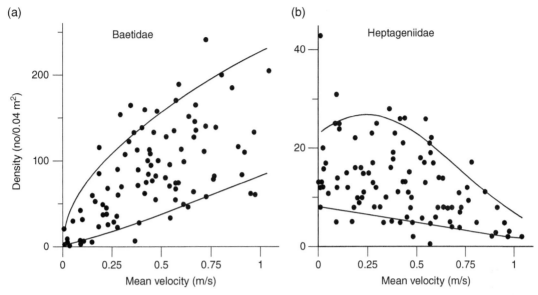

Figure 5.3 Local density of (a) baetid and (b) heptageniid mayflies across the stream bed in relation to near-bed velocity. Data were analyzed statistically as limiting relationship using quantile regression; model lines indicate the 90th and 10th quantiles (upper and lower lines, respectively). Source Lancaster and Downes (2010a) with permission from John Wiley & Sons.

has been used to assess the responses of cottonwoods to reduced floods (Wilding et al. 2014) and in several of applications of ELOHA in the eastern USA (Kendy et al. 2012), but deserves more attention in EFA. Software is available on the web.

5.9.2 Habitat association models

Habitat association models (HAMS) relate the observed presence or abundance of fish or other organisms to habitat variables at the spatial scale of patches, estimate the distribution of these patch-scale habitat features over larger areas, and use these to estimate the expected quantity of habitat for the organisms over the larger areas. With the additional assumption that the relation between habitat and fish abundance will be constant, abundance can also be estimated (Lancaster and Downes 2010a). For example, PHABSIM combines hydraulic models that estimate the amount of area with given values of microhabitat variables such as depth, velocity and so on with AERs reflecting the observed association of

the fish or invertebrate in question with microhabitat variables. Such models are reviewed in Chapter 6, as are models that estimate the fitness of fish occupying a particular patch of habitat. Many kinds of statistical models can be used for HAMs (Ahmadi-Nedushan et al. 2006), but those used for actual environmental flow assessment are typically simple or ad hoc, as with PHABSIM.

5.9.3 Drift-foraging models

Many stream fishes feed on drifting invertebrates. Drift-foraging models can be used to assess habitat quality for these fish in terms of the costs and benefits associated with living there, expressed in units of energy or of growth. Fausch (1984) developed such a model for fish in a flume where he could map flow velocity and drift density from measurements (See Figure 5.1 above). For application to streams, the bio-energetic component of the model must be augmented with a flow model and some source of input on the drift, which could be another model, estimates from data, or assumptions. Drift-foraging models have been used successfully to predict habitat selection (e.g. Gowan and Fausch 2002; Hughes et al. 2003; Hayes et al. 2007), and growth (e.g. Whitledge et al. 2010), and to assess habitat association models (Rosenfeld and Ptolemy 2012; Hayes et al. 2016). The journal *Environmental Biology of Fishes* (97(5)) published a special issue on them in 2014, including a review of practical applications (Rosenfeld et al. 2014). However, the need for information on drift limits the practical utility of these models (Naman et al. 2016).

Drift-foraging models can be components of more complex models. Individual-based models (IBM) incorporating simple drift-foraging models have been used to explore cumulative watershed effects on populations (Harvey and Railsback 2007) or the logical consequences of the behavioral rules built into the models (EPRI (Electric Power Research Institute) 2000; Railsback et al. 2003, 2005). An IBM intended for flow assessment is reviewed in Chapter 6.

5.9.4 Capability models

Models that directly estimate the abundance or biomass of fish or invertebrates as a function of habitat are sometimes called capability models (e.g. Korman et al. 1994), in contrast to suitability models such as PHABSIM. Interest in the use of such models for stream management rose with availability of computers but then declined when it became clear that models fit to data on one or a few streams gave poor predictions on others, and especially on streams in different regions (Korman et al. 1994; Fausch et al. 1988). However, this verdict seems mistaken. Overfitting (using models with too many variables) probably accounted for some of this problem, and spatial-scale issues probably contributed to the problem, as well. Fish select habitat based on features at various spatial scales, so that the importance of fine-scale features can depend on the coarser scale context, and failing to take this into account can lead to poor predictions. Hierarchical modeling (e.g. Deschénes and Rodríguez 2007) seems appropriate for predicting abundance for this reason.

Quantile regression also seems useful for capability modeling (e.g. Knight et al. 2008).

5.9.5 Bayesian networks

Bayesian networks (BNs) are quantitative models with graphical interfaces that resemble familiar "boxes and arrows" conceptual models. Mathematically, they are directed acyclic graphs. However, as implemented with available software, they also have flexible data-management capabilities and algorithms to estimate the probability that some variable will be in a particular state, depending on the state of other variables linked to it through the network. BNs were developed in the field of artificial intelligence, particularly for diagnostic tasks (e.g. what are the probabilities that a patient has one or another disease, conditional on the patient's symptoms and history), but have found application in fields ranging from environmental assessment to criminology to medicine (Marcot et al. 2001; Pourret et al. 2008; Steventon 2008). Applications of BNs to environmental assessments have mostly concerned wildlife, but have been applied to environmental flow assessments especially in Australia (Reiman et al. 2001; Hart and Pollino 2009; Stewart-Koster et al. 2010; Shenton et al. 2011; Chan et al. 2012). Appendix A of Hart and Pollino (2009) provides a good description of BNs, including their limitations. Because the models have simple graphical representations, they have proved useful and effective in group processes, including those involving stakeholders with conflicting interests (Marcot et al. 2006; Steventon 2008). Standard BNs cannot model feedback loops and do not deal well with changes over time, so they are poorly suited for modeling population dynamics (Shenton et al. 2012), but Home et al. (2018) developed a closely related "Conditional Probability Network" model that can model dynamic processes by linking back to the previous time step. Marcot (2012) describes metrics for evaluating BNs.

As an example, Shenton et al. (2011) describe the development and application of a BN model for Australian grayling and river blackfish in the highly modified Latrobe River in Victoria, Australia. The

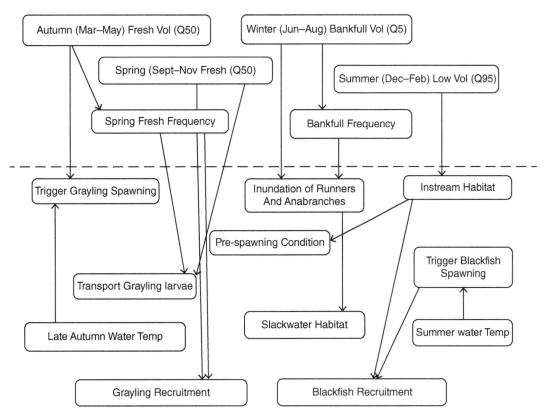

Figure 5.4 The conceptual model for the Shenton et al. (2011) Bayesian Network. Source: Shenton et al. (2011) with permission from John Wiley & Sons.

BN was based on a conceptual model (Figure 5.4) developed by the Victorian eFlows and Monitoring Assessment Program, described by Cottingham et al. (2005) and Chee et al. (2009). Briefly, a working group reviewed the literature and developed draft conceptual models for geomorphic processes, habitat processes and macroinvertebrates, aquatic and riparian vegetation, native fish spawning and recruitment, and water quality. These were circulated to others for comments, revised, and reviewed in a long workshop.

In the conceptual model embodied in the BN, selected aspects of the flow regime drive other aspects of the ecosystem that affect spawning and recruitment, which, together with mortality, determine population levels. Developing the BN model then required dividing each box (node) into discrete states, and then estimating the probability that the node would be in each state, conditional on the state of the other nodes that link to the node in question by the arrows in the conceptual model (the "conditional probability tables"). The states for the flow nodes at the top of the conceptual model were determined by analyzing and adjusting the historical flow regime to emulate natural conditions and other flow "scenarios." Thus, the conceptual model and the conditional probability tables for the light grey nodes are based directly on expert judgement, while the conditional probability tables for the dark grey nodes were estimated from "adjusted" flow data, which also incorporate expert opinion. The heavy dependence on expert opinion underscores the importance of good practice in expert elicitation, discussed above.

Webb et al. (2015a) used a simple BN completed with Netica© software in their study of the effectiveness of releases of water from storage for reducing the amount of terrestrial vegetation in stream channels after a long drought (Figure 5.5). In the figure, the states of the "parent nodes" have been fixed, so the numbers in the "child" node are the estimated probabilities that the vegetation cover will in one or another of the three states of the node (high, medium, or low), given the states of the parent nodes. Thus, instead of simply giving the most likely outcome, the BN quantifies the uncertainty in the prediction.

Bayesian network models are highly versatile. The probability values in the conditional probability tables (CPTs) can be estimated from data, from the output of other models, or from professional opinion, which makes BNs useful for integrating different kinds of information from various sources.

To illustrate "parameterizing" CPTs from data, Figure 5.6 shows a BN for the abundance of brown and rainbow trout in Martis Creek in Nevada and Placer Counties, California, developed using Netica© software and data from 30 years of monitoring by Moyle and his students. Moyle et al. (2011) defined the influence diagram (the boxes and arrows), and the CPTs were derived by the software from the data.

As appears in Figure 5.6, the BN simply summarizes the information in the data, for example that the highest flow came in spring in 27% of years. The probabilities shown for the "child" nodes (those with arrows running to them) are conditional on both the data and the model. To use the model, the user introduces a "finding," say that the mean annual flow was low. This would change the probability of that state to 100%, and the model would adjust the probabilities of the states for trout density accordingly (but not much, in this case, e.g. the probability of "high" brown trout density decreases to 32.3%). Note also

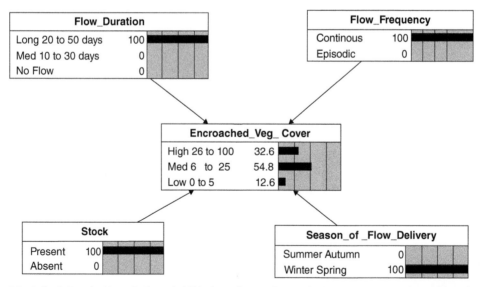

Figure 5.5 A simple Bayesian Network. The probabilities for each state of encroached vegetation cover in the "child" node are given by the conditional probability table for the node, given the values set for each of the "parent" nodes. Source: Webb et al. (2015a), with permission from John Wiley & Sons.

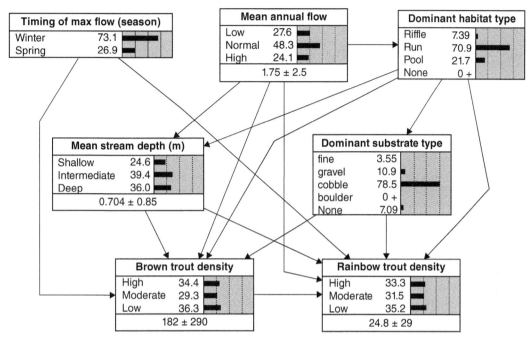

Figure 5.6 A simple Bayesian network showing factors that determine brown trout and rainbow trout density (mostly young-of-the-year) in Martis Creek, California. Conditional probability tables associated with each node (i.e. variable) in the network were derived from data collected from 1979–2008. Source: Moyle et al. (2011).

that the arrows do not necessarily imply causation, only association.

The networks can also include nodes representing decisions and utilities (costs or benefits), in which case they may be called Bayesian decision networks (BDNs). In the current context, the potential states of a decision node might be flow regimes, and the utility nodes might represent power generation, the value of irrigated crops, fish populations or the value of affected fisheries, some measure of biodiversity and so on. Importantly, the BN would not generate point estimates of these utilities, but rather, approximations of the probability distributions of the utilities.

5.9.6 Hierarchical Bayesian models

Hierarchical Bayesian (HB) modeling is a powerful analytical approach that is well suited for ecological problems (Clark 2005; Cressie et al. 2009). Besides

dealing explicitly with uncertainty, HB models fitted to data with a Markov chain Monte Carlo algorithm can be much more complex than standard statistical models, and can include parameters that must be estimated from other variables. Unfortunately, using HB models properly, or even understanding how they work, requires more statistical expertise than is common among workers in EFA (Webb et al. 2017b).

Hierarchical Bayesian models are starting to be used in EFA or closely related studies. For example, Webb et al. (2010b) used hierarchical Bayesian modeling to assess the effects of modified flow regimes on salinity in two rivers in Australia, and on a species of smelt on another. They noted in their summary that:

Inferring the effects of environmental flows is difficult with standard statistical approaches because flow-delivery programs are characterized by weak experimental design, and monitoring programs often have insufficient replication to detect ecologically significant effects. Bayesian hierarchical approaches may be more suited to the task, as they are more flexible, and allow data from multiple non-replicate sampling units (e.g. rivers) to be combined, increasing inferential strength.

In other examples, Bradford et al. (2011) used HB modeling to estimate populations of juvenile salmonids in the lower Bridge River, Grantham et al. (2012) used it to show that the survival of juvenile steelhead in a tributaries to the Russian River in California varied directly with summer flow, and Webb et al. (2015a) used it assess how well periods of inundation remove terrestrial vegetation that establishes in stream channels during severe drought.

Bayesian models incorporate existing, or "prior" information, and use data to "update" the prior information, producing "posterior" information. In Bayesian models, parameters are quantified by probability distributions, and prior distributions must be specified. Prior distributions that are based on existing information are called "informative." In the absence of information, the prior distributions will be flat, or nearly so, and are called non-informative or minimally informative. In either case, the updated or "posterior" probability distributions are proportional to the "likelihood" of the model, times the prior probability distributions. Stated differently, the probability of the model given the data is proportional to the probability of the data given the model (the likelihood) times the prior probability of the model. (Because probability distributions must integrate to one, proportionality is all that is needed.)

Hierarchical Bayesian modeling is also well suited for dealing with classes of streams, and so for ELOHA and other approaches that deal with classes of streams. Hierarchical Bayesian models would assume that the parameters describing some flow–ecology relationship may differ among rivers, but can be treated as samples from distributions. This seems less of an act of faith than to assume that the parameters are the same for all rivers in the class.

Computational requirements limited the use of Bayesian methods until recent decades (Ellison 1996), but computers continue to improve and convenient software is now readily available. McCarthy (2007) provides an accessible introduction to Bayesian modeling in ecology that introduces WinBUGS, a software package, and provides code for example problems (the name comes from Bayesian using Gibbs sampling, a Markov chain Monte Carlo approach). A freeware version, OpenBUGS, is also available, and can be called from the R statistical package. However, like other powerful tools, BUGS should be approached with due caution, and with guidance from someone with a firm grounding in statistics. Clark (2007, p. 456) observed that "Experience tells me that turning non-statisticians loose on BUGS is like giving guns to children." Of course, convenient statistical packages for conventional frequentist statistics have been widely misused as well (Ioannidis 2005; Gelman and Loken 2013), so the problem of misuse is not confined to Bayesian methods. The field of EFA desperately needs good statisticians.

5.9.7 Dynamic occupancy models

Dynamic occupancy models (MacKenzie et al. 2005, 2009) are a new approach to inform EFAs. Like traditional habitat association models, dynamic occupancy models relate the presence and abundance of organisms to conditions in patches of habitat, but can also model population dynamics, and deal with imperfect detection of the organisms in question during data collection. The basic input data are repeated samples of the organisms during multiple sampling "seasons." If applied to a single sampling "season," dynamic occupancy models would be similar to mark–recapture models, except that estimated abundance would be categorical (e.g. none, few, many) rather than numerical, although categories of condition (e.g. breeding, not breeding) can also be modeled. The models are called dynamic because they also estimate transition probabilities from each category to others over time. Habitat conditions are brought into the modeling though regressions similar to resource selection functions. Peterson and Shea (2015) applied such a model to data from three years on 42 fish species in 23 streams in the lower Flint River basin, Georgia.

5.9.8 State-dependent life-history models and dynamic energy budget models

State-dependent life-history models and dynamic energy budget models are highly generalized models rooted in basic biology that provide frameworks for considering the effects of environmental change on the life-history patterns and fitness of animals. State-dependent life-history models and dynamic energy budget models have been applied to various salmonids (e.g. Mangel and Satterthwaite 2008; Satterthwaite et al. 2009; Pecquerie et al. 2011). These models should be useful to EFA because the opportunity for growth provided by the environment is an important consideration. Because the models estimate fitness, they can be used to explore the likely evolutionary changes in life-history patterns resulting from changes in a stream such as those caused by dams, as well as shorter term effects. For example, state-dependent life-history models have been used to consider the effects of conditions in the American and Mokelumne rivers on the relative fitness of resident and anadromous life-histories for female steelhead (Satterthwaite et al. 2009). Although the basic models are highly generalized, and therefore "unrealistic," their base in fundamental biology is an important strength that has been lacking in EFA, and the models can be elaborated for particular streams for which appropriate data are available (e.g. Satterthwaite et al. 2009).

5.9.9 Hydraulic models

Changes in the computer models used to estimate flow for EFA have been as dramatic as changes in measurements of flow. Beginning in the 1970s, 1-D flow models, originally developed for estimating the surface elevation or stage of a river, were applied to EFA, notably in PHABSIM. These were relatively crude affairs that modeled the river as a set of cross-sections, and distributed vertically averaged velocity across the sections by some approximation (Kondolf et al. 2000). Two-dimensional models began to be applied in the 1990s (e.g. Leclerc et al. 1995; Ghanem et al. 1996). These estimate the cross-stream

components of the velocity as well as the downstream component, but again estimate only the vertically averaged flow. A log profile can be tacked onto the vertically averaged flow, but this can be a gross simplification, as shown below and in Chapter 3. One-dimensional flow models are still used in many flow assessment studies, but the trend is clearly toward 2-D modeling, and 3-D models can now be applied to meaningful lengths of natural channels (e.g. Shen and Diplas 2008). However, for a fixed budget, there will be a trade-off between the length of stream that can be modeled and the dimensionality of the model. Lane (1998) gives a description of the physics underlying hydraulic models that should be intelligible to anyone not too put off by equations involving partial derivatives. Some recent papers describe the hydraulic models used as "computational fluid dynamics" (CFD) models, but this seems mainly a matter of fashion.

How detailed the modeling should be will depend on the questions being asked. Shen and Diplas (2008) applied 2-D and 3-D models at two rates of flow to a short reach of the Smith River, Virginia, which is used for spawning by brown trout, and compared the results. At the base flow, some boulders in the study reach were not submerged, but they are deeply submerged at the high flow. When the boulders are submerged, there is upstream flow in the eddies behind the boulders, and a very steep velocity gradient just above the boulders. As would be expected, the 2-D model was unable to resolve these details, which seem to affect redd-site selection by brown trout in this stream. In other cases, such as a salmon spawning riffle in a large river, the velocity field may be homogenous enough horizontally that a 2-D representation is adequate. Whether this is so should be determined on a case-by-case basis, and will depend on the particular questions that the modeling is expected to help answer.

If detailed flow modeling is or will be made available, a relevant question is what to do with the information. There are various ways that it could be used. Most commonly, the information is used as input to another model, such as a habitat association model, that combines the flow information with a simple biological model to calculate some index of habitat. This index is then mapped or plotted as functions of flow; examples are reviewed in Chapter 6. Plotting the index as a function of flow is attractive because flow is normally the subject of dispute, but whether this provides useful guidance about the relationship between flow regimes and future abundance is questionable (Lancaster and Downes 2010a; Shenton et al. 2012). Or, the hydraulic model can be linked to bioenergetics or transport models (e.g. Rosenfeld and Ptolemy 2012; Anderson et al. 2013) to clarify the influence of flow on biotic aspects of habitat. Another approach is to calculate indices of the flow itself, such as the Froude Number, as discussed below. Yet other options would be to use the flow information as input into an individual-based model such as InSTREAM (Railsback et al. 2009), or as input to an expert-based approach such as DRIFT (King et al. 2003). Or, detailed modeling and measurements of the flow field in selected patches of habitat over a range of discharge could be used to help "calibrate" observers for a "demonstration flow assessment" approach.

5.9.10   Hydrological models

Reasonably long gage records are available for many streams, but for most streams they are not. In these cases, some kind of hydrological model will be needed to generate estimated discharge data or flow parameters. This has been a very active area of research, particularly since the International Association of Hydrological Sciences sponsored a 10-year initiative for Prediction in Ungauged Basins from 2003 to 2012. The results are summarized in a book (Blöschl et al. 2012) and reviewed in a long article (Hrachowitz et al. 2013), but many issues remain unresolved, so it is reasonable to expect that work in this area will continue. Alternatively, existing flow records can be extended by correlation with longer records at nearby streams, but it is important to use a method that accounts for similar errors in both records (Helsel and Hirsch 1992).

5.9.11   Temperature models

The physics of the processes that govern stream temperature are well understood, and a variety of models are available for assessing the effect of changes in discharge on temperature. Which models are most appropriate for any given flow assessment will depend on the resources that are available, including time, funding, data and modelers, and on the questions to be addressed. There is not much point to using a detailed model if appropriate input data are not available. Temperature models generally incorporate a flow model, so the possible level of detail in the temperature predictions will be limited by the flow model. Accordingly, conventional water temperature models will not be able to deal with questions about temperature variation within channels, such as thermal stratification in pools at low flows.

5.9.12   Sediment transport models

Having an estimate of the discharge needed to mobilize the bed of the stream is important for EFA for several reasons, including food chain effects (Wooton et al. 1996; Power et al. 2008), flushing fine sediment from spawning gravel (Kondolf and Wilcock 1996; chapter 8), and channel evolution. Sediment transport is also an inherently stochastic process that cannot be predicted in detail. Various models and equations are available for estimating sediment transport, but, as a practical matter, predicting the magnitude of coarse sediment transport is difficult and not highly accurate, largely because transport is a very steep function of the shear stress in excess of that need to initiate transport (Wilcock et al. 2009). Probably the best option for most EFAs is using the BAGS Excel-based software developed for the U.S. Forest Service (Pitlick et al. 2009). This software allows sediment transport at particular sites to be modeled using several equations from the professional literature. In a companion document, Wilcock et al. (2009) provide guidance for selecting appropriate equations and interpreting the output, as well as a first-rate introduction to the topic.

5.9.13 Other uses of models in EFA

Models have other uses in EFA. One is to explore the logical consequences of some set of ideas about the way something works. In a good example, EPRI (Electric Power Research Institute) (2000) explored the consequences of common ideas about fish behavior for habitat selection, and found that the modeled habitat preference varied with discharge. Another use is to embody some such set of ideas in a model, fit the model to data, and assess the ideas in terms of the fit. This works best if the fit of several such sets of ideas are compared (Hilborn and Mangel 1997; Burnham and Anderson 2002). For example, Hill and Grossman (1993) found that a model based on capture rate fit their experimental data on velocity-selection by trout better than a model based on bioenergetic considerations. Models can provide clear summaries of the empirical relations among variables, as exemplified by the various models in Elliott (1994). Models can also provide standards against which to assess data. For example, field data on the growth rate of brown trout has been compared to output from models of the growth of well-fed brown trout to assess habitat conditions in streams (Nicola and Almodóvar 2004, but see Jenkins et al. 1999).

Models can also play an important role in group processes, such as an oversight committee that represents the various interested parties in an EFA. For an example in a related kind of process, the Ecosystem Diagnosis and Treatment (EDT) model has played a significant role in the development of recovery plans for "evolutionarily significant units" of Pacific salmon that listed as either threatened or endangered under the USA Endangered Species Act. This seems odd, because the model was severely criticized by a panel of leading scientists who were recruited by the National Marine Fisheries Service to provide scientific guidance for the overall salmon recovery effort. The EDT model is a complex simulation model, and according to the panel "The inclusion of so much detail may create an unjustified sense of accuracy, but actually introduces sources of inaccuracy, uncertainty, and error propagation"

RSRP (Recovery Science Review Panel (RSRP) (2000, p. 6). Nevertheless, the EDT was used for developing many of the recovery plans. Apparently, the EDT allowed the parties developing the recovery plans to explore the projected consequences of the recovery measures favored by one party or another, and the appeal of this supposed ability overrode any concerns about "inaccuracy, uncertainty, and error propagation" (Williams 2006). However, Bayesian network models can also serve this purpose (Marcot et al. 2006; Steventon 2008), and are more defensible scientifically.

More generally, the proper purpose of biological or ecological models in EFA is to help people think, not to provide answers. As stated by Walters (1986, p. 45): "The value of modeling in fields like biology has not been to make precise predictions, but rather to provide clear caricatures of nature against which to test and expand experiences." Similarly Piketty (2015, p. 70) wrote recently about economics that "Models can contribute to clarifying logical relationships between assumptions and conclusions but only by oversimplifying the real world to an extreme point. Models can play a useful role but only if one does not overestimate the meaning of this kind of abstract operation." To help people think, models have to focus on selected aspects of the problem at hand, which means that other aspects will be neglected. Ignoring this can be disastrous, as exemplified by the economic crisis of 2007–2008. The world of credit default swaps was built on highly sophisticated models that persuaded many intelligent people that the associated risk was negligible, but they failed to recognize that a market based on houses that people could not pay for from their earnings was unsustainable.

5.10 Hydraulic habitat indices

Resource selection functions or habitat suitability criteria are developed from measurements of physical characteristics of the stream at points occupied and not occupied by fish. As an alternative, indices can be estimated directly from measurements or model estimates of characteristics of the flow. For example, the Froude Number is a dimensionless number that has been used as an index of habitat by various authors (e.g. Lamouroux and Souchon 2002; Doyle et al. 2005; Moir et al. 2006). The Froude Number is calculated as velocity divided by the square root of the acceleration of gravity multiplied by depth, but this is vague unless the depth and velocity to be used are specified. Typically, the numbers are calculated from the vertically averaged velocity and the depth at a point in a stream or across a transect (e.g. Jowett 1982), and in the case of a point, a Froude Number of 1 marks an important transition in the physical behavior of the flow. Alternatively, Enders et al. (2009) defined a "bed Froude Number," using as the velocity the velocity measured 10 cm above the bed, and Lamouroux and Capra (2002) calculated a Froude Number using reach-averaged depth and velocity. The physical meaning of the number defined in these ways is unclear. Other dimensionless hydraulic variables, such as the Reynolds Number, have also been used (e.g. Lamouroux and Souchon 2002).

The shear stress on the streambed and the "shear velocity" are other potential indices of the flow field, particularly for benthic organisms. The shear stress is defined as the dynamic viscosity of the fluid times the velocity gradient, and the shear velocity is the square root of the shear stress divided by the density of the fluid. However, at the length scale of benthic organisms, the hydraulic environment on the bed is extremely patchy (Robertson et al. 1997), so these indices should be interpreted with caution.

Crowder and Diplas (2002) suggested using as indices the vorticity of the flow at a point, and especially the integral of the vorticity over a small area, called the circulation, which can be calculated from the output of 2-D or 3-D hydrodynamic models and mapped, together with flow vectors. This facilitates visualization of the structure of eddies in the flow. Although the equations defining circulation may seem daunting to most people involved in EFAs, the basic idea is not that difficult, and modern methods

for visualization of computed fields should help people understand how the flow field changes with discharge.

If flows are measured with velocimeters, other indices such as the "turbulent kinetic energy" (Smith et al. 2006) can be calculated from turbulent fluctuations in the flow. Because flow in natural streams is normally turbulent, the flow at a point in a stream at a given discharge has a velocity that includes downstream, cross-stream, and vertical components that can be decomposed into means and deviations from the mean. Enders et al. (2009) used the standard deviation of stream-wise velocity, turbulent kinetic energy, and shear stress, as well as the bed Froude Number. Harvey and Clifford (2009) used the standard deviation and skewness of the turbulent residuals as well as turbulent intensity. Roy et al. (2010) calculated 14 indices, used a complex statistical analysis to define a smaller set of uncorrelated variables and then partition the fluctuations among six spatial scales. Kerr et al. (2016) report that an index of hydraulic drag that accounts for vertical and lateral turbulence as well as mean velocity is a useful proxy for the energetic costs of holding position.

Fine-scale studies of turbulence should be useful for designing fish passage facilities (Lacey et al. 2012), but it seems too soon to tell whether any of these indices will prove useful for EFA or for predicting the future abundance of fish. Froude Numbers or other indices can be used to develop HSC or as independent variables in resource selection functions, but these would suffer the same problems as others. For example, Enders et al. (2009) found that juvenile Atlantic salmon tended to favor areas with low bed Froude Numbers, but variability among individuals was considerable, and may have resulted from other factors such as food supply that could override any preference for hydraulic habitat. Enders et al. (2009, p. 1825) opined that, "Whereas Atlantic salmon parr seem to react and respond to turbulence on a smaller microhabitat scale, the application of dynamic variables on a reach scale seem to not provide a useful tool for fisheries and habitat managers." Roy et al. (2010) concluded their abstract by noting that "Further research should attempt to link the spatial scales of turbulent flow variability to benthic organism patchiness and fish habitat use." Even at a microhabitat scale, the relation between habitat use and any of the indices should be treated as a hypothesis, particularly because the results may depend on the spatial scale of the measurements (Tullos et al. 2016)

5.11 Hydrological indices

Since EFA is about flow, flow statistics are obvious tools to use, and the relevant question becomes, which ones? Olden and Poff (2003, p. 111) noted that "In recent years the development and application of indices describing hydrological conditions of streams and rivers has exploded in the literature, resulting in dramatic shift from a paucity of indices in the past to the plethora of indices now available," so the choice is not obvious.

Tennant (1976), in a highly cited paper, used data from 58 sites on ten streams in Montana, Wyoming and Nebraska to claim that for most streams, 10 % of the annual average flow was the minimum needed "to sustain short term survival habitat for most aquatic life forms;" 30 % as a base flow would "sustain good survival," and 60 % as a base flow would "provide excellent to outstanding habitat." Environmental agencies in England, Wales, and Scotland now use the flow that is exceeded 95 % of the time (Q95) as a criterion for "special caution" when considering diversions (Armstrong and Nislow 2012), although Armstrong and Nislow argue persuasively that it is too low. Caissie et al. (2015) compared six such methods using flow records from streams in the maritime provinces of Canada, and found that, for their region, methods based on the mean annual flow were generally more protective than standards based on flow exceedance statistics. These methods may seem simplistic, but Caissie et al. (2015) assert that "Based on current knowledge, hydrologically based methods are as good as any other environmental flow approaches (e.g. hydraulic rating and habitat preference methods) In fact, none of the environmental flow methods have been developed on the basis of tested relationships between flow regime alteration and ecological responses."

The Indicators of Hydrologic Alteration (IHA) method (Richter et al. 1996) uses 33 indices, selected with an eye toward the effects of hydropower operations on flow; convenient software for doing the analysis is available from the Nature Conservancy. Olden and Poff (2003) found 171 indices reported in 13 papers, generally describing the magnitude, frequency, duration and rate of change of flow. The indices tend to be highly correlated, and Olden and Poff (2003) found that they could account for most of the variation in the indices using just a few synthetic variables defined by principal components analysis, using data from 420 gages on relatively unmodified streams. These synthetic variables are uncorrelated, which is a virtue for statistical analysis, but they do not seem to be useful aides to thinking, since they cannot be visualized. Using the IHA indices or selecting indices based on the particular situation at hand seems preferable.

As an alternative to indices, Stewart-Koster et al. (2014) have suggested functional linear models, which in effect use the whole hydrograph over some period as the predictor variable in a regression model. This has attractive features, but at the cost of considerable mathematical complexity. And, as with other types of regression modeling, the quality of the result will depend on the quality and quantity of data on the response variable. Whether the method will be useful in practice remains to be seen.

5.12 Conclusions

Tools available for EFA vary widely in nature and in effectiveness. Some are more difficult and expensive to apply than others, so the human and financial resources available for the assessment will affect which should be used. The most effective tools make good use of available information, avoid dubious assumptions, and communicate uncertainty clearly. Bayesian methods are most likely to be optimal when much information is available or for adaptive management. Simpler hydrological methods may be appropriate for initial planning of water projects or when stakes are low. Habitat association models applied at the microhabitat scale should be avoided, especially for territorial fishes.

CHAPTER 6

Environmental flow methods

Summary

Methods of many different kinds have been developed for environmental flow assessment, as well as frameworks within which various methods are combined. The methods can be classified in various ways, such as hydrological, hydraulic rating, habitat assessment, holistic, and combinations thereof; as top-down or bottom-up methods; as sample-based or whole-river methods; as standard setting or incremental methods; as microhabitat, mesohabitat, or whole-stream methods; and as opinion-based and model-based methods. In the past, named methods dealt mainly with fish, although the effects of water projects on other aspects of the environment were often considered at the same time. More recent practice has tended toward including these other considerations into a named framework for EFA. Different methods are appropriate for different situations, but some should simply be abandoned.

6.1 Introduction

There is a plethora of methods for environmental flow assessment: Tharme (2003) counted 207. Doubtless she missed some, and new ones have been developed since. Differences among rivers and streams and the circumstances of the assessments are two reasons for the number of methods. The shortcomings of all of them is another. This review does not try to be comprehensive, but rather discusses examples of the main themes.

Environmental flow methods can be classified in various ways. Tharme (2003) distinguished "hydrological, hydraulic rating, habitat simulation and holistic methods, with a further two categories representing combination-type and other approaches." Methods can also be distinguished as top-down or bottom-up methods, as sample-based or whole-river methods, as standard setting or incremental methods, as microhabitat, mesohabitat, or as whole-stream methods, opinion-based and model-based methods, and in other ways as well. There can be ambiguity between the terms "method" and "model," and methods can have more than one name: Gopal (2013) presents tables that sort them out. None of these classifications is entirely satisfactory; for example holistic methods seem to us to be analytical frameworks for methods rather than methods per se, and we discuss them as such. Here, we briefly describe Tharme's four general categories of methods, and then go into more detail on methods that are, or should be, frequently used, using the

Environmental Flow Assessment: Methods and Applications, First Edition. John G. Williams, Peter B. Moyle,
J. Angus Webb and G. Mathias Kondolf.
© 2019 John Wiley & Sons Ltd. Published 2019 by John Wiley & Sons Ltd.

first three categories from Tharme (2003) to organize the discussion. Then, we discuss four analytical frameworks for environmental flow methods, IFIM, DRIFT, ELOHA, and adaptive management, within which methods may be implemented, and discuss some recent examples in some detail. Linnansaari et al. (2012), Gopal (2013), and Armanini et al. (2015) provide recent and comprehensive review for readers who want more information or different points of view.

6.1.1 Hydrologic, habitat rating, habitat simulation, and holistic methods

Hydrological methods focus on flow per se, and assume that the biology will follow along. For example, in the Maritime Provinces of Canada, 25 % of the mean annual flow has been a traditional criterion for assessing proposed diversions (Caissie et al. 2015). Hydraulic rating methods examine the rate at which stream width, depth, or velocity change as discharge changes, and use inflection points in these curves for criteria. Habitat simulation methods estimate the habitat value of patches of habitat as a function of flow, usually in terms of depth, velocity and substrate size or cover. By summing over patches at each modeled discharge, the methods produce a curve of habitat value over discharge. Usually, these methods combine a hydraulic model with a simple habitat-selection model, particularly when they simulate "microhabitat," but some use more complex models to estimate the potential energetic benefit for drift-feeding fish (usually trout). Holistic methods also deal with the flow regime, but instead of statistics, consider parts of the annual hydrograph that are thought to have important ecological functions: "In a holistic methodology, important and/or critical flow events are identified in terms of select criteria defining flow variability, for some or all major components or attributes of the riverine ecosystem" (Tharme 2003, p. 402). For example, the "functional flow" approach proposed by Yarnell et al. (2015) is a recent addition to this class of methods. This approach was developed for the Mediterranean climate of California, and is based on the idea that

four features of the annual hydrograph provide distinct geomorphic or ecological functions: wet-season initiation flows, peak magnitude flows, spring recession flows, and dry-season low flows.

6.1.2 Top-down and bottom-up approaches

Environmental flow assessment can be approached in terms of how much a flow regime can be modified without causing too much harm (top-down), or in term of what flows need to remain in or be restored to the river to serve specific ecological functions (bottom-up). These can be two sides of the same coin, but the differences can be substantial. Generally, top-down approaches take a more ecological point of view, and bottom-up approaches tend to focus on particular species. For example, the flow assessments for many California rivers have mostly considered flow requirements for various life-stages of salmonids: migration, spawning, incubation, rearing, and so on. However, bottom-up methods can also take a broader view, as with the South African Building Block methodology, discussed below. Bottom-up methods logically require understanding what matters about the flow regime for the stream ecosystem in question. In many instances, particularly when trying to improve an already severely modified flow regime, there is no alternative to acting as if you do have such understanding (e.g. Moyle et al. 1998). The "Natural Flow Paradigm" (Poff et al. 1997) usually provides a basis for top-down methods. It calls for attention to five attributes of the flow regime: magnitude, frequency, duration, timing and rate of change (some authors also add seasonality), and the questions for the assessment become how much each attribute can be modified without causing too much harm.

6.1.3 Sample-based methods and whole-system methods

Hydrologically based approaches deal with entire streams or sections of streams, but hydraulic rating methods and habitat simulation methods models almost always deal with samples of the reach or reaches of stream to which the results will be applied. Generally, the samples are transects or patches of habitat. This raises the question how well the sample of transects or patches represent the whole, or, stated differently, how different would the result be if the method had been applied to the entire stream? This is a question common to many areas of human endeavor, from science to opinion polling to industrial quality control, and methods for addressing it are well worked out. These allow knowledge about the thing to be sampled to be incorporated into the sampling design, often through stratified sampling; methods that also prove for good spatial coverage are appropriate for EFA (Williams 2010a). Although it is not possible to quantify how closely a given sample resembles the whole that it is supposed to represent, it is possible to quantify how close it is likely to be. That is, a probability distribution for the difference between the sample and the whole can be calculated, provided that the sample was selected randomly. Typically, the probability distribution is summarized by standard errors or confidence intervals, which can be calculated using bootstrap methods when traditional statistical formulas do not apply (Williams 1996, 2013). Not only does probability (random) sampling allow for assessing the accuracy of the results, it also gives more accurate results than deliberately selected samples, such as "representative reaches." Statisticians have recognized this since the seminal paper comparing the two methods (Neyman 1934). The failure to follow ordinary scientific norms for sampling is an unfortunately common problem in EFA, and apparently in benthological studies as well (Downes 2010).

6.1.4 Standard-setting and incremental approaches

Any assessment method can be used for setting flow standards, but the intended distinction is between methods that yield flow standards directly as some function of the flow regime, and methods that estimate habitat value as a continuous function of flow, so that the benefit or cost of incremental changes in the flow regime can be assessed. The

Tennant Method and PHABSIM are well known examples of each type. This distinction used to be common, and usually came with the implication that incremental methods were more advanced and better (e.g. Stalnaker et al. 1995). Using Tharme's categories, the distinction is the same as that between hydrological and habitat rating methods on the one hand, and habitat simulation methods on the other.

6.1.5 Micro-, meso-, and river-scale methods

Environmental flow methods generally operate at particular spatial or temporal scales. Hydrological methods and holistic methods operate at the scale of at least a reach of a river, over periods determined by the flow data used. Habitat simulation models such as PHABSIM generally estimate microhabitat at fine spatial scales, but some, such as MesoHABSIM or CASiMiR, operate at intermediate scales, generally the scale of morphological features such as pools and riffles. The temporal scales of habitat simulation models are variable; conceptually, values are calculated for particular rates of discharge, but these can be integrated over time.

Operationally, most microhabitat and mesohabitat models define habitat in terms of measureable physical variables, usually depth, velocity, and substrate or cover, although other variables such as velocity gradient or distance from the bank have been used as well for microhabitat models. As such, they are physical habitat models, and do not address important biological aspects of habitat such as food supply, competition, or risk of predation. Most also operate at a single spatial scale.

6.1.6 Opinion-based and model-based methods

Roughly speaking, methods can be based on expert opinion, the output of numerical models, or a combination of the two. At one time, model-based approaches were thought to be "objective," and so better than opinion-based methods. However, the application of any numerical model involves enough subjective choices that all methods involve considerable expert opinion, so the difference between the second and third of the categories just listed is one of degree, not kind.

6.2 Hydrological methods

6.2.1 The tennant method and its relatives

The Tennant or Montana Method (Tennant 1976) is the best known of this type. The title of Tennant's paper is "Instream flow regimens for fish, wildlife, recreation, and related environmental resources," although it was mostly about fish. Tennant was a biologist for the U.S. Fish and Wildlife Service for many years, and by his account, his method "evolved over the past 17 years from work on hundreds of streams … between the Atlantic Ocean and the Rocky Mountains." However, the ten streams for which he gave data were in the states of Montana (five), Wyoming (four), and Nebraska (one). The method can be boiled down to a table, with flows for October to March in one column, and for April to September in another, and with seven rows ranging from "optimum range" at 60 to 100% of the mean annual flow for both periods at the top, down to "severe degradation" at 10% or less, "excellent" conditions required 30% of mean annual flow in winter, and 50% in the summer. There is also a row for "flushing to maximum," at 200%.

For assessments of any consequence, however, Tennant suggested additional work. Especially for controlled streams where flows could be manipulated, he suggested "photographic regression," with photos taken from elevated points, at relevant levels of discharge. "Pictures may be the best data you will collect for selling your recommendations …" He also suggested that the U.S. Geological Survey be hired to measure depth and velocity at selected transects, so that the biologists could spend their time "on a more complete ecological analysis of streamflow needs," and the discussion and conclusions sections of his paper describe the consequences of various rates of flow for the resources cited in his tile.

The Tennant Method has been widely applied, but does not work well outside the area where it was developed (Caissie et al. 2015), and how well it works within that area in the long term is questionable (Mann 2006). In justifying the simplicity of his method, Tennant described regularities in the relationships among discharge and stream width, depth, and velocity, known to geomorphologists as the hydraulic geometry relations. However, as discussed later in this chapter regarding hydraulic geometry, the regularities probably result in part from biased sampling. Certainly, Tennant neglected variability in flow which is now regarded as critical (Poff et al. 1997; Caissie et al. 2015). Similarly, the "flushing or maximum" criterion seems too low. In the concluding section of his paper, Tennant wrote "Request that maximum flows released from dams not exceed twice the average flow. Prolonged releases of clear water greater than this will cause severe bank erosion and degrade the downstream aquatic environment." This recommendation recognized that dams interrupt sediment transport, but ignored the role that disturbances play in creating habitat.

Hydrological methods are making a comeback. In a recent defense of hydrological methods, Caissie et al. (2015, p. 660) argued that: "The strength of hydrologically based methods lies both in the simplicity of their application and their focus on protecting the hydrological character of rivers as a whole," but they also note that "… it is equally important to protect flow variability in order to maintain some ecological integrity (Poff et al. 1997)." They also argued that:

> … many scientists recognize that there are currently no truly scientifically defensible environmental flow assessment methods, as methods are based on common sense rather than scientific proof and validation (Castleberry et al. 1996; Acreman and Dunbar 2004). This means that hydrologically based methods are as credible as any other methods, provided that they are applied correctly using the best available information and good judgement.

A simple but conservative hydrological approach for general use has been suggested by the Water Footprint Network, a consortium of NGOs, universities, and development agencies. In "The Water Footprint Assessment Manual," Hoekstra et al. (2011, p. 152), noting that ELOHA is labor intensive and expensive, proposed that:

For the time being, the following simple generic rule for establishing environmental flow requirements is proposed:

1 For each month of the year, the mean monthly run-off in developed condition is in a range ±20 per cent of the mean monthly flow as would happen under undeveloped condition, and

2 For each month of the year, the mean monthly base flow is in a range ±20 per cent of the mean monthly base flow as would happen under undeveloped condition.

Richter et al. (2012) proposed a similar presumptive standard for similar reasons, but one based on 10–20 % of unimpaired flows on a daily rather than monthly basis, possibly with a minimum flow below which diversions would not be allowed.

On a technical point, it seems worth considering the shape of the distribution of flows in developing standards for planning. For some regions, the distributions will be approximately normal, but in others they will be highly skewed, so that standards based on medians may be more appropriate than standards based on means. Means are also more affected by extreme values, so unless the flow record is long, estimates of the means are likely to be more uncertain than estimates of medians.

The potential problem with preliminary flow recommendations of this sort is that they may be hard to change in light of new information (Tom Annear, cited in Armanini et al. (2015)). However, planning will proceed using some kind of flow standards, so offering an adequately conservative, or precautionary, standard seems sensible. It would also be helpful if these indices were routinely calculated in any assessments for which the necessary data are compiled. The calculations are easy, and it would be instructive to learn how the indices compare with the results of other methods, over a range of streams.

6.2.2 Indicators of hydraulic alteration (IHA)

The IHA or Range of Variation method is based on the sensible idea that flow regimes should be based not just on measures of the central tendency, but also on measures of variation (Poff et al. 1997). The method was developed by Brian Richter and colleagues with the Nature Conservancy (Richter et al. 1996, 1997, 1998), and software for the analysis is available online from the Nature Conservancy. The method calculates 32 statistics on the magnitude, duration, timing and rate of change of flow from daily time series, both for a base period, usually unimpaired or existing conditions, and the proposed conditions. The differences between the two sets of statistics provide a measure of the hydrological effects of the project, but the ecological effects remain to be assessed. Some rule of thumb about the maximum allowable change could be used, but the real strength of the approach is to direct attention to the various aspects of the flow regime that need to be considered. However, some of the statistics seem intended to address rapidly varying flow resulting from diversions to hydropower generators, and may not be relevant for streams where that is not an issue. The available software makes calculating and graphing the statistics easy, so the main task required is to develop the input data. Even for systems with long gage records, this may be non-trivial. Conditions in many watersheds will have changed gradually over the period of record, and reconstructing a record of unimpaired flow flows can be difficult enough on a monthly basis, if the flow is significantly affected by diversions. Reconstructing daily flows generally requires considerable estimation. Use of the IHA should be routine in environmental flow assessments where developing the input data is practicable, but it should be used to provide food for thought, not as a source of answers.

6.3 Hydraulic rating methods

Hydraulic rating methods deal with the rate at which stream width, depth and velocity change as discharge changes, and use inflection points in these curves for criteria. Thus, for example, the wetted width of a stream typically increases rapidly as discharge increases from zero, but then the increase slows. The shape of this curve will vary from transect to transect, but an average curve can be calculated from a sample of transects. Then, if the wetted width and discharge are made dimensionless by dividing by mean values, data from streams of different sizes can be plotted together and fit with a curve, as in Figure 6.1, so that generally applicable recommendations could be developed. This is perilous, unless scatter around the curve is taken into account. However, the curve does show a connection between hydrologic and habitat rating methods: you can use a habitat rating approach to get the criteria for a hydrological method.

One approach to developing flow standards uses kinks in the width–discharge curves, based on the assumption that habitat value is roughly proportional to width. However, from a long-term field study, Armstrong and Nislow (2012) showed that growth potential for juvenile Atlantic salmon and brook trout increased linearly with discharge in spring, summer, and autumn, contradicting that assumption.

Another hydraulic rating approach makes explicit use of hydraulic geometry relations. Despite the obvious variability among and along rivers, various authors have reported that the regularities that seem to exist in hydraulic geometry can be used to advantage in EFA (e.g. Jowett 1998; Lamouroux and Jowett 2005; Rosenfeld et al. 2007; Booker 2010). Generally, these papers refer back to a classic paper (Leopold and Maddock 1953) showing that the variation with discharge in stream width, depth, and velocity at a station could be approximately described by simple power functions, and that the same variation at stations along a stream could be described by other simple power functions:

$$w = aQ^b, d = cQ^f, v = kQ^m$$

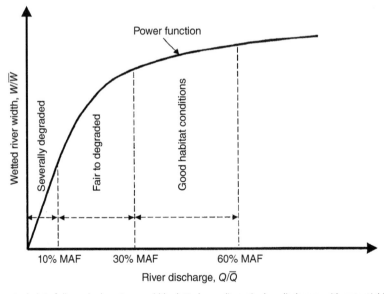

Figure 6.1 Conceptual plot of dimensionless stream width plotted over dimensionless discharge, with potential flow criteria shown by dashed vertical lines. Source: Caissie et al. (2015) with permission from John Wiley & Sons.

where w, d, and v are width, depth, and velocity, and Q is discharge. By continuity, since $Q =$ width $\times$ depth $\times$ velocity, the sum of b, f and m, and the product of a, c and k, equal 1. Jowett (1998, p. 451) reported that "Hydraulic geometry can be used to indicate whether hydraulic conditions approach a 'threshold' such as a minimum acceptable depth or velocity, thus predicating the need for more extensive habitat survey and analysis." Rosenfeld et al. (2007, pp. 765–766) found with some qualifications that "… hydraulic geometry relationships performed reasonably well for predicting optimal flows …" although they actually meant optimal flows as predicted by PHABSIM.

The reported success of hydraulic geometry methods may seem strange, given the variation in most stream channels that is apparent to anyone who has walked along them, and the variation in the hydrologic geometry exponents reported by various authors (Hatfield et al. 2003). Fonstad and Marcus (2010) described the along-stream variation in stream width and depth using data from transects measured every 100 m over many kilometers of stream. The along-stream variation in width and depth is much greater than the along-stream change predicted by downstream hydraulic geometry. That is, the widths and depths predicted by hydraulic geometry are so highly smoothed that their practical utility for EFA is doubtful.

Biased sampling may explain why predictions from hydraulic geometry seem to be useful. The data for the hydraulic geometry studies have been from transects used for rating stream gages, or from 1-D PHABSIM studies. The rating transects are deliberately located in sites with as simple a flow field as possible. PHABSIM transects are also almost always deliberately selected, in part to avoid flow fields too complex for the hydraulic models, and it seems likely that investigators either deliberately or unconsciously select transects that seem to them to represent the central tendency of the stream or habitat type being sampled. In addition, Jowett (1998) averaged transects over multiple study sites, further suppressing variation. In short, using hydraulic geometry to determine fish–habitat relationships is likely to lead to results that have little relationship to the real stream in which the organisms live. Hydraulic geometry may have a place in data-sparse situations, but it should be used with great caution.

6.4 Habitat simulation methods

The term "habitat simulation" as used by Tharme (2003) refers to a group of methods (or models) such as PHABSIM that Lancaster and Downes (2010) called habitat association models (HAMS), but we include as well two other methods that combine hydraulic models with biological models to simulate habitat value as a function of discharge: drift-foraging models, and individual-based models.

Habitat simulation models generally predict how fish will be distributed over habitats, and such models are called species distribution models (SDMs) in the ecological literature. Species distribution modeling is a very active area of research, largely because of concerns about how climate change will affect distributions of species. Generally, the level of statistical sophistication is much higher in the SDM literature than in the EFA literature, and the SDM literature gives much more attention to evaluating models, so people working on habitat simulation models would do well to pay attention to it. Elith and Leathwick (2009) provide a thoughtful review.

Until recently, the accuracy of the hydraulic modeling available for habitat simulation methods was questionable (Kondolf et al. 2000). Poor hydraulic modeling is still often used, but much better modeling has become available. In principle, models could resolve the complexities in the flow field to very fine spatial scales, but increasing demands for computational power and accurate definition of the stream bed with increasing detail in the modeling impose practical limits (see Tonina and Jorde (2013) for a good discussion of relevant hydraulic models). Three-dimensional (3-D) models have been used for academic papers, but models currently in use for EFA are 1-D or 2-D. How well hydraulic models predict the flow field depends largely on the complexity of the stream bed, and so of flow, and on how well the bed topography is described in the model. Thus, simulations of the flow over relatively smooth salmon spawning gravels in large rivers should to be more accurate than simulations of more complex situations. Recently published reports of tests of 2-D model predictions with independent data give coefficients of determination in the range of 0.65–0.95 when there was detailed information on the bed topography (Chapter 7). In our view, however, the biggest problems with habitat simulation models are biological, as discussed below.

6.4.1 Habitat association models

Microhabitat models The Physical Habitat Simulation system, PHABSIM (Bovee 1982; Bovee et al. 1998) is by far the best-known and most widely used of these models (Souchon et al. 2008). Together with its derivatives, such as the River Hydraulics and Habitat Simulation system, RHYHABSIM (Jowett 2002) and the Riverine Habitat Simulation system, RHABSIM (Payne 2005), PHABSIM has been applied around the world in many situations to assess the habitat for fishes, guilds of fishes, invertebrates, and even for riffle-inhabiting ducks (Kondolf et al. 2000). PHABSIM was developed with 1-D hydraulic models, but 2-D models are increasingly used, and River2D, a 2-D hydraulic model supported by the University of Alberta, includes PHABSIM as a module (some authors, e.g. Gard 2014, use the name River2D instead of PHABSIM). However, the basic approach has remained remarkably stable since it was first developed.

The biological models in the commonly used HAMs reflect associations or correlations between habitat variables and the density of the organisms of interest; Lancaster and Downes (2010) call them abundance–environment relations (AERs). The AERs traditionally used in PHABSIM and related methods are simple univariate curves, usually called "suitability curves" or "preference curves," depending on the way they are derived, one for each habitat variable considered. Usually these are depth, mean velocity and substrate/cover, but sometimes others

variables are used, as well. The curves take values ranging from 0 to 1, and are usually derived from observations of habitat values at positions occupied by fish (see Figure 6.3), either with or without adjustment for the availability of habitat with such values.

One way or another, depending on the hydraulic model used, the bed of the reach of stream to which the model is applied is divided into cells or "tiles," and, for a given discharge, a depth and water velocity are assigned to each tile by the hydraulic model. The values for other microhabitat variables, usually substrate or cover, are taken from field data. To calculate a composite suitability for each tile, the values of the AER curves are combined, usually by simple multiplication, producing another number between 0 and 1. Sometimes, another statistic, such as the geometric mean of the values, is used instead, but the result is again between 0 and 1. Then, the area of each tile is multiplied by its corresponding suitability, and these "weighted" areas are summed over the reach. Repeating this process at different rates of discharge yields a curve of habitat value over discharge. The process is detailed in Box 6.1, copied from Williams (2010a).

The procedure described in Box 6.1 for calculating a composite suitability assumes that the various microhabitat variables influence habitat selection independently, which seems unlikely, and statistically better alternatives are available that avoid this assumption (Ahmadi-Nedushan et al. 2006). Of these, logistic regression has been most commonly proposed (e.g. Thielke 1985; Knapp and Preisler 1999; Guay et al. 2000; Beakes et al. 2014). The model of Guay et al. is examined in detail in Chapter 7, but these alternatives are seldom used in actual assessments.

The most commonly used HAMs, such as PHABSIM, are microhabitat models, but others, discussed below, operate at the mesohabitat scale. For reasons described in Chapter 7, and in Railsback (2016), we think that the use of HAMs at the microhabitat scale should be abandoned. There are several basic problems with microhabitat HAMs. First, we agree with Giger's (1973, p. 103) comment, which

Box 6.1   PHABSIM

PHABSIM is a collection of hydraulic and biological models used to assess the value of habitat in a stream as a function of discharge for a particular species or life stage. PHABSIM operationally defines and estimates a suitability, S, for a species or life stage, and uses S ($0 \leq S \leq 1$) to weight the area of the stream, yielding a statistic called weighted usable area (WUA). Conceptually, S varies continuously over the surface of a stream, and is defined operationally in terms of "microhabitat" variables; usually these are water depth, velocity, and substrate size or cover, but sometimes other variables such as distance to cover or velocity gradient. Substrate and cover are estimated by field surveys, usually as categorical variables, but water depth and velocity are estimated over a range of discharge with a hydraulic model. The biological models normally used to calculate S for depth and velocity are curves called habitat suitability criteria (HSC), which vary between 0 and 1 as a function of one of the microhabitat variable at points or small areas in the stream. For categorical variables such as substrate, a suitability value is assigned to each category. The sampling and modeling issues involved with developing habitat selection models are well covered in Manly et al. (2002), and the discussion applies to HSC as well.

The hydraulic models used in PHABIM are usually one-dimensional (1-D), and are used to estimate depth and velocity at points along transects perpendicular to the flow. Increasingly, however, two-dimensional (2-D) models are being used. With these, areas of the stream are divided into many small cells or tiles, and depth and velocity are estimated for each. At a given discharge, the values of the suitability curves for each point or cell are combined, often by simple multiplication, to calculate WUA. Details are complicated by the large number of options available in PHABSIM, but in the general case, with the 1-D models, the river reach of interest is represented by a set of transects, and each transect is assumed (e.g. through "weighting factors") to represent some fraction of the total reach. One of several hydraulic models is used to estimate water depth and velocity at usually 20 or more points along each transect. The values at the points are assigned to "cells" with nominal areas calculated from the weighting factors and the distance between the points on the transect. WUA for the transect at a given discharge is estimated by:

$$\widehat{WUA}_t = \sum_{i=1}^{n} \alpha_i \hat{S}_i$$

where the summation is over the n cells on the transect, α_i is the nominal area of the ith cell on the transect, and, for the default option,

$$\hat{S}_i = \hat{S}_{v,i} \ \hat{S}_{d,i} \ \hat{S}_{s,i}$$

and $s_{v,i}$, $s_{d,i}$, and $s_{s,i}$ are the values of the HSC for the species or life stage in question for velocity, depth, and substrate/cover for the ith cell, again at the given discharge; the "hats" indicate that the terms are estimated from samples. Repeating this process at each transect over a range of discharge and interpolating results in curves of WUA over discharge, and summing over these gives a composite curve of WUA over discharge for the reach. Usually, the sum is normalized by stream length, so the final results are reported as WUA per length, which has a dimension of length. The process is similar with 2-D models, except in this case the α_i are real areas that are determined by the grid of the hydraulic model. The composite curves are the basic product of PHABSIM, although they can be used to assess flow regimes rather than specific levels of flow by combining the curves with a hydrograph to produce times series of WUA. According to Annear et al. (2004, p. 149), "…the primary value of PHABSIM is its ability to identify trade-offs between streamflow and hydraulic habitat …"

pre-dated the development of PHABSIM: "Micro-habitat information presently consists of limited physical data not easily related to food and shelter requirements. Neither can the data be used in a spatial context." Microhabitat HAMs also depend on two ideas that were common when PHABSIM was developed in the 1970s: that stream ecosystems are basically equilibrium systems, and that stream ecosystems are structured primarily by abiotic factors. Thus, the abundance of cohorts would be controlled by equilibrium ecological processes and by physical habitat through density-dependent processes, so that if a stream were "fully seeded," the number of fish would reflect environmental conditions, making modeling instream flow relationships fairly simple. Such ideas have since been abandoned (e.g. Power 1990; Mangel et al. 1996). Other basic problems were discussed in Chapters 1 and 5. As nicely summarized by Armstrong and Nislow (2012, p. 530):

> PHABSIM includes three assumptions regarding the biology of fish. First, that local abundance reflects habitat quality; second, that preference is consistent across discharges and third, that fish are free to move in response to change in habitat with discharge. Each of these assumptions may be invalid (Armstrong 2010). Indeed, a conceptual modeling analysis from first principles (Railsback et al. 2003) suggests that the basic structure of PHABSIM is flawed.

It is fair to ask, "If PHABSIM is as bad as we claim why does, it continue to be used?" There are several reasons:

- PHABSIM was developed, and for many years promoted, by scientists at the U.S. Fish and Wildlife Service, and later, after a transfer of relevant personnel, by the U.S. Geological Survey. Thus, PHABSIM seemed to have an official stamp of approval.
- PHABSIM seems to provide managers with a rational basis for making decisions about modifications of flow regimes; the failure to account for statistical uncertainty in PHABSIM results contributes to this illusion.

- Managers in some areas have come to expect that PHABSIM will be used.
- Training in the use of PHABSIM has been readily available.
- There are consultants and agency staff who have built their careers around it, and resist change.
- There is an associated technical jargon that seems scientific, makes the method opaque to outsiders, and provides a language for negotiating for more or less water in streams.
- A convenient alternative has been lacking.

Although other microhabitat HAMS are superior in various ways to PHABSIM, they share most of the basic problems. An example is examined in Chapter 7.

Mesohabitat models

Mesohabitat models such as MesoHABSIM (Parasiewicz 2001, 2007; Vezza et al. 2014) are habitat association models developed at the spatial scale of habitat types, or mesohabitats. They have been used mainly in the northeastern USA and Europe. According to Parasiewicz (2007):

> Meso-scale units can be defined as areas where an animal can be observed for a significant portion of their diurnal routine, and it roughly corresponds with the concept of "functional habitat" (Kemp et al. 1999). Because of the natural mobility of fish, observation at the meso-scale is less affected by coincidence than at the micro-scale and can be expected to provide relatively meaningful clues about an animal's selection of living conditions (Hardy and Addley 2001).

MesoHABSIM was designed in part to avoid the extensive field work required by PHABSIM. Briefly summarized, the steps in MesoHABSIM are:

- Map habitat types at four or five different flows spanning the range of management interest;
- Measure abundance and habitat attributes in all habitat types in "representative" sites;
- Develop AERs relating the presence or abundance of fish to habitat attributes in the habitat types;
- Estimate the distribution of the habitat attributes in the habitat types at the different levels of

discharge, accounting for changes in the boundaries of the types;

- Calculate expected abundance at the four or five flows and interpolate between them to estimate future abundance as a function of flow.

The implementation of MesoHABSIM seems more thoughtful and methodologically advanced than most implementations of PHABSIM, but still suffers problems such as deliberate selection of sample sites. Parasiewicz and Walker (2007) published a comparison of fish abundance with the habitat index used by MesoHABSIM, with data collected a year after the data used to develop the model. The relationship was statistically significant, but this is a weak test, and there was a great deal of scatter in the data. Moreover, the data showed high variability in the number of fish collected by electroshocking at 220 locations where habitat data for the HPI were also collected. Accordingly, the sampling error associated with applying the method to the whole reach would further weaken the relationship. Parasiewicz and Walker (2007) also applied PHABSIM and another model, HARPA, and tested them in the same way; the relationships between fish abundance

and the indices calculated by those models were not statistically significant. Importantly, each of the three models provided different guidance to managers. Overall, however, MesoHABSIM still suffers from the basic flaws of habitat association models, and if used, should be applied along with other methods, or in the context of strong adaptive management.

CASiMiR (Computer Aided Simulation Model for Instream Flow Requirements) is another meso-habitat model based on habitat suitability curves, called "membership functions," and was developed in Europe for assessing diversions for hydropower (Mouton et al. 2013). Instead of numerical criteria, CASiMiR uses qualitative categories such as low, medium and high to describe habitat quality, and combines the curves for the various habitat variables using "fuzzy" logic. This allows a particular value of a habitat variable to have some degree of membership in one category and a complementary degree of membership in the next category. Thus, a given velocity, for example, might have an 80% membership in medium habitat value and a 20% membership in high habitat value (Figure 6.2). Then, the curves are combined following the rules of fuzzy logic, which

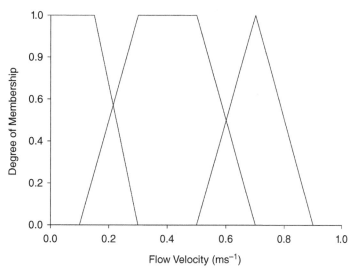

Figure 6.2 An example of membership functions for low, medium, and high habitat value. Source: redrawn from Noack et al. (2013) with permission from John Wiley & Sons.

avoids the assumption that organisms respond independently to habitat variables.

CASiMiR membership functions can be developed from expert opinion or from data (Mouton et al. 2013), as with PHABSIM suitability curves, but it retains the basic problems of other habitat association models, and it is not clear that CASiMiR membership functions work better than suitability estimated with logistic regression. CASiMiR comes with jargon associated with fuzzy logic that may appeal to some, and the partial memberships have considerable appeal. However, Bayesian Network models, discussed in Chapters 5 and 9, have essentially the same feature, as well as other advantages, and seem a better alternative.

6.4.2 Bioenergetic or drift-foraging models

Drift-foraging models simulate the potential energy gain for fishes that feed on invertebrates drifting in the current (Grossman 2014). The basic idea of these models is simple. Drift-feeding fish swim against the current to stay in one place and let the stream carry drifting invertebrates within striking distance; then they make a quick foray to capture the prey, and return to their original position. There is an energy cost to swimming against the current, and to the forays, which increases exponentially with the velocity of the current. The fish gain energy from the captured prey which initially increases with water velocity as the current brings more prey within reach, but then decreases as the prey drift too rapidly for the fish to catch. Conceptually, these factors define a cost curve and a benefit curve, and the models are based on the idea that fish should select holding positions that maximize the difference between these curves (Figure 6.3).

Kurt Fausch developed the first such model (Fausch 1984), followed by Hughes and Dill (1990) and Hill and Grossman (1993). These models substantiated the thinking of Chapman (1966) and other biologists of the time about the behavior of drift-feeding salmonids. Drift-foraging models have become common in the literature over the following decades, and the

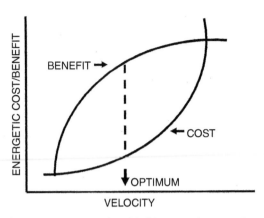

Figure 6.3 A conceptual model of the energetic costs and benefits for a drift-feeding fish as functions of water velocity. Source: Grossman (2014) with permission from Springer.

journal *Environmental Biology of Fishes* published a special issue on the behavior and ecology of drift foraging in 2014 (vol. 97, p. 5). Readers with a particular interest in drift-foraging models can find much good information in its articles, and also differences of opinion, or at least in emphasis, among the authors.

Drift-foraging models come in two main kinds. Individual-based models (IBMs) deal with virtual fish using virtual habitat, and allow the virtual fish to select habitat, grow, reproduce, and otherwise behave (including feeding on benthic prey). These models can be used to explore many biological questions besides growth and habitat selection. Net energy intake (NEI) models are more habitat-centered, and assess microhabitats in terms of the potential energy benefit to fish occupying the habitat. Conceptually, these models estimate a map over the habitat of the potential net energy intake for a particular kind of fish, given assumptions about the density and energy content of drifting invertebrates. With some additional assumptions, these maps can be used to estimate the number of fish that can be supported by the habitat, and how they should be distributed (e.g. Hayes et al. 2007). Both kinds of model can be used in two ways for EFA; one is to inform thinking about how changes in flow regimes may affect drift-feeding fish, and the other is to predict the effects of a particular change in flow regime in a particular reach of stream. For reasons discussed previously and reiterated below, we think they are useful for the first purpose, but not the second. We discuss NEI models first, and then IBMs, but various considerations apply to both.

Net energy intake models

As just noted, NEI models need to account for the energy cost of holding position against the current, the ability of the fish to detect and capture drifting prey, and the number and energy content of invertebrates passing within striking distance. These bioenergetic factors must be simulated, measured, or simply assumed. Different models focus on different factors or sets of factors, and the complexity of the models varies.

As well as a bioenergetics model, most NEI models include a hydraulic model that simulates the flow field over the habitat, although Urabe et al. (2010) simply estimated the flow field from measurements across closely spaced transects. The accuracy of the flow simulation does not matter so much if the model is intended simply to inform thinking, but clearly matters a lot for predicting what will happen in a particular stream; for that purpose, collecting independent flow data with which to test the hydraulic model is critically important. Although their purpose was more general, Wall et al. (2016) did this, and reported coefficient of determination (R^2) of 0.58 for measured and predicted depth, and 0.64 for velocity. Other authors have reported better accuracy (Chapter 7), but doubtless at the expense of more detailed data on the topography of the streambed. Given a fixed budget, more detailed topography will cover a smaller area of the channel.

The importance of the energy cost of swimming for NEI depends strongly on several factors. One factor is the length of the fish, which is intuitive if swimming speed is scaled in units of body lengths per second. For larger fish, such as adult trout, the cost of swimming may increase slowly enough over the relevant range of water velocity that it can be treated as a constant (Grossman et al. 2002).

However, for smaller fish, such as young-of-the-year salmonids, swimming costs increase more rapidly (Rosenfeld et al. 2014). Turbulence also increases the energy cost of swimming, and again, smaller fish will be more affected. Hence, more detailed hydraulic modeling may be necessary for smaller fish. Another factor is the quality of the velocity shelter available to the fish. Recall the quotation from Bachman (1984, p. 9) in Chapter 1 about adult brown trout holding position next to a stone: "While a wild brown trout was in such a site, its tail beat was minimal … indicating that little effort was required to maintain a stationary position even though the current only millimeters overhead was as high as 60–70 cm s^{-1}." In such cases, what matters is more likely the ability of the fish to detect and capture prey, which also decreases with flow velocity.

Whether, or in what circumstances, it is necessary to calculate the energy cost of swimming, seems to be an open question. Grossman (2014) found that the velocity at which the capture success curve declined most rapidly (Figure 6.4) was a good predictor of the velocities occupied by rainbow trout and rosyside dace in Coweeta Creek, North Carolina, and Grossman et al. (2002) found the same for rosyside dace and several other species of minnow in the same stream. However, Grossman (2014) noted that this model had not been tested elsewhere, and Rosenfeld et al. (2014) describe both the swimming cost function and the capture success function as key sources of uncertainty for drift-foraging models.

For the benefit curve, the abundance and energy content of the drifting invertebrates clearly matters. Most NEI models assume a constant flux of drift, although some account for depletion of the drift by predation within the simulated habitat (e.g. Hayes et al. 2007), and the models can be modified to make the drift concentration a function of flow (Hayes et al. 2016). Just how to do this is unclear. As part of an application of the concept of response lengths to EFA, Anderson et al. (2013) modeled the influence of flow on drift, and predicted that drift density should decrease with discharge. However, Hayes et al. (2016) found empirically that drift density

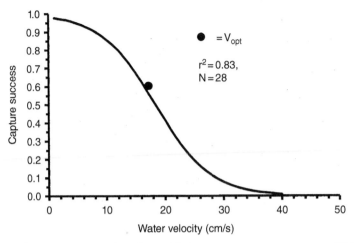

Figure 6.4 Success rate of prey capture versus velocity for rosyside dace at summer water temperatures (18C). Source: Grossman (2014) with permission from Springer.

increased with discharge. Dealing with the drift and other sources of food probably is most difficult part for models that aim to predict what will happen in particular streams (Fausch 2014). In a recent review of the literature on invertebrate drift, Naman et al. (2016, p. 1299) observed that: "Overall, the controls on drift-foraging consumer–resource coupling remains a fundamental gap in our understanding of trophic dynamics in streams, as well as the ability to accurately assess trophic interactions and habitat quality with drift-foraging models."

Individual-based models

InSTREAM is the main IBM that has been used for EFA. It is a complex and sophisticated model for stream-dwelling salmonids that has provided the basis for many publications by Steve Railsback, Bret Harvey and colleagues, and is available at http://www2.humboldt.edu/ecomodel/instream.htm (accessed October 2018). As described in the abstract of Railsback et al. (2009):

InSTREAM is a simulation model designed to understand how stream and river salmonid populations respond to habitat alteration, including altered flow, temperature, and turbidity regimes and changes in channel morphology. The model represents individual fish at a daily time step, with population responses emerging from how individuals are affected by their habitat and by each other (especially via competition for food). Key individual behaviors include habitat selection (movement to the best available foraging location), feeding and growth, mortality, and spawning. Fish growth depends on prey availability and hydraulic conditions. Mortality risks due to terrestrial predators, piscivorous fish, and extreme conditions are functions of habitat and fish variables. Field and analysis techniques for applying InSTREAM are based in part on extensive analysis of the model's sensitivities and uncertainties. The model's software provides graphical displays to observe fish behavior, detailed output files, and a tool to automate simulation experiments.

The model's original purpose was to address one of the most difficult general problems of impact assessment for stream-dwelling trout: understanding how alteration of habitat affects populations of animals that actively adapt to habitat change by moving. InSTREAM can predict how trout populations respond to changes in any of the inputs that drive the model, especially flow, temperature, turbidity, and channel morphology. InSTREAM can also predict how populations respond to changes in ecological conditions such as food availability or mortality risk. Because InSTREAM provides an observable virtual ecosystem, it is also a useful tool for addressing many basic ecological research questions.

Importantly, InSTREAM models the effects of flow on the growth and reproduction of the modeled fish over time, not just on which microhabitats the fish should select, and so predicts the response of populations to the flow regime. It can also model benthic feeding, or a mixed strategy of drift and benthic feeding (Harvey and Railsback 2014). As an "observable virtual ecosystem," InSTREAM has proved valuable; for example, an early version (EPRI (Electric Power Research Institute) 2000) showed that, given reasonable rules about how salmonids respond to their environment, we should expect that habitat selection will vary with discharge, contradicting a fundamental assumption of PHABSIM. Having such a tool for testing ideas is a tremendous advantage.

Whether InSTREAM can usefully predict the effects of management on the abundance of fish in real streams is much less clear. InSTREAM is a simulation model, and as such an application of InSTREAM is a thought experiment (Schnute 2003). As complex as it is, the virtual ecosystem described in InSTREAM is enormously less complex than real ecosystems. Railsback et al. (2009) are forthcoming about these simplifications, for example, at p. 10:

> Trout have additional adaptive behaviors that we have chosen not to represent mechanistically in this version of InSTREAM, because doing so does not seem necessary to meet the model's purposes. These behaviors include variation in diel activity patterns (feeding vs. hiding); allocation of energy intake to growth, energy storage, or gonad production; year-to-year spawning effort.

Similarly, the amounts of food available in the drift and on the bottom are treated as constant in time and space. Some of the behavioral rules coded into the program, such as the rule that fish do not feed at night, are unrealistic (Armstrong 2010; Bradford and Higgins 2001). Spawning habitat is described only in terms of area. In the current version, flow is represented by a 1-D hydraulic model. This leaves the model open to the criticism of complex models generally by May (2004, p. 793): "It makes no sense

to convey a beguiling sense of 'reality' with irrelevant detail, when other equally important factors can only be guessed at." The Recovery Science Review Panel, convened by NOAA Fisheries to provide guidance for salmon recovery efforts coast-wide, was sharply critical of the Ecosystem Diagnosis and Treatment (EDT) model, which is now widely used in Washington and Oregon, largely on that basis: "The inclusion of so much detail may create an unjustified sense of accuracy, but actually introduces sources of inaccuracy, uncertainty, and error propagation" Recovery Science Review Panel (RSRP) (2000, p. 6). Applied to particular streams, InSTREAM is best used as a source of hypotheses, rather than as the basis of a function that relates flow regime to population abundance.

Two more recent versions of the model, modified for anadromous fish, have been described: InSALMO (Railsback et al. 2013) and InSALMO–FA (Railsback and Harvey 2013; Railsback et al. 2014). InSALMO was applied it to juvenile Chinook salmon (*Oncorhynchus tshawystcha*) in Clear Creek, a tributary of the Sacramento River in California, and InSALMO–FA was applied to juvenile steelhead (*Oncorhynchus mykiss*) in the same stream. Here, FA is for facultative, because there can be anadromous and non-anadromous members of the same population, and whether a particular individual is anadromous may depend on environmental conditions.

InSALMO–FA nicely illustrates one of the problems with simulation models. According to Railsback et al. (2014, p. 1271): "The purpose of inSALMO–FA is to help predict and understand how river management actions such as altered flow and temperature regimes and habitat restoration projects affect the spawning and juvenile rearing life stages – especially smolt production – of facultative anadromous salmonids" However, although the model can be calibrated to predict reasonable numbers of virtual smolts, it assumes that the virtual fish start out the same propensity to migrate (although not at the same time), and chose a resident or anadromous life history depending on their early growth. But, there is now strong evidence for genetic variation among juvenile

O. mykiss in their response to environmental conditions (Pearse et al. 2014; Phillis et al. 2016). Thus, in a mixed population of resident and anadromous *O. mykiss*, some juveniles will have a much stronger propensity to migrate than others. As a result, the calibrated model must have gotten the right result for wrong reasons, which is all too easy to do with a complex simulation model.

6.5 Frameworks for EFA

Often, individual methods used in EFA are put into some king of framework when they are applied to particular situations, and the spatial scale of the application is an important consideration in developing or selecting a framework. In the USA, environmental flows are usually set for individual streams, and frameworks such as the Instream Flow Incremental Methodology (IFIM) have been developed accordingly. In other situations, as in the Murray–Darling basin in Australia, environmental flows are being set regionally, and regional frameworks such as the FLOWS methodology (DEPI 2013) have been developed.

6.5.1 Instream flow incremental methodology (IFIM)

The IFIM is the framework most commonly proposed for EFA in the US. The IFIM was developed by an interagency group in the late 1970s, largely in response to a surge in applications for small hydropower plants following the energy crisis of that decade. As described by Annear et al. (2004, p. 187):

> The Instream Flow Incremental Methodology (IFIM) is a modular decision support system for assessing potential flow management schemes. This method quantifies the relative amounts of total available habitat available throughout a network of stream segments for selected flow regimes. It was designed to prescribe instream flow regimes that result in no net loss of total habitat, or to develop mitigation plans to compensate for habitat potentially lost as a result of proposed flow management.

> … The IFIM is composed of a library of linked analytical procedures that describe the spatial and temporal features of habitat resulting from a given river regulation alternative. The unique feature of IFIM is the simultaneous analysis of habitat variability over time and space. This methodology is composed of a suite of computer models, manuals, and data collection procedures that address hydrology, biology, sediment transport, and water quality (see Stalnaker et al. 1995 and Bovee et al. 1998). Several studies have demonstrated the relation between usable habitat and fish populations (Orth and Maughan 1982; Nehring and Miller 1987; Bovee 1988; Jowett 1982; Nehring and Anderson 1993). …

The IFIM and PHABSIM were developed simultaneously, and the terms IFIM and PHABSIM are frequently confused, in part because there is not a crisp definition of IFIM, and in part because the terminology has not been consistent over time. This was particularly true in the 1980s, when articles in the professional literature that were really about PHABSIM came out under titles such as "Evaluation of the instream methodology for recommending instream flows for fish" (Orth and Maughan 1982), or "A critique of the instream flow incremental methodology" (Mathur et al. 1985). However, Bovee et al. (1998) begin by stating:

> The Instream Flow Incremental Methodology (IFIM) is a decision-support system designed to help natural resource managers and their constituencies determine the benefits or consequences of different water management alternatives. Some people think of IFIM as a collection of computer models. This perception is understandable because IFIM is supported by an integrated habitat simulation and analysis system that was developed to assist users in applications of the methodology. However, IFIM should be considered primarily as a process for solving water resource allocation problems that include concerns for riverine habitat resources. (p. 1.)

Orth (1987, p. 171) similarly notes that "(T)he IFIM process includes evaluation of effects of incremental changes in stream flow on channel structure, water quality, temperature, and availability of suitable microhabitat in order to recommend a flow regime that will maintain existing habitat conditions." However, he adds that "The physical habitat component (PHABSIM; (citation omitted)) is the most frequently used component, often to the exclusion of other components." Orth is not the only one to suggest that in practice IFIM usually boils down to PHABSIM: Shirvell (1986) said flatly that "IFIM is PHABSIM." EPRI (Electric Power Research Institute) (2000), in a review of instream flow assessments in Federal Energy Regulatory Agency processes, reports that although PHABSIM was the method most commonly used, the IFIM process was rarely implemented.

IFIM as initially proposed and as described by Bovee et al. (1998) is nevertheless worth considering, partly because it represented an attempt to take a comprehensive approach to assessing instream flow needs, and partly because it illuminates the orientation of the developers of PHABSIM. Bovee et al. (1998) begin with a broad perspective, depicted in Figure 6.5, showing concern for water quality and channel structure as well as microhabitat. Another figure (Figure 6.6) draws attention to errors in observations and modeling that should be taken into account. However, it is easy to suspect that these figures were not designed for careful consideration, since Figure 6.5 depicts an endless loop. In addition, the "observed values" and "system conceptualization" boxes on the right side of Figure 6.6 have nothing to do with the output of the real system, the "parameterization" box is in the wrong place, and there should be a directed linkage from the "system conceptualization" box to the "system representation" box.

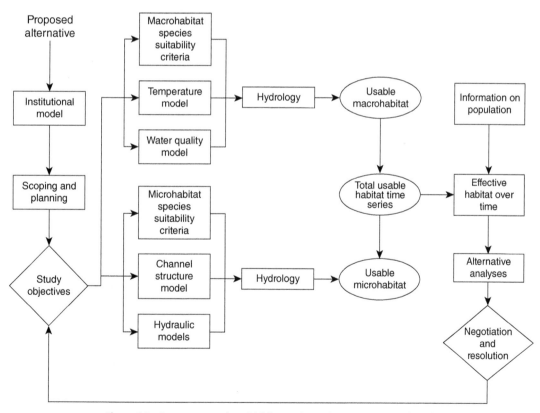

Figure 6.5 Components and model linkages of IFIM. Source: Bovee et al. (1998).

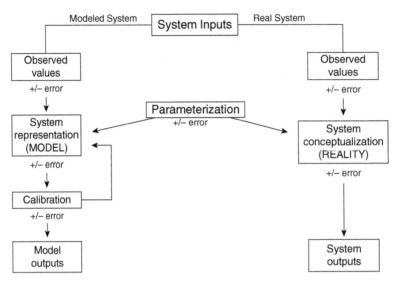

Figure 6.6   Schematic of sources of error in modeling. Source: Bovee et al. (1998).

The fundamental concern permeating Bovee et al. (1998) is with successfully negotiating environmental flow agreements. More attention is given to negotiation than to biology. Much of the document concerns negotiations per se, and this emphasis carries over into the discussion of the technical aspects of the work. For example, in explaining the need to define a "currency" for such a negotiation, Chapter 2 states (p. 35): "... it takes many more data and vastly greater modeling expertise to predict changes in fish populations than it does to predict changes in habitat availability. If a stakeholder insists the currency be fish populations (or worse, economic values of the fishery), you can expect a long and arduous study, with no guarantee of being able to measure the currency at the end of the study."

Regarding habitat suitability criteria, Bovee et al. (1998) note that judgment-based criteria "... are as valid in an application of IFIM as data-based criteria, if they are supported by a consensus of opinion of the stakeholders" (p. 80). The same concern with agreement appears in a discussion of the necessary number of transects, under the heading Schedules and Budgets:

The geographic coverage of PHABSIM sites, the amount of replication, and the level of detail are the most controllable factors in regulating time and cost estimates. These factors collectively distill down to the number of transects that will be used to represent the segment. Although there is no fixed formula to determine the exact number of transects required for every mesohabitat type, the average number of transects used to describe single channel mesohabitat sites usually ranges from two to six (two for the most uniform mesohabitats and five or six for the most complex). This estimate is based on our reviews of many PHABSIM studies conducted over the past two decades, including our own. ... Issues related to geographic coverage, number of replicates, and transect density can often be addressed by staging a field trip for the stakeholders. The purpose of the field trip should be to obtain consensus regarding the approximate numbers of transects needed in each site for planning purposes. (p. 53).

Similarly, "IFIM is not designed to produce the 'one best answer'. The best answer is whatever the consensus of stakeholders says it is" (p. 93). However, Bovee et al. (1998) say nothing about negotiating a monitoring program to assess the consequences of the consensus agreement. In short, the objective that the developers of IFIM and PHABSIM have in mind is negotiating agreements about instream flow standards. Whether the standards actually achieve the intended biological result seems beside the point. In any case, using two to six transects to represent a mesohabitat will result in massive sampling uncertainty (Williams 1996, 2010b).

6.5.2 Downstream response to imposed flow transformation (DRIFT)

DRIFT is perhaps the most advanced of the holistic methods developed by scientists in South Africa. King et al. (2003) provide a good description of DRIFT, and summarize it as follows in their abstract:

… DRIFT's basic philosophy is that all major abiotic and biotic components constitute the ecosystem to be managed; and within that, the full spectrum of flows, and their temporal and spatial variability, constitute the flows to be managed. The methodology employs experienced scientists from the following biophysical disciplines: hydrology, hydraulics, fluvial geomorphology, sedimentology, chemistry, botany and zoology. …

DRIFT is a structured process for combining data and knowledge from all disciplines to produce flow-related scenarios for water managers to consider. It consists of four modules. In the first, or biophysical module, the river ecosystem is described and predictive capacity developed on how it would change with flow changes. … In the third module, scenarios are built of potential future flows and the impacts of these on the river and riparian people. [The second and fourth modules concern subsistence users of river resources and economic issues such as mitigation and compensation costs.] …

DRIFT should be run in parallel with two other exercises which are external to it: a macro-economic assessment of the wider implications of each scenario, and a Public Participation Process whereby people other than subsistence users can indicate the level of acceptability of each scenario.

DRIFT is also based on the principles that water-allocation decisions must be made quickly, even in the face of scant information, but the decisions should be precautionary and implemented adaptively. Although

there is much attention in DRIFT to all aspects of the problem of making flow allocations, the discussion below is limited to the biophysical module.

Work for the biophysical module begins with selection of "representative sites." These are selected deliberately, to have "the highest proportion of natural features, because these provide good clues to flow–ecosystem relationships" (King et al. 2003, p. 623), with the usual practical constrains of access, cost, etc. Unfortunately, this means that there is no statistical basis for applying the results to the rest of the river, and is a weakness of the approach. The second step is the development of a daily flow record, preferably for 30+ years, which is analyzed in a manner something like IHA, but with ten statistics rather than 32 (Figure 6.7). Hydraulic modeling of the study sites then translates the discharge levels into stage, water velocity, and so on, to allow assessments of their biological effects. It appears that both 1-D and 2-D approaches have been used. At this point a set of flows regimes is picked for analysis, and scientists from each discipline use methods of their choice to assess the effects of the flow regime on the relevant environmental factors, in terms of each of the ten aspects of the flow regime. Assessments are made in terms of tendency (toward or away from natural conditions) and severity, using a categorical scale (none, negligible …, critically severe), which is supposed to reflect ranges of percentage changes. What is strikingly different from unusual practice in the northern hemisphere, and what makes the approach truly holistic, is the way these assessments are integrated (King et al. 2003, p. 629):

This is best done in a workshop environment, so that all can understand the predicted changes. Typically, for each flow reduction, the geomorphologists first describe the anticipated changes to the physical environment, followed by aquatic chemists outlining chemical and thermal changes, and then the vegetation specialists describing shifts in aquatic and riparian plant communities. At this stage, the predicted changes to the environment of the fauna have been described, allowing the fish and invertebrates (and any other faunal) specialists to record their predictions.

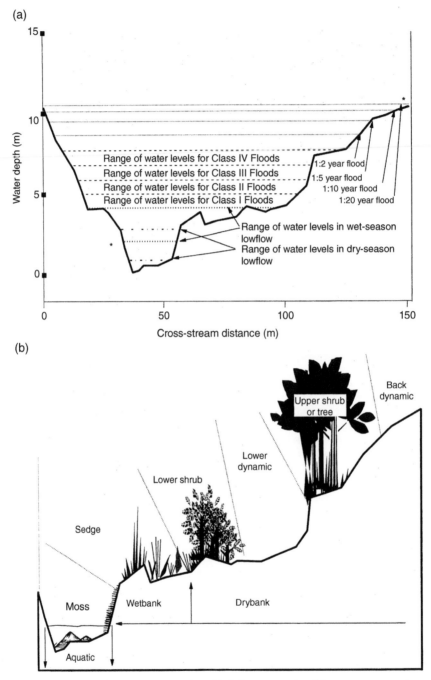

Figure 6.7   Linkage of hydrological data to river cross-sections: (a) discharge ranges for all flow components and corresponding water depths; (b) expanded detail of one bank, with zones of riparian vegetation and other features which can be linked to different water levels. Source: King et al. (2003) with permission from John Wiley & Sons.

Excel-based software has been developed for DRIFT (Arthington et al. 2007), but it still depends heavily on expert opinion, and so the willingness of competent scientists to participate. Overall, DRIFT was a major advance, although it could be improved by incorporating Bayesian methods to make the blending of data and opinion more transparent and formal. However, like any other approach, DRIFT can be abused; DRIFT was reduced to poorly supported suitability curves in an application to the Kishenganga River between India and Pakistan (Waters and Bailly 2013).

6.5.3   Ecological limits of hydraulic alteration (ELOHA)

ELOHA is a new framework for EFA, announced by Poff et al. (2010), that has been under development for some time by an international group of scientists, with strong support by the Nature Conservancy, a major conservation group. ELOHA was intended for developing flow standards for streams within a region, rather than a single stream. It has been applied widely, for example in a dozen regions of the USA (Kendy et al. 2012). Because ELOHA is relatively new and is getting considerable attention, we review it at some length.

As summarized by Poff et al. (2010, p. 147):

The ELOHA framework includes the synthesis of existing hydrologic and ecological databases from many rivers within a user-defined region to develop scientifically defensible and empirically testable relationships between flow alteration and ecological responses. These relationships serve as the basis for the societally driven process of developing regional flow standards. This is to be achieved by first using hydrologic modelling to build a "hydrologic foundation" of baseline and current hydrographs for stream and river segments throughout the region. Second, using a set of ecologically relevant flow variables, river segments within the region are classified into a few distinctive flow regime types that are expected to have different ecological characteristics. These river types can be further subclassified according to important geomorphic features that define hydraulic habitat features. Third, the deviation of current-condition

flows from baseline-condition flow is determined. Fourth, flow alteration–ecological response relationships are developed for each river type, based on a combination of existing hydroecological literature, expert knowledge and field studies across gradients of hydrologic alteration.

Scientific uncertainty will exist in the flow alteration–ecological response relationships, in part because of the confounding of hydrologic alteration with other important environmental determinants of river ecosystem condition (e.g. temperature). Application of the ELOHA framework should therefore occur in a consensus context where stakeholders and decision-makers explicitly evaluate acceptable risk as a balance between the perceived value of the ecological goals, the economic costs involved and the scientific uncertainties in functional relationships between ecological responses and flow alteration.

The ELOHA framework also should proceed in an adaptive management context, where collection of monitoring data or targeted field sampling data allows for testing of the proposed flow alteration–ecological response relationships. This empirical validation process allows for a fine-tuning of environmental flow management targets. The ELOHA framework can be used both to guide basic research in hydroecology and to further implementation of more comprehensive environmental flow management of freshwater sustainability on a global scale.

Thus, the main steps are: (i) build a hydrologic foundation; (ii) use this to develop a classification for stream segments; (iii) determine levels of hydrologic alteration within each class; (iv) develop hydrologic–ecological response relationships for each class of streams. Then, (v) use these relationships to develop flow standards, in a social process that takes account of competing needs for water; and (vi) apply these standards in the context of adaptive management.

Hydraulic alteration–ecological response relations are central to the approach. In many cases, it can be easy to identify relationships between aspects of the flow regime and responses of fish populations or ecosystems, such as the high flows needed in some streams for migration by salmon. In other cases, it may take years of study to identify such relationships. For example, whether very low summer flows in the Eel River in California are preceded by winter flows high enough to mobilize the gravel streambed can determine whether the stream ecosystem will develop in ways that promote the growth of cyanobacteria, some of which are toxic, during the following low-flow season (Power et al. 2015). Without long-term research at a university field station, this relationship would not have been discovered and explained.

It is probably useful to distinguish different forms of these relationships. One kind relates some flow attribute directly to a response through a causal chain, as in the Eel River example just discussed. Another kind is relates the flow attribute to an index or rating, as in the Michigan approach, discussed below. The professional literature is the main source of information on flow–ecology relationships, but generally these should be treated as hypotheses when they applied to particular streams or sets of streams. Relationships can also be discovered empirically by searching through data sets on flows and ecological conditions in streams, as in the Queensland example discussed below. However, this sort of "data dependent" analysis carries large risks of identifying spurious relationships.

Applications of ELOHA vary considerably; for example, some use computationally intensive approaches for classifying stream segments and looking for ecological responses to changes in flow, while others use existing, simpler classification systems and make more use of expert knowledge. In our view, the results are mixed, with the simpler applications being more successful, provided that well-qualified experts are involved. However, ELOHA does not seem to be working for its intended purpose of providing an economical way to develop regional flow standards for data-sparse regions (Hoekstra et al. 2011; Richter et al. 2012). Two examples are discussed below, one from Queensland, and one from the State of Michigan, USA. Although project in Michigan began before ELOHA was announced, it fits well in the category, and has been cited as an application of ELOHA by Kendy et al. (2012).

ELOHA in southeast queensland

A major test of ELOHA in southeast Queensland, Australia., described in Arthington et al. (2012), Rolls and Arthington (2014), and Mackay et al. (2014), occupies the computationally intensive end of the gradient just described. The hydrologic foundation was developed from 60 stream gages within the region, with an average of 25 years, and at least 15 years of data within the period 1975–2000, to describe historical (recent) conditions. Any gaps in the daily flow records were filled by regression modeling or interpolation. A reference data set, representing pre-development conditions, was generated from by output of an integrated water quality and quantity model.

Flow regimes in the area are highly variable, and many of the streams are intermittent. Thirty-five hydrologic metrics reflecting the magnitude (23), timing (5), frequency and duration (4), and rate and frequency of change (3) were calculated for each record at the "nodes" used in the model, or from stream gage records. These in turn were used to develop stream classifications for the reference and historical records separately, using the Mclust package in the R statistical package. These resulted in six reference and five historical classes. Then, the extent of hydrologic alteration between the reference and historical records was estimated using the Gower metric. Several exploratory analyses were also done, but these seem intended to explore the results, not affect them.

To assess the ecological effects of changes in flow regimes, data were collected on riparian vegetation, aquatic vegetation, and fish, at 44 sites along 22 reaches reflecting the major hydrologic gradients in southeast Queensland. Sites were sampled three times at approximately four-month intervals, and the results were used to estimate the ecological response to hydrologic alteration at the sites.

The main findings of the project seem to be that overall there had been relatively little change in the flow regimes or in the ecology of the study sites, that some undammed streams have been altered by land use and abstractions, and that different dams had different effects on the flow regimes downstream, depending on how much water they impound and how they are managed. This is thin broth, which may be explained by several factors.

Unfortunately, the flow records used are too short to characterize the highly variable flow regimes adequately, so the project lacked a firm hydrologic foundation. This may help account for the meager results. Arthington et al. (2012) decided that 15–25 years was "sufficient to calculate metrics representative of the long-term values," citing Kennard et al. (2010a), who used a resampling approach to estimate how the variance of 102 hydrologic indices decreased as the length of flow record increased. Kennard et al. (2010a, p. 150) reported that:

> Time series long enough to capture rare events are typically uncommon and consequently compromises must frequently be made in which the benefits of increased accuracy are balanced against the availability or spatial coverage of such long-term data. The results of the present study suggest that improvements in accuracy generated by inclusion of records of increasing length typically decrease in magnitude after 15 years.

However, the length of record needed to estimate metrics representative of long-term record in a stream must depend on how variable the flow is, so recommending a single number seems strange, and on analysis turns out to be unjustified. The analytical approach used in Kennard et al. (2010a) is flawed in several ways, resulting in underestimates of the uncertainty in flow metrics from streams with variable flows (Williams 2017).

Moreover, many of the Queensland streams go dry, or nearly so, every year. Very low flows are notoriously hard to model, because the uncertainty in the model results is large relative to the flow, particularly for alluvial streams, where interactions between the surface flow and groundwater are important. Third, the methods used to look for hydrologic alteration–ecological response relationships do not seem powerful as Bayesian methods, such as used by Webb et al. (2015a). Finally, the application appears to be disconnected from actual management of the streams in question, which could result in a lack of focus.

ELOHA in michigan

As a party to the Great Lakes Charter of 1985 and the Annex to the charter of 2001, Michigan was obligated to develop a water management system that would prevent further "adverse impacts" to water-related resources from new water withdrawals. Accordingly, in 2006, Michigan adopted legislation requiring development of an integrated assessment model to determine the potential for new withdrawals to reduce flows "such that the stream's ability to support characteristic fish populations in is functionally impaired," plus development of regulations to avoid such impairment. A Groundwater Conservation Advisory Council (GWCAC), comprising representatives of major interest groups, was established to oversee the development of the model by scientists, mainly from the Michigan Department of Natural Resources and the University of Michigan (working group). Various reports and journal articles describe the model and its development (e.g. Steen et al. 2008; Hamilton and Seelbach 2011; Zorn et al. 2011, 2012; and others cited therein).

Michigan has a moist continental climate with annual average precipitation of ~0.7 to 1.0 m. The area is strongly influenced by Pleistocene glaciation, and the surficial geology varies from exposed bedrock to coarse grained morainal deposits, so the permeability of subsurface is highly variable spatially. According to Zorn et al. (2012), "many Michigan streams are low-gradient, sand and gravel, hydrologically stable environments with temperatures governed principally by groundwater inputs that can be quantified as base-flow yields." The streams support over 60 species of fish.

The approach developed by the working group is based on what are essentially habitat suitability curves, using landscape-scale habitat variables: basin area, July mean temperature, and "base flow yield," (quantified as median August streamflow divided by catchment area), instead of the usual microhabitat measures of depth, velocity, and substrate. As such, it suffers many of the same conceptual problems as other methods using habitat suitability criteria. However, the selection of variables seems well founded; earlier work had shown that clusters of Michigan fish species were spatially distributed in a meaningful way when plotted on axes of catchment area and low-flow yield (Zorn et al. 2002), and Brenden et al. (2008) used regression tree analysis to develop a stream segment classification based on catchment area and mean July temperatures. Moreover, the curves are used in an interesting way, described below, and were developed from a great deal of data on fish distributions and abundances in Michigan. The data were collected mainly by the Michigan Department of Natural Resources, and also by the U.S. Forest Service and the University of Michigan. Michigan also has accomplished fish biologists who have worked in the area for many years, and so was well situated to meet the challenge posed by the legislation.

The Michigan approach uses a stream segment classification based on stream size (streams, small rivers, large rivers) and water temperature (cold, cold transitional, cool, warm), using class limits developed by expert opinion. There are no cold, large rivers in the state, so the system includes 11 classes, into which fit ~9 000 stream segments of varying length, defined by consolidation of ~34 000 segments in the state in the USGS National Hydrography Dataset, again using expert opinion.

The data used were: fish density at 1720 electrofishing sites and 158 rotenone sites distributed around the state, July mean temperatures estimated for each segment by regression and kriging with data from 830 stations; base-flow yield estimated from nearby stream gages or by a regression model using gage data and landscape variables such as surficial geology and topography, using data from 147 stream gages (Zorn et al. 2012); and catchment area from GIS data.

The working group developed a model for estimating the response of the individual species of fishes in each segment to reductions in flow, using fish density data for over 30 species, and the three habitat variables mentioned above: catchment area, July mean temperature, and base flow yield. For each species, they fit normal curves to the habitat data from the 20 % of the sample sites with the highest density for the species, and assumed that the means represented optimal conditions. The group then discretized the curves into bins each one standard deviation wide, which were given scores of 4, 3, 2, 1, and 0. Thus, for example, sites with a catchment area within half a standard deviation of the mean (for the top 20 % of sites) were given a score of 4 for that variable, sites with areas between 0.5 and 1.5 standard deviation of the mean got scores of 3, etc. The three habitat scores were then combined for the species using the minimum value for each triplet. Note that the response curves are not based on actual changes in the landscape variables, but rather the distributions of the fishes over the variables in data from the sampling sites; these were assumed to be equivalent.

Curiously, this process was done separately for 33 species in the northern part of the state and 43 species in the southern part, with a great deal of overlap among the species, "...to control for some of the spatial variation ..." (Zorn et al. 2012). Why the optimal physical habitat for a species should vary spatially is unclear. Most likely, the Working Group's decision to treat the areas separately reflects spatial variation in fish communities, and underscores the conceptual muddle of suitability curves based on physical habitat.

A definition of "characteristic fish populations" is needed to assess whether a stream's ability to support such populations would be impaired. The definition was developed by assessing densities of fishes at several hundred electrofishing sites as function of the composite scores, where the density for each species was scaled by dividing by the median density of the species over all sites, to make the scaled densities of more and less abundant species comparable. The scaled densities for the species were then averaged to get a composite density for the site. These scaled composite densities are close to linearly related to the scores, which was taken as a kind of test of the method, although the same was not necessary true for individual species. After inspecting these densities, composite scores of 2 and 3 were selected as defining characteristic and "thriving" fish assemblages, and the "model" described above was used to define a curve showing the response of the characteristic fish assemblage to withdrawals of water. Curves were also developed for the thriving fish assemblage, and used to indicate the risk of an adverse resource impact.

The GWAC, a political body, decided that a flow reduction resulting in a 10% decline on the characteristic species curve would constitute an "adverse resource impact," as shown by the point labeled "c" in Figure 6.8. However, there is naturally a good deal of uncertainty associated with this estimate, so that more careful assessment of the potential harm from a proposed withdrawal is appropriate as the anticipated proportional reduction in flow increases. To address this, the

working group tentatively defined two other points on the "proportion of flow removed" line in a similar way. A vertical line dropped from the intersection of the 10% decline line and the "thriving species" curve identifies one point on the "proportion of flow removed" axis in Figure 6.8, and another point was similarly defined by drawing a horizontal line showing a 20% decline, and then a vertical line dropping from its intersection with the thriving species curve. This resulted in three zones of effect less than an adverse resource impact: a "safe" zone, and then two zones increasing potential for harm. The final Michigan legislation used this zone concept, but defined the zones using narrative rather than graphical criteria.

The model can be used to simulate the effects of diversions on fish in the streams because the scores for base flow yield and July mean temperature will change as water is diverted. The change in the scores will not be continuous, but can be smoothed by

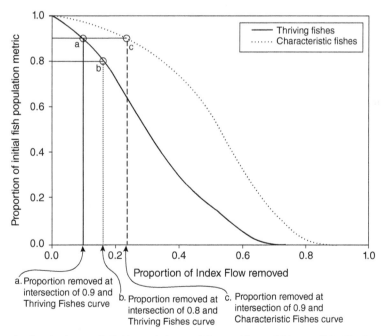

Figure 6.8 Curves showing the decrease in thriving and characteristic assemblages of fishes, and identification of three zones of increasing potential for harm. Source: Hamilton and Seelbach (2011).

plotting the scores as the flow is decreased by 10, 20, 30 %, etc. and drawing lines through these points.

The Michigan approach was developed in response to a legal mandate by accomplished scientists who have worked in the area for some time, and had a good many data to draw upon. Professional judgment played a major role in developing the method, but it seems likely that their judgments were as good as could be expected in the circumstances. However, the approach has shortcomings, several of which are described in the citations given above. For example, Zorn et al. (2008, p. 20) state that:

> One weakness of our regional-scale HSI (habitat suitability index) approach is that although we were able to describe the central tendency of fish–base flow relationship for each river type, we also identified considerable variation around this for individual species and sites. Our simple model is useful for setting regional flow standards but certainly does not account for all the complex mechanisms that affect fish abundances at specific sites. Our approach, which involves looking across spatial gradients of [base flow yield] reduction, lacks the resolution and mechanistic insight that can come from studies tracking responses of individual streams through time. Site-scale models can bring greatly enhanced accuracy and mechanistic realism; however it is impractical to do enough of these to provide comprehensive regional coverage. Also, the required assumptions about modeled habitat changes indexing actual fish population changes have rarely been validated, thus the empirical confidence in many HSI studies is low. However, correspondence between model scores and measured fish densities in this study provides empirical support for our HIS-based approach. As both regional and site-scale modeling approaches have somewhat complementary strengths and weaknesses, we recommend conducting a targeted series of site-scale studies across a range of Michigan river types for comparison with the regional results.

Moreover, the method is based on relating changes in abundance to changes in features of physical habitat that were estimated with models. However, Brenden et al. (2007) compared modelled stream habitat features with the same features measured in the field, and cautioned that:

If model-predicted local habitat data are included in analyses meant to elucidate fish assemblage–habitat relationships, then the results from our research on Michigan and Wisconsin streams suggest that conclusions regarding observed environmental gradients or directional relationships between local habitat features and individual species or assemblage metrics should only be considered as tentative.

It seems likely that applications of ELOHA could be improved by using Bayesian methods, which quantify uncertainty in the assessment, and provide a formal and transparent way to incorporate expert judgement. Moreover, hierarchical Bayesian modeling would allow for variation in relationships between flow and ecological responses among the different streams in each class of streams.

6.5.4 Adaptive management

Adaptive management has been a popular recommendation for environmental flow assessment since Castleberry et al. (1996). According to Poff et al. (2006, pp. 164–165) "Ideally, the ELOHA framework should be used to set initial flow standards that can be updated as more information is collected in an adaptive cycle that continuously engages water managers, scientists and stakeholders to 'fine tune' regional environmental flow standards." Linnansaari et al. (2012, p. v) wrote that: "Regardless of the type of framework to be established, it is fundamental that the established environmental flow standards are preceded and followed by a controlled monitoring program and the possibility to refine the environmental flow regime standards by adaptive management in an iterative process."

Adaptive management has nevertheless been difficult to apply effectively. Probably the greatest obstacle is mental; good adaptive management requires a more scientific mind-set than is common among regulators, agency managers, stakeholder advocates, and even many fish biologists. In our experience, they tend to focus more on reaching decisions than on learning. However, an exemplary applications in Australia is described below.

A conceptual model from Healey et al. (2008) provides a good overview of adaptive management, which is presented as a cyclical process in which even the understanding of the problem and the goals of management can change in light of new information (Figure 6.9). Performance criteria, to keep the assessments from becoming post-hoc rationalizations, are an important but uncommon element.

It is common to distinguish active adaptive management, or actual management experiments, from passive adaptive management, which is essentially observational (e.g. Gregory et al. 2006). An example of active adaptive management applied to environmental flows was described by Failing et al. (2004) and Gregory et al. (2006), and also in Chapter 7. Briefly, a dam on the Bridge River, British Columbia had existed with no flow releases for about 40 years, although inflow from tributaries below the dam provided some flow to most of the study reach. The proposed experiment consisted of releases at four rates of discharge for several years each, with an active monitoring program. Unfortunately, only two of the treatments were implemented, so learning was limited (Bradford et al. 2011).

Litigation regarding environmental flows in the lower American River in California produced an example of passive adaptive management (Castleberry et al. 1996). The court decision set did not mandate flow releases from a dam, but did prohibit diversions from the reservoir by a municipal utility district unless interim flow standards were met below the dam. The judge recognized the uncertainty regarding the flows needed to achieve the intended level of environmental protection, and so ordered the parties to cooperate in studies to clarify what the flow standards should be. One of us (JGW) was appointed to supervise these studies.

A major issue in the controversy concerned rearing flows for Chinook salmon. One side argued that lower flows in the spring would result in faster growth of juvenile Chinook, while the other argued that lower flows would result in slower growth. Water temperature in the spring rearing season varies inversely with flow, so the argument was really about the relation between water temperature and

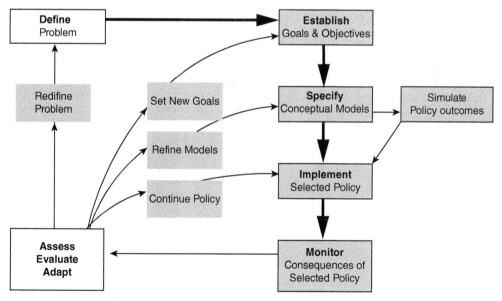

Figure 6.9 Conceptual model of the adaptive management cycle. Note that "policy" as used here may include taking some action. Source: Healey et al. (2008).

growth, which depends strongly on the amount of food available. With enough food, juvenile Chinook grow rapidly at temperatures that would be harmful if less food were available. Because there is considerable variation in spring flow and temperature in the lower American River regardless of management, the relation between temperature and growth could be clarified by sampling juvenile Chinook and determining their growth rates by otolith microstructure analysis. Although the monitoring program ended prematurely, the preliminary results strongly suggested that growth was faster at lower flows (Williams 2001).

Active adaptive management is difficult because: (i) it is politically difficult to arrange; (ii) ecosystem responses take time, so the experiments do as well; (iii) other and uncontrolled factors also affect the system of interest, subjecting flow experiments to unexpected problems. A 12-year adaptive management program on the San Joaquin River in California failed to give clear results (Dauble et al. 2010), probably because of such confounding factors. Although learning takes longer with passive adaptive management, it seems much easier to do, and effective learning is still possible provided there is enough variation in the key driving variables, as in the American River example just discussed, and the monitoring program is well designed. In these cases, passive adaptive management seems the preferred approach. This does not preclude taking a scientific approach; sciences such as geology are primarily observational. Rather, whether the adaptive management is active or passive, the challenge is to do it well. The selection of good response measures can be critical; in the American River case just discussed, learning depended on measuring attributes of individuals (growth rates from otoliths), rather than just population attributes such as abundance. Individual-based metrics tend to be more informative (Osenberg et al. 1994)

6.5.5 Evidence-based EFA

An "evidence-based" approach to EFA, modeled on approaches from evidence-based medicine, and contrasted with "experience-based" methods, has been developed in the Australian state of Victoria over the last decade, and an application of the approach is described by Webb et al. (2015a). The approach was developed for streams for which environmental flows and a monitoring program had already been defined using an experience-based method, so the evidence-based approach can be regarded as a rigorous way to apply passive adaptive management to environmental flows. The approach was developed by research scientists, particularly at the University of Melbourne, and includes structured reviews of the scientific literature, structured expert elicitation, monitoring data collected with standardized methods over large spatial scales, and Bayesian methods to combine these different kinds of information to assess the ecological response to environmental flows.

The more experience-based method currently used in Victoria was described by Hart and Pollino (2009) as follows:

> … The FLOWS methodology establishes a series of environmental objectives for each river reach as the basis for developing the final environmental flow recommendations. The ecological objectives are linked to a particular environmental asset, such as native fish, water quality, stream channel form or macro-invertebrates. The water requirements of the environmental asset are linked to particular flow components, which include low and high flows for summer and winter, freshes and cease-to-flow ….
>
> A technical panel is employed to determine the objectives and flow components required in each reach of the river. The panel also determines the volume, timing, duration and frequency that are associated with each flow component. …

Thus, FLOWS is one of several holistic methods that were precursors to ELOHA, but with less emphasis on stream classification.

Developments in water politics in the twenty-first century in Australia, plus the standard problems of insufficient data or funding for monitoring, motivated development of the evidence-based approach (Stewardson and Webb 2010). Under the Australian constitution, the power to manage water resides with the states rather than the commonwealth government. However, despite the evident need for better basin-wide management of the Murray–Darling revealed by the Millennium Drought, the states in the basin could not agree on a new plan. To break this impasse, the commonwealth government resorted to its obligations under international environmental treaties, such as the Convention on Biological Diversity, to pass the Water Act of 2007. This in turn required that the basin-wide management provide for environmental protection (Skinner and Langford 2013). The Water Act of 2007 established the Murray–Darling Basin Authority (MBDA), and mandated development of a Basin Plan that was adopted in 2012 (Hart 2016).

Across the Murray–Darling basin, meeting the sustainable limit to extractions required by the Water Act will require reducing annual diversions from the rivers by up to 2750 Gl, which is equivalent to a constant flow of over $87\,m^3\,s^{-1}$. Even though this allocation of water to the environment will be reduced proportionally with allocations to out-of-stream uses in dry years, the allocation is serious business for a drought-prone region where many rivers often go dry. Political resistance to the reduced diversions created considerable pressure to show that estimates of environmental needs are reliable, and to develop more defensible methods for estimating such needs. In particular, the dependence on expert opinion in methods such as the FLOWS approach seemed vulnerable to legal challenge (Webb et al. 2015a).

During the drought, invasion of stream channels in Victoria by terrestrial vegetation became a major problem. Webb et al. (2015a) describe the evidence-based approach using the example of an analysis of the efficacy of channel inundation for reducing the amount of terrestrial vegetation. The approach is generally applicable, using Bayesian modeling to combine information from systematic reviews of the scientific literature, expert opinion derived from structured questions, and monitoring data, to predict the ecological responses to environmental flows. The steps in the approach, as described in Webb et al. (2015a) and shown in Figure 6.10, are:

- Develop evidence-based conceptual models relating flows to ecological responses using systematic literature reviews, and formulate the conceptual models as a Bayesian Network (BN) models;
- Quantify the linkages in the conceptual models using expert opinion in a structured process that allows for filling in the conditional probability tables for the BNs,
- Collect relevant monitoring data across wide spatial scales using compatible methods;
- Translate the conceptual models into process-based mathematical expressions of the relationship between flow and ecological response, in the form of Bayesian hierarchical models; and
- Use the BN models to develop prior probability distributions for the Bayesian hierarchical models.
- With the hierarchical Bayesian model, update the prior distributions with data from the monitoring program to predict the ecological response.

Besides duration and frequency of inundation, the model included the season of inundation, slope of the bank, and the presence or absence of livestock, as covariates. In this case, the model estimated means and standard deviations of the expected terrestrial vegetation cover at 27 sites on seven rivers for four different inundation conditions. The estimated effectiveness of inundation varied considerably among the deliberately selected sites, from very high at Broken River site to very low on the Yarra River site, although the standard deviations of the estimates were small (Figure 6.11). As more monitoring data become available, these estimates can be updated, or, if it is indicated, the conceptual and numerical models can be revised. However, because the sites were selected deliberately, the results cannot be applied rigorously to other reaches of the rivers. Although not used for the analyses of flow and terrestrial vegetation in Webb et al. (2015a), stream classifications developed with Bayesian mixture modeling (Webb et al. 2007; Kennard et al. 2010b) could be used as needed, and not all the steps listed above are strictly necessary. According to Webb et al. (2015a, p. 513):

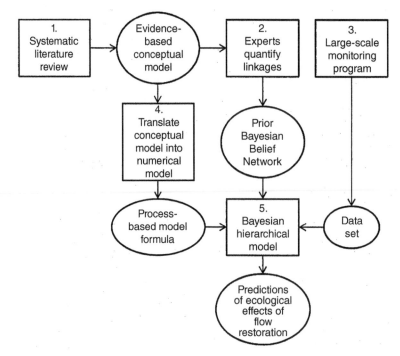

Figure 6.10 Work and information flow for the framework presented in this paper. Boxes are tasks and correspond to the step numbers listed below. Circles are outputs from each step. Source: Webb et al. (2015a), with permission from John Wiley & Sons.

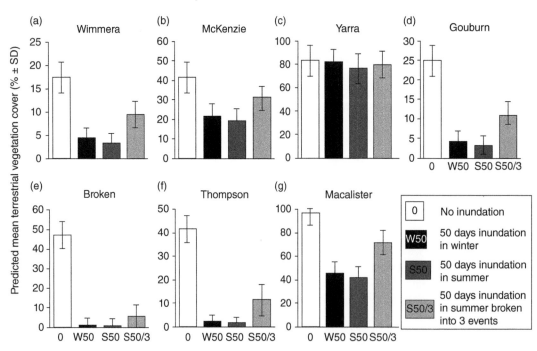

Figure 6.11 Predicted cover by terrestrial vegetation for four treatments at six of 27 sites on seven rivers; error bars show standard deviations. Source: Webb et al. (2015a), with permission from John Wiley & Sons.

We believe that the core of the approach lies in the Bayesian hierarchical analysis of data collected over large scales. It could be argued that if one already had a very strong idea of what processes to model and/or hypotheses to test, then the systematic review component is not required. Similarly, if one is very confident in the quality and precision of the data collected, then the effort involved in conducting expert elicitation to develop informative prior distributions may have marginal benefits.

Long-term collaboration among state and catchment managers, consultants, and researchers was a critical, and time-consuming, aspect of developing and applying evidence-based EFA (Webb et al. 2014). For example, the approach uses Bayesian methods, and the capability to do Bayesian modeling resides mostly in research institutions. For researchers to be wholeheartedly involved, however, requires that they be able to publish their work, since publications are the coin of the academic realm. Thus, managers needed to accept that publishing is a legitimate need. Perhaps as an unintended consequence, various aspects of the approach are well described in the literature. (e.g. Chee et al. 2009; Webb et al. 2007; Stewardson and Webb 2010; Webb et al. 2010a, 2012, 2014, 2015a, b; de Little et al. 2013). On the other hand, academics need to understand the practical difficulties with implementing monitoring programs, and accept that data will be less than ideal. The elements of models, and Hierarchical Bayesian models, are discussed in Chapters 5 and 9.

6.6 Conclusions

Methods for EFA range from very simple to very complex. Environmental flow assessment is a fundamentally difficult problem to which complex modern methods of statistical modeling can be usefully applied, but the value of any method cannot be judged by its complexity. Methods and models are not self-executing, and any can be misused. Methods should be selected to match the issues involved in the assessment and the resources available for it, but Bayesian methods will be optimal in many cases. Habitat association models such as PHABSIM that operate at the microhabitat are popular in some countries but are fundamentally flawed and should be abandoned.

CHAPTER 7
Good modeling practice for EFA

Summary

The proper use of models requires a good understanding of the model, the data at hand, the thing being modeled, the questions being asked, and the reasons for the modeling. Although many kinds of models are useful for EFA, considerations that apply to most or all are described, as are some that are specific to statistical models. If possible, models should be tested. The normal purpose of testing models should be to increase or decrease confidence in the model for a given purpose, not to yield a binary verdict such as valid or not valid. Models of habitat selection by juvenile Atlantic salmon are used to illustrate main points.

7.1 Introduction

The use of models is ubiquitous in environmental flow assessment (EFA), so understanding the selection, implementation, and testing of models is important. However, that understanding is complicated by the great variety in the models that are or could be used. A major assessment is likely to involve: a hydrological model to estimate future and perhaps reference discharge; a hydraulic model to simulate the depth and velocity of the flow in some reaches at a given discharge; a water temperature model and perhaps a water quality model to simulate temperature and other water quality variables, given a flow regime; and biological models to predict the response of aquatic organisms to the flow regime. There are many differences among these types of models, but some considerations apply to all.

Models are best used in EFA to help people think, not to provide answers. Models are not representations of the real world; rather, they are representations of selected aspects of the world, typically described by data that are thought to be relevant for the question being addressed. It follows that proper use of models requires a good understanding of the model, the data at hand, the thing being modeled, and the questions being asked. Developing that understanding is a major challenge, and typically requires a joint effort by people with different areas of knowledge. Using model results without this understanding is perilous.

The physics of water flow and temperature are well understood, so the complexity and accuracy of flow and temperature models are limited mainly by the input data and computer power, which are real constraints. The situation is quite different with biological models. As discussed in Chapter 1,

Environmental Flow Assessment: Methods and Applications, First Edition. John G. Williams, Peter B. Moyle,
J. Angus Webb and G. Mathias Kondolf.
© 2019 John Wiley & Sons Ltd. Published 2019 by John Wiley & Sons Ltd.

ecological systems are enormously more complicated than physical systems. Ecological systems have many weakly interacting elements that are often poorly known or unknown, with often non-linear relationships among them, as well as a few more strongly interacting elements and better-known relationships. Modeling ecological systems, or major elements of ecological systems, therefore needs different approaches and even more attention to uncertainty.

Environmental flow models generally fall into two broad categories: simulation models and estimation models. Essentially, simulation models are thought experiments (Starfield 1997; Schnute 2003), in which parameter estimates often come from the literature or from expert opinion. Most models used in EFA are simulation models. If the subject of the model is well understood, as with hydraulic models, then simulation models may be able to make reliable predictions. There is much more uncertainty with most biological systems, however, so simulation models of these systems are mainly useful for showing the consequences of the way we think the systems work, rather than for predicting how they actually work. Estimation models, on the other hand, are inherently statistical, with parameter values obtained by fitting the model to data. They answer the question, how consistent are the data are with the model? They can also be useful for making predictions, at least for the situations from which the data came, and the uncertainty in the prediction can be estimated. Despite these differences, various issues are relevant to both types of models. Therefore, this chapter summarizes some standard guidance about modeling practice, with emphasis on biological modeling, discusses testing models; and then reviews some examples from the literature.

7.2 Modeling practice

Many authors have described good modeling practice and the steps in the modeling process. See, for example: Jakeman et al. (2006) for environmental models generally, Hirzel and Le Lay (2008) for wildlife habitat suitability modeling, Blocken and Gualtieri (2012) for hydraulic (computational fluid dynamics) models, Chen and Pollino (2012) for Bayesian Network models, Grimm et al. (2014) for ecological simulation models, and Rose et al. (2015) for models of the effects of ecosystem restoration on fish. Other examples are cited in these papers. Despite the different fields of emphasis, and how finely the authors divide the steps in the process (ten in Jakeman et al. 2006, 31 in Rose et al. 2015), there is considerable agreement among authors on the main points, which are nicely summarized by Chen and Pollino (2012) as follows (slightly modified):

- *Clearly define the purpose of the model and the assumptions underlying the model;*
- *Thoroughly evaluate the model and its results;*
- *Transparently report the whole modelling process, including its formulation, parameterization, implementation, and evaluation.*

Unfortunately, despite the general agreement in the literature regarding good practice in modeling, actual practice often falls short, as noted by Jakeman et al. (2006, p. 612):

> In summary, the modeling process is about constructing or discovering powerful, credible models from data and prior knowledge, in consort with end-users, with every stage open to critical review and revision. Sadly, all too often in reality it is the application of a predetermined model in a highly constricted way to a problem, and to the social dimensions of which the modeler is unaware.

Rather than tell modelers what to do, like the papers cited above, we offer a set of questions that anyone associated with the modeling process can ask.

7.2.1 What are the purposes of the modeling?

Without a clear statement of the purposes or objectives of the EFA, it will be impossible to judge whether models can satisfy them, so clarity is critical. Broadly speaking, the purpose usually will be to predict the ecological or geomorphic responses to a contemplated change in the flow regime. Because of the complexity of ecological systems, however, firm predictions are not a realistic objective. Instead, the better approach is to acknowledge uncertainty, and to make learning about how the system responds one of the objectives of management. Many assessments take a narrower objective, however, such as predicting the effect of the project on habitat for one or a few species that people care most about. These species may be described as indicators of ecological conditions generally, to try to bridge the gap between the two kinds of objectives, but such claims seem too convenient and should be examined critically before they are accepted.

Unfortunately, the real objectives often differ from the stated objectives. Most EFAs occur in the context of disputes over the allocation of water between instream and out-of-stream uses, so the actual objective of parties in an assessment may be to minimize or maximize the legal, regulatory or financial obstacles to approval of a water project. Or, particularly for regulatory agencies, the actual objective may be to provide an answer that allows a decision to be made, rather than to provide an accurate answer that acknowledges uncertainty. When an assessment must be done, and higher-ups do not understand the need for accepting uncertainty and using an adaptive approach, this result is likely: the staff will be expected to provide clearer guidance for a decision than the complexity of ecosystems allow. This disconnect between the nominal and real objectives for assessments has been a major drag on the practice of EFA and the development of better methods, and a major hazard for the environment.

7.2.2   How should you think about the natural system being assessed?

The best you can do is model the way you think things work, so having a good conceptual model or models of the natural system in question is crucial for determining what numerical models to use and how to use them. Developing one or more conceptual models should be done early in the assessment. Models can only deal with selected features of the natural system, so the conceptual model should identify these and the relations among them, and should describe the spatial and temporal scales of different aspects of the system. Numerical models implicitly describe conceptual models, and these should be consistent with the stated conceptual models. If it is conceptually unclear how the numerical models under consideration relate to the conceptual model(s), then different numerical models or approaches should be used instead. The conceptual model should be subject to change as the assessment progresses and new information becomes available, but the temptation to change to conceptual model to fit an available numerical model should be resisted.

7.2.3   What data are or will be available, and how good are they?

Good data are often unavailable for environmental flow assessments, and this limits the types and resolution of the models that can be usefully applied. There is not much point to using a fine mesh in a 2-D hydraulic model without commensurate topographic data for the stream bed, or to using a detailed stream temperature model if air temperature data have to be estimated from measurements at some regional airport. Similarly, there is not much point to modeling the flow in a small part of a stream with expensively high accuracy, if the assessment has to cover the whole stream. Testing a model with poor or sparse data can also be problematic, as discussed in Chapter 9.

How well things can be measured should also be considered. Estimates of biological variables, such as fish abundance, are likely to be approximate, and data on variables that may seem to be well measured, such as water depth and velocity at points occupied by diver-observed drift-feeding fish, may reflect biased sampling, as discussed below. Finally, data may include errors, from equipment malfunctions or human error. Careful attention to the data is critical for good modeling.

7.2.4 How will the available budget be distributed over modeling efforts, or between modeling and data collection, or between the assessment and subsequent monitoring?

Assessments have finite budgets, so there will be competition for funding for various aspects of the problem and their corresponding modeling efforts. Getting a good allocation of funding over all efforts will be unlikely unless there is clarity on the purpose of the assessment and a good conceptual model of the overall problem. Even then, unforeseen factors will likely make changes in the assessment necessary, so maintaining a contingency fund is generally prudent.

Particularly for assessments concerning proposed dams or other essentially irreversible decisions, there is an argument for going all-out on the assessment, especially if revenue from the project will provide funding for follow-up work. In developed countries, however, assessments are likely to concern changes in the operation of existing projects, as when a permit for a hydroelectric project is up for renewal. In these cases it may make sense to base initial modifications on simple models, such as a Bayesian network model parameterized with expert opinion, and to allocate the bulk of the available funding to data collection over time; i.e. to a monitoring program and adaptive management. This assumes that money may become available for additional modeling once the data are in hand, leading to reconsideration of the modifications. Assessments are more likely to deliver good value for the money in the context of adaptive management.

7.2.5 How will the uncertainty in the results of the modeling be estimated and communicated?

Models deal only with selected aspects of the world, so applying model results to the world inherently entails uncertainty. This makes estimating and communicating the uncertainty in the results important for EFA. With estimation models, this can usually be done with confidence intervals or various other statistics, although these may understate the uncertainty if the assumptions of the model are not met. With simulation models, such as hydraulic models, the uncertainty can be estimated by comparing simulated and observed data, or by Monte Carlo methods, with repeated runs using draws from statistical distributions for key parameter values (e.g. Wilcock et al. 2009).

Dealing honestly with uncertainty is fundamental to good practice in modeling for EFA. Francis and Shotton (1997) described six types of uncertainty associated with fisheries management that also apply to EFA: process, observational, model, estimation, implementation, and institutional uncertainty. The first four of these are relevant to modeling. "Process uncertainty" arises from natural variability. Even if it were somehow known that on average a given reach of a river with a given flow regime could support some number of fish, the actual number that could be supported in a particular year would deviate from that average. No amount of study can reduce this uncertainty, which is large for ecosystems (Mangel et al. 1996). "Observational uncertainty" arises from measurement and sampling errors. Putting more effort into the observations can make them more accurate, although beyond some point the extra effort will not be worthwhile. "Model uncertainty" reflects incomplete knowledge. In science, models are often formulated as equations or sets of equations, but in a broad sense anyone who thinks he or she understands how something works has a model of it. However, ecological processes are so complicated that no one understands them completely, and models of such processes are necessarily incomplete. Moreover, models are typically most useful, especially for prediction, when they focus on a few main processes, and ignore others. However, the need to simplify limits the realism of the model. Values for the variables that adapt a model to a particular situation must be estimated; typically these are estimated from observed data, so "estimation uncertainty" results both from observational uncertainty and process uncertainty (errors in measurement and chance variation). Less obviously, the formulas used to calculate parameters from observations often imply models, so estimation uncertainty also derives from model uncertainty. For example, the traditional method for measuring the vertically averaged water velocity at a point in a stream assumes an idealized velocity profile that is often absent in natural channels. Beale and Lennon (2012) ended a review of uncertainty in species distribution models with language that applies as well to other models used for EFA:

> We argue that it is just as important to quantify uncertainty in model predictions as it is to make the predictions themselves, yet the importance of prediction uncertainty is rarely emphasized. By correctly identifying the existence and sources of uncertainty, even if in the worst case predictions are impossible, advice and management actions will be better informed than if based on false certainty.

7.2.6 How will the model and model development be documented?

Reports of modeling for EFA should describe clearly and honestly what was done, and why, and how the results help meet the purposes of the EFA. Generally, technical details that will be of interest mainly to specialists should go into appendices. Too frequently, the main bodies of reports contain mind-numbing details that seem intended to persuade naïve readers that the modeling was very scientific and rigorous, or at least that clients have gotten a lot of work for their money, but serve mainly to obscure more relevant points. How much information should be reported will depend on the model and the particular situation, but as a basic principle, it should be possible for someone else to reproduce the analysis.

7.2.7 How will the models be tested?

Models can be used with more confidence if they are properly tested in the application at hand, and when that is feasible, tests should be part of the assessment. In some cases, for example with models of the geomorphic response to a change in flow or sediment supply, the temporal scale inherent to the problem will preclude tests of the complete model, but partial tests, for example of the hydraulic component of

the model, may still be possible. Particularly with complex simulation models, simply checking that the computer code actually does what it is supposed to do is an important part of testing the model.

7.2.8 How good is good enough to be useful?

Models will never give perfect predictions, so there will always be room for disagreement whether the glass is half empty or half full, even if the uncertainty around model predictions is reasonably well known. One logical criterion is whether the uncertainty in the model results would affect a management decision (Williams et al. 1999), but the answer also depends on how the modeling results will be used. For example, uncertain temperature modeling results may tell you that you cannot predict whether a contemplated change in flow will push water temperature beyond some biological threshold, which could be a very useful thing to learn.

7.2.9 Who will use the results of the modeling, and how will they be used?

Models for EFA are normally used to provide input to decisions, and models will be most useful if they are understandable, or at least acceptable, to the people making the decisions. People making the decisions will often lack a good understanding of either the models or the things being modeled, and modelers will often lack a good understanding of the problems faced by managers, so an important aspect of an EFA is simply educational. Involving key parties and regulatory staff in the process of developing and implementing the models probably is the best way to do this (Webb et al. 2014).

7.2.10 Do you really need a model?

Other considerations may render some question that could be addressed by modeling irrelevant. In most instances, environmental flows will be balanced against other considerations such as water supply, power generation, recreation or flood management. Even within the EFA itself, factors such as water temperature, water quality, sediment transport, riparian vegetation, geomorphic flows, (Chapter 8) and habitat for other species or life stages normally will be considered as well, and may determine the eventual decision. For an example from our experience, flow releases from Folsom Dam into the lower American River during the fall provide spawning habitat for Chinook salmon. The American River is a substantial tributary to the Sacramento River that runs through the city of Sacramento, and the main reservoir on the river is operated for both flood control and water supply. Chinook salmon spawn and rear below the most downstream dam, so that the relation between flow and spawning habitat seems important. However, the reservoir must be drawn down to a defined level by the middle of the spawning season (mid-November) to provide flood protection for the city of Sacramento. Drawing the reservoir down too far will result in reduced flows in the winter and spring if the following winter is dry, affecting rearing habitat for juvenile Chinook, and increasing releases for spawning will increase this risk. Reducing flows after salmon begin spawning, if rains do not come, may strand redds. Finally, a limited amount of cool water deep in the reservoir may be available to speed the fall cooling of the river, which is typically too warm for egg survival when salmon first appear on the spawning grounds, and high spawning flows followed by a dry winter will result in less cool water being available the following fall. Given these constraints, microhabitat-based predictions of the relation between flow and spawning habitat may have limited influence on management of the river, even when providing spawning habitat for Chinook salmon is an objective of management. The point is that models should not be used simply because they can be, or because it is conventional to use them. Instead, the reasons for using the model in each case should be carefully thought through and articulated.

7.3 Behavioral issues in modeling for EFA

Modeling involves many subjective choices about what to model and how to model it. As one consequence, the motives of those involved with the modeling deserve more attention than they have received. In a recent review, Hämäläinen (2015, p. 244) remarked that: "There are many studies related to uncertainties in environmental modelling and impact assessment (for a review see, e.g. Refsgaard et al. 2007). There is also literature on uncertainties due to model structure error (Refsgaard et al. 2006) but very few studies on the uncertainties related to the skills and behavior of the modeler."

Motives affect modeling for EFA in various ways, even if modelers are not consciously influenced by unstated interests. People are social animals, and the groups to which we belong influence our thinking. Members of a group may reach consensus on modeling some issue, and then reinforce each other's reluctance to reconsider the consensus in light of new evidence. Or, we may give too much weight to the opinions of people who we know and like, or who are on the same side of a conflict. There is also a natural tendency to stick with the familiar, so people who are familiar with one kind of model will tend to use it, even if the choice is poor. Modelers also tend to fall in love with their models, especially if they have worked hard to develop them or to develop skill using them, and their affection can blind them to the model's shortcomings. And, abandoning one model in favor of another is in a sense admitting error, and will raise questions about prior work done with the first model.

Bad models will be used if other models are not available (Hilborn and Mangel 1997), and, for the reasons just discussed, once bad models are established, a kind of inertia can keep out improvements or better approaches. For example, traditional PHASIM combines a hydraulic model with suitability curves for each microhabitat variable considered in the application. Typically, these curves are developed separately, and then some assumption is needed about how influential each curve should be in evaluating habitat, with equal weighting as the default. Logistic regression allows for analyzing the variables together, so that the weighting is derived from the data. This more defensible approach was described by Thielke (1985) and demonstrated by

Knapp and Preisler (1999), and a direct comparison by Guay et al. (2000) showed that logistic regression gave a better fit to the data. Nevertheless, the PHABSIM community has been reluctant to embrace this improvement, and when United States Fish and Wildlife Service (USFWS) (2013) did use logistic regression, it did so to define suitability curves for each variable separately, still leaving the problem of how to weight them, even though both Knapp and Preisler (1999) and Guay et al. (2000) were cited! The heavy hand of tradition seems the most plausible reason for this.

Unfortunately, "who pays the piper calls the tune" applies as strongly to EFA as to anything else, particularly when allocations between instream and consumptive uses of water are at stake. Most people involved with modeling in such situations politely pretend that everybody is just arguing about scientific issues, but only the dullest believe it. Hubert Morel-Seytoux (2001, p. 320) was once asked to explain how modelers for opposing sides in a controversy used the same groundwater model to reach contradictory conclusions. He found that neither side had made a mistake, and reported that "I have no doubt the hydrologists were competent. In fact they knew *very well* what parameters to choose and what assumptions to make in order to obtain results that would meet their clients' desire." Similarly, one of us once listened to an argument between consultants for a power company and staff of regulatory agencies about the appropriate shapes of habitat suitability curves for a PHABSIM study, where it was obvious to all, but unsaid, that the real argument was about more or less water being diverted to hydropower generators. Unfortunately, environmental flow assessments rarely lack this dynamic.

Behavioral problems overlap, because people tend to rationalize away the conflicts between what they should do and what they feel compelled to do. Few people study environmental science or modeling with the ambition of representing special interests or applying flawed methods, but many of them end up doing so, and are then tempted to rationalize what they have done. This is a human thing to do, and there is little point to blaming people for being human, but we should anticipate this behavior and try to counter it with critical analyses of the resulting work.

7.4 Data-dependent activities in developing estimation models

The process of developing conceptual and numerical models entails many choices about just what to model and how to model it. Even for established methods such as PHABSIM, such choices include the shape of the suitability curves to be used and the method by which they will be combined (e.g. arithmetic or geometric product). The question arises how much these choices should be informed by the data to which the model will be fit.

Looking for interesting patterns in data is central to the ordinary practice of science, and a search on "exploratory data analysis" on Google Books leads to many books by statisticians on how to do this effectively; John Tukey's 1977 book by that title is a classic. Almost always, however, the data at hand are only a sample of many similar sets of data that could be collected, and the real interest is in patterns in the population from which the data sets are drawn. Accordingly, exploratory data analysis can mislead by finding patterns that exist in the sample, but not in the population, and applying statistical tests to determine whether such patterns are meaningful is a mistake.

The situation is more complicated in the context of developing models, where experts offer conflicting advice. On the one hand, in *Model Selection and Inference* (1998) Burnham and Anderson take a clear position:

Examples of data dredging include the examination of crossplots of all the variables or the examination of a correlation matrix of the variables. These data-dependent activities can suggest apparent linear or nonlinear relationships and interactions *in the sample* and therefore lead the investigator to consider additional models. These activities should be avoided, as they probably lead to over-fitted models with spurious parameter estimates and non-important variables as regards the *population*. The sample may be well fit, but the goal is to make a valid inference from the sample to the population. This type of data-dependent, exploratory data analysis might have a place in the earliest stages of investigating a biological relationship and should probably remain unpublished ....

On the other hand, in *The Ecological Detective*, Hilborn and Mangel (1997, p. 40) advise:

Plot your data. Get to know them by using standard computer graphics routines to fit various curves (linear, polynomial, logarithmic, exponential). When there are two or more variables plot the data in many ways and look for correlation. Think about plausible functional relationships.

Even without classical data dredging, there are normally enough options in developing and applying models to data (the "garden of forking paths," or "researcher degrees of freedom") that bogus claims of statistical significance or goodness of fit are easy to get (Gelman and Loken 2013, p. 10):

In this garden of forking paths, whatever route you take seems predetermined, but that's because the choices are done implicitly. The researchers are not trying multiple tests to see which has the best p-value; rather, they are using their scientific common sense to formulate their hypotheses in reasonable way, given the data they have. The mistake is in thinking that, if the particular path that was chosen yields statistical significance, that this is strong evidence in favor of the hypothesis.

Similarly, even when the scientific hypothesis to be tested is stated in advance, the statistical hypothesis actually tested will be different, and there may be many possible statistical hypotheses for the investigator to choose from (Gelman and Loken 2013). Zuir et al. (2010) propose "a protocol for data exploration to avoid common statistical problems" that calls for formulating the biological hypothesis as the first step in the study, but this does not seem to

resolve "researcher degrees of freedom," and clearly is not helpful for situations when conducting the study turns up unexpected and interesting results.

So what can be said?

First, even in laboratory science, data-dependent activities in statistical modeling are such a problem that filing (pre-registering) a detailed description of how the data will be collected and analyzed is regarded as best practice (Ioannidis 2005). This is more restrictive than it may sound, since pre-registration almost requires that you replicate an earlier study, and so obtain a data set that is independent of the one used in developing the study and analysis. This makes the study confirmatory rather than exploratory. Correspondingly, when it is feasible, testing with independent data will generally be the best course for models used for EFA. However, in cases such as a model fit to historical data on fish abundance and flow in some stream, testing with independent data will not be feasible.

Second, the plausibility of the result matters. Ioannidis (2005, 696) notes that "As has been shown previously, the probability that a research finding is indeed true depends on the prior probability of it being true (before doing the study), the statistical power of the study, and the level of statistical significance (citations omitted)." Nuzzo (2014) provides a clear explanation why this is so, giving support for the adage that extraordinary claims require extraordinary evidence. If applying a model produces a surprising result, the method and its application should get extra scrutiny before it is accepted as a basis for action.

Third, it helps to show that modeling results are robust to minor changes in the model or the data, that applying the model to other data sets gives the same or similar results, or that other kinds of evidence also support the results.

Finally, the farther you are from the gold standard of pre-registration, and the more data-dependent your analysis, the less you can claim about your results based on p-values. In particular, for exploratory studies, p-values and confidence intervals really only describe the data analyzed, and do not provide a basis for inference (Berry 2017). For applying an established model to a particular situation, such as an assessment of the effects of a flow modification on a well-studied species, something similar to pre-registration seems appropriate. However, this gold standard may be impossible to meet. This is especially true in academic work where the objective is developing new methods or new insights, and time and money are constrained by the terms of research grants. If so, the best you can do is the best you can do, and you should not let the best be the enemy of the good. However, you should be honest about how you got your results, and qualify them accordingly. For testing models, however, as opposed to developing them, p-values derived from data-dependent methods are simply unsuitable.

Unfortunately, articles based on data-dependent analyses have been influential in EFA, especially regarding PHABSIM (Moyle et al. 2011, Appendix A). Jowett (1982) is perhaps the most egregious, and exemplifies the practice objected to by Burnham and Anderson (1998) in the quotation above. Jowett screened a large number of variables, including nine variations of WUA, for correlation with the abundance of large brown trout from in New Zealand rivers. Then, he selected from these "significantly" correlated variables to build regression models for trout abundance, and got the strongest relationship with a model with eight predictor variables including WUA for benthic invertebrates and adult brown trout, together with a fudge factor to deal with streams that were too warm for brown trout. On that basis, he claimed that "This study demonstrates that WUA is an important determinant of adult brown trout abundance, refuting one of the major criticisms of IFIM." Of course, correlation does not show causation, and "… in a world with a large number of unrelated variables and no clear a priori specifications, uncritical use of standard methods will lead to models that appear to have a lot of explanatory power" (Freedman 1983). To be fair, Jowett was not alone; other fisheries scientists, e.g. Wesche et al. (1987), committed similar statistical sins, and evidently the reviewers of these papers were blind to

the problem, as well. At the time, so were scientists in many other fields. Nevertheless, Jowett (1982) is still cited as showing that fish abundance is positively related to habitat indexed by WUA (e.g. Locke et al. 2008; Rosenfeld and Ptolemy 2012).

In a recent example illustrating how "forking paths" can lead to spurious findings of statistical significance, Sabo et al. (2017) found significant associations between features of the flow regime in the Mekong River at a mainstem gage site and catch in a well-documented fishery. However, the fishery is not on the Mekong, but on a tributary with unusual hydrology that connects the Mekong to a large natural lake and adjacent floodplain. Water flows from the Mekong to the lake during the wet season, and back to the Mekong during the dry season, so the seasonal hydrograph at the lake lags that in the Mekong by more than a month. The fishery targets fishes migrating from the lake and floodplain to the Mekong as the water level in the lake declines. Nevertheless, Sabo et al. claimed that the features of the Mekong River hydrograph were "drivers" of the catch. They then designed a flow release schedule for proposed hydroelectric dams across the Mekong that they claimed would increase catch in the fishery. One of the features, the magnitude of the monsoon flood pulse, is known to be important (Halls et al. 2013), but the others are not.

There are several other problems with the proposed flow regime (Halls and Moyle 2018; Williams 2018), but the relevant point here is that the Sabo et al. (2017) catch data do not match catch data in an authoritative report on the fishery (Halls et al. 2013), or in an earlier report by one of the Sabo et al. authors (So et al. 2007), although the patterns of the data are somewhat similar (Figure 7.1). The catch data are estimates based on complex stratified sampling described in detail by Halls et al. (2013), and it seems likely that Sabo et al. (2017) failed to take proper account of the stratifications. Thus, the significant associations are with the wrong numbers! This has practical significance; the proposed flow regime was published in a major journal, and could be used to justify constructing the dams.

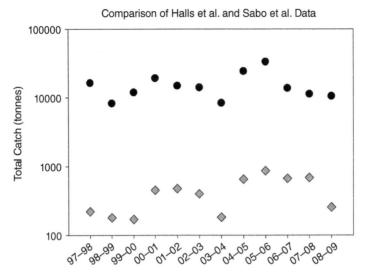

Figure 7.1 Comparison of estimates of the seasonal catch in the Dai fishery from Halls et al. (2013, black circles) and Sabo et al. (2017, grey diamonds). Although the trends in the data are somewhat similar, there are clear differences, and the magnitudes are far off. Note log scale. Source: John Williams.

7.5 Sampling

Once there is a decision to apply models, various issues arise about how to do so. There may be relevant data available, but generally more must be collected. This raises questions about where to collect data, and how to do it.

7.5.1 General considerations

Sampling is an important and generally neglected part of EFA. Except for hydrological models, data used to apply EFA models to specific situations are almost always collected by some kind of sampling, raising the questions what, where, and how to sample. Again it is important to be clear about objectives. As noted by Cochran (1977, p. 5), "A lucid statement of the objectives is most helpful. Without this, it is easy in a complex survey to forget the objectives when engrossed in the details of planning, and to make decisions that are at variance with the objectives."

Samples are collected in order to make inferences about the thing being sampled, and random or probability sampling provides the firmest ground for doing so. Sampling is a well-developed part of statistics, and good general texts that provide comprehensive discussions are available, for example Cochran (1977), Jesson (1978), or Thompson (2002), as well as books on sampling for ecological studies, such as Manly and Navarro Alberto (2015). Williams (2010a, p. 442) discussed sampling for environmental flow assessment, and summarized the main points in his conclusion:

> To be scientifically credible, [environmental flow methods] based on samples should incorporate probability sampling, and report interval estimates for whatever index of habitat the method produces. The sampling design should provide for a balance between sampling efficiency and spatial balance. The GRTS design, which is increasingly used in water quality monitoring, seems well suited for EFMs as well. All streams are unique, so knowledge of the stream and of the assessment method to be used are critical for developing a good sampling design, and results will be most satisfactory if a statistician familiar with sampling is involved in planning the study.

In a bit more detail, the "sampling universe," the thing to be sampled, and to which model results can properly be applied, should be clearly defined. In the context of models for EFA, this may be simply a reach of stream, or it may be all the streams in a region, or something in between. Or, it could be a particular species or life stage of fish within a set of streams, etc. "Sampling units," such as short reaches where depth and velocity will be estimated with a hydraulic model, or reaches within which fish will be collected by electroshocking, must also be defined, along with a "sampling frame" within which the sampling units can be randomly selected. The essential feature is that all potential sampling units have a knowable, non-zero probability of being selected. Then, an efficient sampling plan should be developed, generally one with good spatial balance, which allows for getting a good return of information from the investment of sampling cost. For EFA, the "generalized random tessellation stratified" or GRTS design (Stevens and Olsen 2004) is likely to be the best choice.

7.5.2 Spatial scale issues in sampling

Issues of spatial scale should be considered in developing and executing sampling plans: "There is growing awareness that patterns of habitat use by animals cannot be isolated from issues of scale" (Crook et al. 2001, p. 525). With fish, for example, there are different sampling methods to choose from: diver counts, electrofishing, seining, etc. These methods differ in their inherent spatial scales, and also in attributes such as potential for errors in identification or detection.

Frequently, the interest will be in the habitat features associated with the presence or abundance of organisms, and this can depend on the spatial scale of the observations. For example, Douglas (1995, cited in Cooper et al. 1998) investigated the relationship between the proportion of the channel with cover and the density of juvenile rainbow trout. At the scale of individual riffles, there was no relationship, but there was a strong relationship at the scale of reaches, which contained multiple riffles. Habitat features can be measured at points, or within

patches, usually called quadrats in sampling, and the size of the quadrat can affect the result. The substrate size apparently selected by several species of darters changed depending on the size of the quadrats used (Welsh and Perry 1998). For two species that occur between rocks in the stream bed, the size of the substrate selected increased as the quadrat size increased from 5 to 25 cm, but for a third species that occurs on top of rocks, the substrate size decreased over the same size range. For other species that occur over sand, the substrate size did not vary with quadrat size.

Nested sampling designs and hierarchical modeling allow for making inferences about habitat selection or habitat quality at multiple spatial scales (Deschénes and Rodríguez 2007). For example, in a nested sampling study of habitat use by juvenile steelhead, Cavallo et al. (2003) conducted snorkel surveys at three spatial scales in the lower Feather River in California: coarse (25 km), intermediate (300–500 m), and fine (25 m). They found that at the coarse spatial scale, the longitudinal position was the most important factor affecting the presence of smaller (<100 mm) steelhead. At the intermediate scale, most fish were observed in glide or riffle habitat, but all steelhead <80 mm were observed within approximately 2 m of shore. Depth and cover explained most habitat selection within that two-meter strip. Spatial scale issues are described further below, in a review of studies of models of habitat selection by juvenile Atlantic salmon. How to deal with habitat selection at multiple spatial scales in sampling plans is an issue that needs more attention, but it is clear that knowledge of the biology of the species of interest is critical for developing a good sampling design.

7.5.3 Cleaning data sets

Once data are collected, data sets should be checked for errors before it they are used. Electronic data loggers make it easy to collect various kinds of data, and avoid much human error. However, errors can still result from electronic, mechanical or human problems: a battery can run low, a water temperature sensor can be exposed to air if flow drops far enough, a technician can click on the wrong button while collecting or processing the data, etc. For example, recently the US Geological Survey technician who services the stream gage near where one of us lives selected the wrong conversion factor for the current meter used to measure flow for calibrating the gage, which resulted in an inexplicable increase in flow during the dry season. The USGS has a protocol for checking data, and in the course of this the technician realized his error and corrected the flow record, but only after he had been alerted to a potential problem.

Checking data sets is tedious and time-consuming, so there is always a temptation to neglect it; this should be resisted. EFA generally involves many kinds of data, and each kind is apt to have its own sources and types of error. Plotting the data is often an effective approach for detecting errors. For example, for time series data, plotting each data point against the adjacent one makes unusually large changes, or periods of no change, easy to see. These will flag either errors or unusual conditions that should be investigated.

7.6   On testing models

7.6.1   The purpose of testing models

All models are simplifications that deal only with selected aspects of the world, as emphasized by the saying that "all models are wrong, but some are useful." Thus, the normal purpose of testing models should be to increase or decrease confidence in the model for a given purpose, not to yield a binary verdict such as valid or not valid.

7.6.2   Why testing models can be hard

Showing that a model gives accurate predictions does not mean that it is a reasonable representation of reality, and choosing between two models that give equivalent predictions can be difficult, as shown by an historical example described by Oreskes (1998). The Ptolemaic model of the universe, with the earth at its center, gives good predictions of the position of the heavenly bodies as seen from the earth. The Copernican model also gave good predictions, but sixteenth-century astronomers favored the Ptolemaic model because the Copernican model seemed to fail a test. Because they underestimated the size of the universe, the astronomers tested the Copernican model by looking for changes in the angular position of stars that would be expected if the earth rotated around the sun, but did not find them. The risk of making such glorious errors in testing models for EFA is small, but the risk of making more pedestrian errors in testing models is large; however, it can be reduced by testing models in as many ways as possible.

7.6.3   The problem with validation

Oreskes et al. (1994) argued persuasively that validity and verity are not properties of models, so models cannot be validated or verified. The theoretical ecologist Richard Levins (1966, p. 430) said much the same, although in different language: "A mathematical model is neither an hypothesis nor a theory. Unlike a scientific hypothesis, a model is not verifiable directly by experiment. ... The validation of a model is not that it is 'true' but that it generates good testable hypotheses relevant to important problems." Even if a species distribution model does a good job of predicting the probability that some species will be present or absent in a particular patch of habitat, that does not make the model valid in the sense that logic can be valid. Consider the situation where a species distribution model is validated on one stream, but is not validated when tested at a different time, or on a different stream. What does it mean if a model is both validated and not validated? This is not a hypothetical case; for example, Girard et al. (2003) found that a species distribution model developed with microhabitat data collected under clear conditions did a good job of predicting fish distributions on clear days, but not on cloudy days.

From another point of view, the problem with validation is that it turns what should be a matter of degree into a binary decision; at some point, a decision must be made whether the model works well enough to be useful, but past that, the attention should be on how well the model works in various circumstances. As an example of poor practice, consider the use of validation in a recent report by the United States Fish and Wildlife Service (USFWS 2013, p. 18):

> The criterion used to determine whether the model was validated was whether the correlation between measured and simulated velocities (for intercept equals zero) was greater than 0.6. A correlation of 0.5–1.0 is considered to have a large effect (Cohen 1992). The model would be in question if the simulated velocities deviated from the measured velocities to the extent that the correlation between measured and simulated velocities fell below 0.6.

A plot of simulated data with a correlation coefficient of ~0.62 shows how much error this criterion would allow in a "validated" model (Figure 7.2).

It can be argued that validation and verification have limited, technical meanings that are well understood by modelers, but experience shows that these technical meanings often suffer "meaning creep"

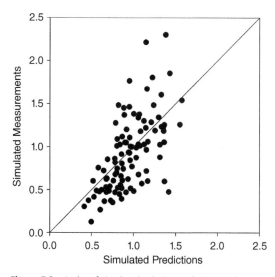

Figure 7.2 A plot of simulated velocity predictions and simulated measurements, created from random numbers with multiplicative errors, with a Pearson product moment correlation of 0.619. Source: John Williams.

toward standard usage. For example, Gard (2010, p. 124) wrote that "It is the standard practice in instream flow studies to assume that all independent variables are equally significant and to have compound suitability calculated as the product of the HSI of the independent variables (Bovee 1996, p. 120). This assumption has previously been tested and validated in the peer-reviewed literature (Vadas and Orth 2001)." Actually, the abstract of Vadas and Orth (2001) says that "Across all guilds, depth was consistently the most important factor in habitat selection," but the point here is that Gard was using "validated" in the general sense, and not as a term of art among modelers.

7.6.4 The limited utility of significance tests

Tests of statistical significance should have a limited role in selecting or testing models, although they have been frequently used. For example, Dunbar et al. (2002, p. iv) used significance tests to "verify" the flow models in their study: "For the model verification, a chi-squared test was used to test the significance of differences between measured and modelled values. In all but one case (a marginal failure for the high calibration flow on the River Senni) the model predictions passed at the 5% confidence level." Similarly, Thomas and Bovee (1993) used chi-square tests to assess where habitat suitability criteria were transferable. Recently, Lamouroux and Olivier (2015) claimed that their habitat models predicted changes of fish populations, based on p-values less than 0.05 for some of their study reaches. However, a significant result in such tests means only that the model fits the data considerably better than a specified null hypothesis, such as no effect; this does little to instill confidence in the results as a guide to management (Williams et al. 1999). From a different point of view, significance tests, like validation, turn a question of degree into a yes or no question, as with the test reported by Girard et al. (2003, p. 1390): "The predictions of the hydrodynamic model were considered valid if the 95% confidence intervals (CI) of slope and the intercept of the relationships between predicted and measured values included unity and zero, respectively."

Misuse and misunderstanding of p-values are common enough that the American Statistical Association recently published a statement explaining what they are and what they do and do not mean (Wasserstein and Lazar 2016). The six principles given by the statement are quoted below; note especially the last:

1 p-values can indicate how incompatible the data are with a specified statistical hypothesis;

2 p-values do not measure the probability that the studied hypothesis is true, or the probability that the data were produced by random chance alone;

3 Scientific conclusions or policy decisions should not be based only on whether a p-value passes a specific threshold;

4 Proper inference requires full reporting and transparency;

5 A p-value, or statistical significance, does not measure the size of an effect or the importance of a result;

6 By itself, a p-value does not provide a good measure of evidence regarding a model or hypothesis.

7.6.5   Tests should depend on the nature of the method being applied

There are various types of models used in environmental flow methods (EFMs), so models should be evaluated in different ways. There are at least three complementary ways to test the models. The first, applicable to all models, is to assess the plausibility of the model, the assumptions embedded in it, and their implications; do they make sense in the context of EFA? Tests of plausibility can be complex, and not simply based on subjective judgement. For example, simulation models can be very useful for testing whether the assumptions of some other model or class of models are plausible. In a report for the Electric Power Research Institute (EPRI (Electric Power Research Institute) 2000), Railsback used an individual-based model to show that, given reasonable rules for behavior by trout, microhabitat

selection will vary with discharge. This undercuts a basic assumption of PHABSIM and related models. Similarly, Railsback et al. (2003) used a newer version of the same model to identify reasons that the observed density of juvenile salmonids may not reflect habitat quality.

For models that make quantitative predictions, say of abundance or where fish should be found, a second way to test the model is to assess the fit between the model results and data. There are various ways to quantify the goodness of fit, discussed below. Models should be tested at the spatial scale at which they operate; for example, habitat models that estimate "microhabitat" cell by cell (or tile by tile), should be tested at the spatial scale of the cells.

The third way is experimental: to modify the flow in some stream, and monitor to determine whether the system behaves as predicted by the model. This is easy to say, but the response usually takes time, and money is seldom available for the necessary monitoring; unexpected or unlikely events can confound the "experiment," and it may be unclear what response was expected. For example, if some metric of habitat changes as predicted, but the abundance of fish does not, how should we judge the model?

7.6.6   Models should be tested multiple ways

Using only one criterion to test models increases the chance of a spurious result, and creates a temptation to choose the criterion that puts the model in the best or the worst light. Using multiple approaches reduces these risks, and also increases understanding of the model. There are various statistical measures of goodness of fit, for example, and graphical comparisons with data or other models, or graphical investigation of the residuals from a fit, can also be informative. These are discussed below. With methods that combine two distinct kinds of models, say hydraulic models and abundance–environment relations, each should be tested separately as well as tested in combination.

7.6.7   The importance of plausibility

As noted in the discussion of data-dependent activities in modeling, the plausibility of a research finding matters, and the same is true for the assumptions embedded in models. Discussing model selection, Burnham and Anderson (1998, p. 8) noted that "*If a particular model (parameterization) does not make biological sense, it should not be included in the set of candidate models*" (emphasis in the original).

In most applications, EFMs incorporating species distribution models are used with the implicit assumption that occupancy indicates habitat quality, so that protecting occupied habitat or providing habitat similar to occupied habitat will benefit the species. This may be true, but it may not; reasons that this assumption may fail are well known (Fretwell 1972; Van Horne 1983; Garshelis 2000; Railsback et al. 2003). For example, with territorial animals, subordinates may be forced to use habitats that they would otherwise avoid, or habitat conditions that an organism would prefer may not be available in the study area. Or, if food is scarce, an animal may forage in energetically profitable but hazardous locations that it would avoid if food were more abundant. Or, a species may occupy both habitats that support positive population growth and habitats that do not (source and sink habitats). In such a case, evaluating habitats based on the characteristics of sink habitats would mislead. Or, the organism may be responding to variables not included in the model, or to variables that are in the model, but at a different spatial scale from that inherent in the model. Even if there is a plausible biological linkage between the habitat as modeled and the fitness of the animal of concern, the degree of plausibility should be critically assessed.

7.6.8   The importance of testing models with independent data

Models should be tested against independent data, i.e. data not used in developing and calibrating the model. Ideally, the independent data should be collected separately, preferably at a different time or place, so that the generality of the model can be assessed. When using independent data is not practicable, the model can be repeatedly fit to randomly or systematically selected subsets of the data, and then tested with the rest ("cross-validation"). The evaluation is in terms of the distribution of the repeated fits, which can be summarized by standard errors or confidence intervals. With these recursive methods, however, you are working with only one sample, and other samples could give substantially different results.

7.6.9   The quality of the data limits the quality of the tests

Even if good methods are used for testing a model, the quality of the tests will be limited by the quality of the data, so the quality of the data should be considered carefully before the results of the test are relied upon: were the selection of sampling locations and the equipment used appropriate? and so on. Unfortunately, getting good data is usually difficult. If the data were collected by others, or for a different purpose, extra care is in order.

7.6.10   The importance of replication

Factors that regulate fish populations are complex and variable, so it is unrealistic to expect that any model will give good results for all applications, and models may give good results by chance. It follows that tests or evaluations of models or methods should not happen only once. Rather, assessments should be based on as many applications of the model as is feasible. With novel models, this may be only one, but the probability that a good result is a consequence of bias or chance is substantial (Ioannidis 2005), so the result should not be given too much weight until the test is replicated by others. This caution is standard in laboratory science, where things are much better controlled than in the field, and is yet more appropriate for EFA.

7.6.11   Models should be tested against other models

Even poor models will be used if better models are not available, as emphasized by Hilborn and Mangel (1997). If there is no convenient alternative, showing that a model gives poor results may not affect its use, particularly if using the model has become conventional. Thus, showing that a model is "generally accepted" can be a weak endorsement, and comparisons of different models, showing that one gives more reliable or plausible results than the other are generally more useful than testing a single model against some standard. For example, Rosenfeld and Ptolemy (2012) and Hayes et al. (2016) showed that relative to models predicting the net rate of energy intake, PHABSIM seriously underestimated flow needs of drift-feeding salmonids.

Model selection with some criterion such as the Akaike Information Criterion (AIC) or the Bayesian Information Criterion (BIC), discussed in Chapter 5, is a formal way of testing models against other models, but this can be done informally as well. In an informative example, Turgeon and Rodríguez (2005) fit logistic regression models and classification tree models to data on juvenile Atlantic salmon and microhabitat in two nearby reaches of a stream in Quebec, and compared the results. [This is the first of a set of papers on habitat selection by juvenile Atlantic salmon, that we discuss at some length.] Each reach was 75 m long and 15–20 m wide. "The models aimed at predicting presence or absence of salmon at location, either active or at rest, on the basis of habitat features at the location." Accordingly, assuming that habitat selection indicates habitat quality, the model results could be used in approximately the same way as habitat suitability curves for assessing flow regimes, although the authors expressed no such intention.

Working at mid-day in June or August, divers located fish (50–160 mm), watched them for several minutes, estimated their size, distance from the bottom, and whether they were feeding or resting, and marked their location. Later, measurements were made of depth, velocity (0.15 and 0.4 m from the bottom, substrate size, instream or overhead cover variable (boulder, large wood, submerged or overhead vegetation, undercut bank, broken water surface), and distance from the edge of the stream. The same variables were measured at a similar number of randomly selected locations without fish. The data were

standardized by subtracting the mean and dividing by the standard deviation. For each reach, models were fit for both active and resting fish, using data from that reach.

As candidate variables for the models, Turgeon and Rodríguez started with all those measured, including interaction and quadratic terms, to allow for non-linear effects. Then, using a somewhat complex stepwise selection procedure, they eliminated most of the variables for the logistic regression models, and all but a two or three for the classification tree models.

For the logistic regression model, no microhabitat variable was selected for all four models, and for the active fish, only depth was selected for both! Given that microhabitat for drift feeding fish is typically defined in terms of depth, velocity, and substrate, this is a striking result. For the resting fish, only substrate and boulders were selected for both models. Depth had by far the largest coefficients for the models of active fish, and the presence of boulders had the largest coefficients for resting fish (Table 7.1). Since the data were standardized, the absolute value of the coefficients shows their relative weights in the models.

As with logistic regression, no variable was selected for all four classification tree models. Only depth was selected for the both models for active fish. The presence of boulders was selected for both models of resting fish, and depth was selected for one of them (Figure 7.3). No other variables were selected.

Turgeon and Rodríguez (2005, p. 547) tested each of the models against data from the other reach, using numerical metrics and prediction maps, and concluded that: "The results suggest that [logistic regression] and [classification trees] are suitable but not equivalent tools for modeling distribution of juvenile Atlantic salmon. For both types of models, accuracy (predictive power) was high. Values of performance measures were high in calibration trials and declined in validation trials. ..."

Comparing the variables selected in the models for the two reaches raises questions. For example, for the classification tree model of fish at rest, depth is

Table 7.1 Results of logistic regression – McFadden's ρ^2 is a modification of R^2 for logistic regression that typically takes lower values than R^2.

	Active		At rest	
Model term	Reach 1 ($n = 182$)	Reach 2 ($n = 152$)	Reach 1 ($n = 127$)	Reach 2 ($n = 131$)
Constant	0.807	−0.673	−1.656	−1.701
Depth	3.544	3.959	0.638	—
Velocity at 40 %	−0.075	—	—	1.155
Distance to bank	0.854	—	0.673	—
Substratum particle size	—	—	0.622	1.008
Rock > 20 cm	0.610	—	1.816	1.721
Depth² (Velocity at 40 %)²	—	−1.756	—	—
	−0.497	—	—	−0.862
Substratum × Depth	—	—	−0.722	—
McFadden's ρ^2	0.52	0.45	0.47	0.50

Source: Turgeon and Rodríguez 2005 with permission from John Wiley & Sons.
All models were globally significant at p < 0.0001. McFadden's ρ^2 is reported for each model also.

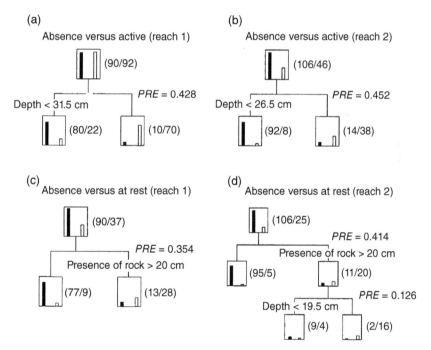

Figure 7.3 Classification tree models for predicting presence versus absence for active (a: reach 1; b: reach 2) and resting fish (c: reach 1; d: reach 2). Vertical bars represent the frequency of absence (black) and presence (white) at each node. Splitting values and proportional reduction in error (PRE) values are given on the branches of the trees. Absence/presence numbers for each node are given in parentheses. Source: Turgeon and Rodríguez (2005) with permission from John Wiley & Sons.

selected as a variable for reach 2 but not for reach 1, whereas for the logistic regression model, depth is selected for reach 1 but not reach 2. This seems strange. However, the large coefficients for depth for active fish in the logistic regression models are consistent with the classification tree results. Intuitively, variables that play similar roles in the different models seem more likely to be biologically important, instead of reflecting chance variation in the sample of data to which the model is fit.

Fitting the models to the combined data for the two reaches would be an obvious next step toward exploring which variables are more important, and recursive methods, similar to the bootstrap, can be used for a deeper analysis (Breiman 2001). Suppose, for each of the four cases shown in Figure 7.3, many random resamples were drawn with replacement from the original data sets, and the classification tree model was fit to each resample, producing a "random forest" of models. Then, the variation among the trees could be used to assess the importance of the different variables, by sorting out those that are selected most frequently for from those that are selected less frequently. A similar approach could be used for the logistic regression models. If used thoughtfully and transparently, this approach seems powerful, although testing the models with independent data from other reaches would be better. In any case, the study provides a useful caution against simply assuming that microhabitat should be defined in terms of depth, velocity, and substrate, and shows that fitting models to data may not be straightforward.

7.7   Experimental tests

7.7.1   Flow experiments

Flow experiments can test whether a change in a flow regime results in the biological response that the EFM leads you expect, and so are the most powerful tests of EFMs, but doing them well is difficult. Two good flow experiments that tested PHABSIM have been described in the literature. One, by the Michigan Department of Natural Resources,

manipulated the discharge in a small trout stream, as described by (Nuhfer and Baker 2004). In the other, the provincial power company in British Columbia (BC Hydro) began but aborted the application of adaptive management in the Bridge River, below a large dam. Nevertheless, work that was completed allowed for another test of PHASIM (Bradford et al. 2011; Failing et al. 2004).

Hunt Creek is an extremely stable, groundwater-fed stream in Michigan that supports a large population of small (mostly <18 cm) brook trout. Nuhfer and Baker (2004) monitored populations by electrofishing in the spring and fall for five years in a proposed 0.6 km treatment reach and in upstream and downstream control reaches. Then, using a by-pass, they diverted 50, 75, and 90 % of the summer flow around the treatment reach, for two years for each treatment; bulkheads with traps blocked movement into or out of the treatment reach. They estimated WUA with 63 transects and suitability indices for diurnal (feeding) and nocturnal (resting) young of the year, yearlings, and older fish. As summarized in the abstract, the study found "generally insignificant or inconsistent relationships between WUA and population parameters such as abundance, survival, and growth, suggesting that PHABSIM was poorly suited for predicting biological impacts of water diversions from low gradient brook trout streams ..." They also observed that many fish changed their behavior during the larger diversions, moving from riffles into pools, and forming schools. The flow reductions had remarkably little effect on either the growth or abundance of fish, which Nuhfer and Baker attribute to the shift in habitats, and to brook trout being well adapted to summer drought conditions. These findings are consistent with those of Xu et al. (2010).

As an experimental rather than an observational study, Nuhfer and Baker (2004) is particularly strong. The authors considered the possibility that their results may reflect poor estimates of WUA, and indeed this is plausible (Williams 2010b). However, Nuhfer and Baker (2004, p. 12) argued that "... if these efforts (63 transects) were not sufficient to characterize habitat in a 600 m stream reach, then the labor required for adequate model projections would be prohibitive for most resource agencies."

The Bridge River drains glaciated terrain in southwestern British Columbia, with a mean discharge of about 100 m³ s⁻¹ and a pronounced snowmelt peak. Formerly, the river flowed eastward through a steep gorge to join the Fraser River. The gorge made an attractive site for a dam, and one was built there in 1960, drowning considerable habitat for salmonids upstream from the gorge. Except for spills, the entire flow of the river was diverted to a hydroelectric plant on an adjacent river, although groundwater and tributaries provided inflow reaching about 1 m³ s⁻¹ at the confluence with the Yalakon River (mean discharge 4.4 m³ s⁻¹) about 3 km below the dam. Even with these low flows, the lower Bridge River supported good populations of various salmonids, raising the question whether more flow would support more fish, and discussion of releases from the dam from the dam began in the 1980s. Eventually, a 1999 out-of-court settlement of litigation provided for a release averaging ~3 m³ s⁻¹ with a late spring peak, until a Water Use Plan was approved by the provincial government and the Canadian Dept. of Fisheries and Oceans. 3 m³ s⁻¹ does not seem

much water for such a large river, but as Bradford et al. (2011) explain, the confined channel in the gorge was unsuitable for rearing with the natural high flows, and was used by salmonids primarily for migration to lower-gradient reaches farther upstream with better habitat.

Negotiations for a Water Use Plan developed an experimental design with releases of 0, 1, 3, 6, and 9 m³ s⁻¹ planned, using pre-1999 monitoring for the 0 release, and the court-ordered flow for the 3 m³ s⁻¹ release, but for unrelated reasons the experiment was not fully implemented, and releases have remained at 3 m³ s⁻¹. Nevertheless, extensive monitoring and a PHABSIM study were conducted. Because the pre-1999 flows were approximately the optimum flow predicted by a PHABSIM study in part of the channel (Figure 7.4), the 0 and 3 m³ s⁻¹ monitoring results for that reach could be used to assess the PHABSIM predictions.

The PHABSIM results predict that fish abundance should decline when the flow in Reach 3 increased to 3 m³ s⁻¹, but it stayed about the same, except for juvenile Chinook salmon. Thus, the experiment supports PHABSIM only for Chinook salmon, and Bradford et al. (2011) noted an alternative explanation

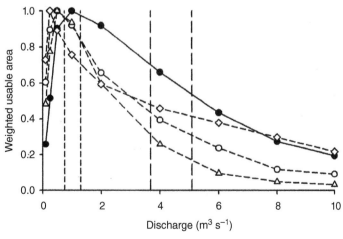

Figure 7.4 Curves of WUA over discharge for Reach 3, Bridge River, from a study by Triton Environmental for BC Hydro. Symbols are: filled circles, age-1 rainbow trout; open circles, age-0 rainbow trout; diamonds, coho salmon; triangles, Chinook salmon. Source: Bradford et al. 2011, with permission from John Wiley & Sons.

for this result. The releases increased water temperature during the Chinook salmon incubation period, resulting in faster development and emergence as early as December, instead of March, probably resulting in higher post-emergence mortality. On the other hand, the wetted area of the river in Reach 3 increased about 30% after releases began, indicating some reduction in average habitat quality, probably because flow velocity in the thalweg tripled to ~0.6 ms^{-1}.

7.7.2 Behavioral carrying-capacity tests

Morhardt (1988, cited in Zorn and Seelbach 1995) proposed "behavioral carrying capacity" as a way to test the utility of WUA as an indicator or habitat quality. For such a test, a section of stream is blocked in such a way that fish can leave but not enter, and then the section is deliberately overstocked. The idea is that fish will leave the section until the population reaches carrying capacity. Then, if WUA is a good indicator of carrying capacity, the density of the remaining fish should be proportional to the WUA of the section, and if the test is repeated at various flow conditions with different amounts of WUA, the relationship between WUA and carrying capacity can be tested. The advantage of this approach is that multiple tests can be conducted relatively quickly, but this comes at the cost of the "behavioral carrying capacity" assumption. Zorn and Seelbach (1995) applied such a test with eight repetitions to smallmouth bass in a millrace channel in Michigan where flow could be controlled, and found that that the relationship between WUA and the density of fish remaining was negative and non-linear.

7.7.3 Virtual ecosystem experiments

A common approach to EFA is to observe habitat variables at spots in a stream where fish are or are not observed, and to develop a statistical model, such as the Turgeon and Rodríguez (2005) models described above, that predicts where fish will be or will not be found. Then, the model is used to calculate the relationship between flows and the amount of habitat of the type that fish prefer, with the assumption that the

will be a direct relationship between the amount of such habitat and the future population of fish.

Railsback et al. (2003) tested that assumption, and so also methods that incorporate the assumption, using a complex individual-based model (IBM) of a virtual population of trout that was programmed to select the best available habitat in the virtual stream, subject to a variety of more or less realistic constraints. "Best" was defined in terms of "expected maturity," a measure of fitness, and the constraints included such things as territorial behavior and limits to the area over which each fish selects habitat. The model ran on a daily time-step, and fish suffered mortality each day with a probability based on their physiological condition and habitat, so that each day fish could reassess their situation and move if better habitat became available. Railsback et al. (2003, p. 1580) summarized their results in their abstract:

There was no strong relation between fitness potential and the density of fish in the IBM; cells where fitness potential was high but density low were common for all age classes, and fitness potential was not proportional to density. This result was consistent at high and low abundance and high and low overall habitat quality. We developed a statistical model of trout density observed in the IBM as a function of the four habitat variables that vary among cells. We then tested the ability of modeled mean density to predict population response to habitat changes resulting from stream flow modification. Modeled density partially explained population response to flow, but only at flows near the flow at which the density model was developed, and not for groups (e.g. juveniles) experiencing strong competition for habitat. Modeled density predicted population response opposite that observed for age-0 trout and incorrectly predicted response of all age classes to major changes in flow. These results make sense if habitat selection is understood as an emergent property of (i) the mechanisms by which habitat affects fitness, (ii) habitat availability, (iii) population abundance and size structure, and (iv) how individuals compete with each other. We identified eight reasons why animal density may not reflect habitat quality and several inherent limitations of habitat selection modeling.

7.8 Testing models with knowledge

When new models are being developed, one reasonable test of the model is simply whether the results are consistent with prior knowledge. For example, Bélanger and Rodríguez (2002) developed a dynamic turnover model based on the assumption that local movement is a good indicator of habitat quality; fish should move to good habitat more often than to poor habitat, and leave poor habitat more often than good habitat. The model was applied to three sites in the same river system, using mark-recapture data. One site supported brook trout, while the other two supported both brook trout and Atlantic salmon. The model showed higher equilibrium abundance for Atlantic salmon in riffles and for brook trout in pools, as expected from previous studies. Of course, passing such a test does not mean that the model is ready for general use, but any time a model gives results at variance with prior knowledge, it should be carefully examined. More importantly, the observed abundances varied strongly over the 65-day study period, "suggesting that movement-based parameters may be more stable than measures of abundance for evaluating salmonid habitat" (p. 155).

7.9 Testing hydraulic models

Hydraulic models used for EFA are increasingly 2-D, although 1-D models are still used; 3-D models are not yet in common use, although the potential for them is clear (Tonina and Jorde 2013). The details of testing will depend on the kind of model, but general principles apply to all; the discussion below assumes 2-D models. There are several published comparisons of the results of 1-D and 2-D modeling for NHMs, such as Benjankar et al. (2015). These show that the differences can be substantial.

For numerical habitat models, testing a hydraulic model against independent data is important but seldom done (Wright et al. 2017). Generally, these models are calibrated in terms of water surface elevation, but are used to estimate flow velocity and depth averaged over small areas defined by the model grid. Accuracy will be affected by approximations in the model, by errors or approximations in the input data describing the shape of the channel, by errors in measurements of discharge in the calibration data, and by the calibration process itself. It should be standard practice to present scatter plots and other graphical depictions of observed versus simulated values, and comparisons should also be summarized with statistics. Bennett et al. (2013) described a broad range of statistical and graphical summaries that can be used, and also a five-step approach for characterizing model performance that seems appropriate for hydraulic models. As summarized in their abstract:

> A five-step procedure for performance evaluation of models is suggested, with the key elements including: (i) (re)assessment of the model's aim, scale and scope; (ii) characterisation of the data for calibration and testing; (iii) visual and other analysis to detect under- or non-modelled behaviour and to gain an overview of overall performance; (iv) selection of basic performance criteria; and (v) consideration of more advanced methods to handle problems such as systematic divergence between modelled and observed values.

Willmott et al. (2015) and Duveiller et al. (2016) reviewed several indices of agreement that seem applicable to hydraulic models and make conflicting recommendations, so which index is preferable is still under discussion, and may depend on the context. In the meantime, calculating and reporting several seems advisable, and, since these will be unfamiliar to most readers, also reporting better-known statistics such as a correlation coefficient seems advisable.

As an example of good practice, May et al. (2009) provided a scatter plot showing the mean and local bias (Figure 7.5), and also the associated correlation coefficient, presumably the Pearson product moment correlation. Similarly, Wall et al. (2016) showed scatterplots of observations for both the calibration data and for independent data, with coefficients of determination of (R^2) of 0.48 for depth and

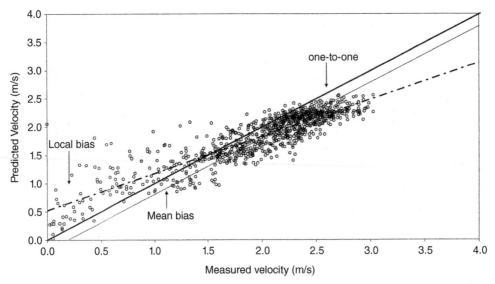

Figure 7.5 Scatter plot of predicted versus measured velocity in the Trinity River, California, USA, with a correlation coefficient of 0.87. Data were collected during floods for a study of bed mobility at salmon redds, which accounts for the high velocities. Source: May et al. (2009) with permission from John Wiley & Sons.

0.64 for velocity with the independent data. Some authors (e.g. Wright et al. 2017) show histograms of deviations from model predictions, but such a plot would mask the systematic bias (over-prediction at lower discharge and under-prediction at high discharge) revealed in Figure 7.5.

When hydraulic models used for EFA are tested against independent data, the results can be surprisingly poor. For example, the coefficient of determination (R^2) between measured and predicted velocity reported by Guay et al. (2001) was only 0.17. The spatial density of the input data is an important factor. Gard (2009) compared the results of PHABSIM using a traditional transect-based hydraulic model and River2D, a hydraulic model commonly used for NHMs. However, the correlation coefficients (r) for observed velocities and depths and those simulated by River2D were only 0.23 and 0.46 (Gard 2010). Gard (2010) attributed these poor results to sparse input data (2.58 points per 100 m^2), and reported much higher correlations (0.64–0.82) for observed and modeled velocities

using 40 or more input data per 100 m^2. This seems consistent with results described by Tonina and Jorde (2013). Harrison et al. (2011) reported coefficients of determination (R^2) of 0.85 at base flow and 0.74 at bankfull discharge in a recently reconstructed but still relatively simple channel. May et al. (2009) reported correlations of 0.87–0.96. As shown by their scatterplot, however, there was still non-trivial error, with over-predictions at velocities below about 1.5 $m^3 s^{-1}$, and under-predictions at higher velocities.

7.10 Testing EFMs based on professional judgement

Expert opinion is ubiquitous in EFA, obviously so in methods such as Demonstration Flow Assessment, and implicitly in other methods. When models are used, there is always judgment applied in deciding which models are selected and how they are implemented (Kondolf et al. 2000). One virtue of Bayesian

Network models is that they can combine expert opinion, data, and modeling results or data in a transparent and quantified way.

Some of the points made above, for example about plausibility, apply as well to expert opinion, and opinion about things such as the distribution of organisms or their abundance can be similarly tested against data, although in some cases it may take time to accumulate the necessary data. For example, as the result of litigation, Putah Creek in the Central Valley of California has a flow regime based largely on the testimony of one of us (PBM). Moyle's "Inland Fishes of California" is a standard reference work on the subject, and with the students in his field class he had sampled Putah Creek for over 15 years, so his status as an expert was well supported. Importantly, Moyle et al. (1998) described their expectations for the flow regime in sufficiently specific terms that when most features of the recommended flow regime were established by a court decision, and sampling the fishes continued, it was possible for Kiernan et al. (2012) to show that the expectations were realized. Since the flow regime mimicked the natural flow regime, the example lends support to the natural flow regime paradigm.

Hydrologically based methods are really a class of expert opinion methods, as suggested above, in that expert judgement goes into picking some level of flow that is calculated from flow records for gaged streams, and from some kind of regional regression equations for ungaged streams. Suppose, for example, that someone decides that the minimum bypass flow past a diversion in some stream be set at 20 % of the mean annual natural flow. Then, one can ask how well the mean annual natural flow is known, which may not be very well if the flow record is short, and, one can ask whether 20 % of the estimated annual natural flow actually provides the intended protection of the stream ecosystem, provided that the intended benefits are described in useful detail. There is also the question whether the estimated 20 % criterion was met, but that is more a test of management of the stream than of the assessment.

7.11 Testing species distribution models

Methods such as PHABSIM and mesohabitat simulation (MesoHABSIM) are built on species distribution models, although various other names for these models are used as well, sometimes with more specific meanings: examples include resources selection functions (e.g. Manly et al. 2002), habitat-association models (Lancaster and Downes 2010a), and numerical habitat models (Guay et al. 2000). Generally, the models fit the observed distribution of the organisms of concern to environmental variables that describe their habitats, and then inferences are made about how the organisms will fare when habitat changes. The change can be through time, as with climate change or when a stream's flow regime will be changed, or over space, as when such models are used to identify areas for reintroduction of rare species, selecting areas for restoration, etc. The assumptions underlying the models are questionable, as shown by the virtual ecosystem experiment described above. The models are nevertheless widely used, so we describe ways to test them.

How to model species distributions is an active area of work. There is a large and rapidly expanding literature on testing species distribution models in ecology and wildlife management, and many numerical criteria have been proposed for the purpose, perhaps because none work best in all cases; for example, some may work better for common species, others for rare species. Vaughan and Ormerod (2005), Mouton et al. (2013), and probably others provide reviews. Unfortunately, this literature is mostly ignored by people working on EFA, who tend instead to use tests of statistical significance, ad hoc criteria, or visual examination of graphs or maps instead (e.g. Freeman et al. 1997; Hardy et al. 2006; Bondi et al. 2014). Beakes et al. (2014) and Peterson and Shea (2015) are exceptions that use more appropriate criteria.

Species distribution models are usually tested in terms of their ability to correctly predict presence or absence of organisms in patches of habitat (goodness of fit), quantified by some statistic. However, this is at best a partial test of their reliability, and there are problems with the various statistics (Vaughan and Ormerod 2005). Besides goodness of fit, other concerns are generality, or transferability (Vaughan and Ormerod 2005; Wenger and Olden 2012); how imperfect detection may affect results (MacKenzie et al. 2002; Peterson and Shea 2015); and how plausible the model assumptions or results are in terms of current knowledge. Finally, the utility of the models for the intended purpose should be considered, not just predictive power; a model predicting that fish will be in water will be accurate, but not useful.

7.11.1 Goodness of fit

Probably the most familiar statistic for goodness of fit is the coefficient of determination, written as r^2 or R^2. However, as is common with statistics, there are caveats. For example, adding variables to an estimation model will always yield a higher R^2 for a given data set, but R^2 does not take account of the number of variables in the model. An "adjusted R^2" can be used, but other approaches for comparing estimation models, such as using the Akaike Information Criterion (AIC) or the Bayesian Information Criterion (BIC), seem preferable (Gruebber et al. 2011). For simulation models, there are correlation coefficients and various indices of similarity (Willmott et al. 2015; Duveiller et al. 2016), as well as graphical comparisons.

For species distribution models, goodness of fit can be summarized in a 2×2 table showing the numbers or percentages of correct or incorrect predictions, and various indices of performance can be defined in terms of the elements of this table, labeled as shown below. Mouton et al. (2013) give the formulas for seven such accuracy statistics in terms of a, b, c, and d, plus their sum, n: percentage correctly classified, sensitivity, specificity, normalized mutual information, odds ratio, Kappa, and true

skill statistic. For example, the percentage correctly classified is $([a + d]/n) \times 100$.

a, # of true positives	b, # of false positives
c, # of false negatives	d, # of true negatives

Another approach involves calculating what is called the receiver operator characteristic curve, and calculating the area under the curve (AUC). As the name suggests, this approach was developed by engineers (for assessing radar detection), but is now commonly used in other fields. The AUC gives an expected value of 0.5 for random guesses, and 1.0 for a perfect fit. Unfortunately, all of these measures have shortcomings (Vaughan and Ormerod 2005; Lobo et al. 2008; Beale and Lennon 2012), so none is suitable for all situations.

Graphs or prediction maps are useful for assessing goodness of fit, but because of quirks in human perception, they can be misleading. For example, perception of the vertical distance between two lines can be compromised if parallel lines are slanted; people normally perceive the orthogonal distance between the lines instead of the vertical distance. Similarly, with prediction maps, there is a tendency to see observations in low-suitability habitat that are close to high-suitability habitat as confirming the method used to estimate suitability. Using numerical indices along with prediction maps or graphs is good practice.

7.11.2 Prevalence

One problem for the performance indices just described is that the prevalence of the organism of interest matters; it is easy to make a good guess whether a species will occur in a patch of habitat if it occurs almost everywhere or almost nowhere. When habitat is divided into many small patches, as when a 2D hydraulic model is used to define patches, a large percentage will be vacant unless the fish are abundant, as noted by Al-Chokhachy et al. (2013) for bull trout and Beakes et al. (2014) for juvenile Chinook salmon. This affects the indices of performance described above, and makes them less suitable for use with numerical habitat models, if the availability of habitats is estimated using the hydraulic model. Guay et al. (2000) and Beakes et al. (2014) tried to avoid this problem by comparing their occupied sites with a similar number of randomly selecting unoccupied sites, but this introduces sampling uncertainty regarding the unoccupied sites. On the other hand, if unoccupied sites are sampled in areas out of the normal range of the fish in question, the test can be biased. For example, sampling unoccupied sites in pools when the fish of interest lives in riffles may improve the apparent predictive power of the model. Using a number of these indices seems like good practice if any are used. Vaughan and Ormerod (2005) describe a bootstrap approach for improving the performance of the accuracy statistics, but recommend instead using some model-selection criterion such as the AIC or the BIC.

7.11.3 Imperfect detection

False negatives and false positives can occur in the data, as well as in model predictions, with serious consequences for fitting and testing models. With some sampling methods, such as electrofishing, the chance of misidentifying fish may be negligibly small, so that detection means that the sampling site is occupied. However, the converse does not hold, and there may be a good chance that a site is occupied even if the fish is not detected. With other sampling methods, such as snorkeling, the chance of misidentifying a fish may not be negligible, particularly if water clarity is marginal and the species in the area are not easily distinguished. Or, detection may be better in some sites or habitats than others, or under different conditions, as is most obvious with fishes such as juvenile salmonids that are small enough to burrow into the stream bottom. Most studies using species distribution models ignore imperfect detection (Kellner and Swihart 2014), but unless detection is very good or does not vary across sites, this can compromise the results (Rota et al. 2011; Lahoz-Monfort et al. 2014).

Studies described below show that the effect can be substantial. Statisticians have long been conscious of this problem, but it is largely ignored in EFA.

Although imperfect detection cannot be avoided, it can be estimated using an approach somewhat like mark–recapture estimates of populations, with multiple samples taken during a sampling period (MacKenzie et al. 2002, 2009; Peterson and Shea (2015)). This at least allows for estimating the uncertainty in the model results that arises from imperfect detection. Or, it may be possible to define a useful "detectability function" by which detectability declines with distance, turbidity, etc. Detectability needs much more attention in EFA.

Of course, tests of models will not work if investigators simply refuse to accept the results. For example, consider the following from a report by Dunbar et al. (2002) "Further validation of PHABSIM for the habitat requirements of salmonid fish," at p. 4:

> Various summary indices of time-varying physical habitat were plotted against population densities, but no strong relationships were found. It is clear from this work that there are no simple relationships of this sort, probably due to the many other factors that affect fish densities/population numbers. Nevertheless, failure to find simple relationships should not in any way reflect on the validity of PHABSIM itself. Such relationships can only be defined under controlled conditions, where all factors are held constant other than physical habitat. Worldwide, several suitable experimental facilities exist.

7.11.4 Spatial scale and other complications

A series of papers on species distribution models for juvenile Atlantic salmon illustrate many of the points made above. The *Centre Interuniversitaire de Recherche sur le Saumon Atlantique* (CIRSA) maintains a field station on the Sainte-Marguerite River, a largely undisturbed river in the Canadian province of Quebec northeast from Montreal. Some of the papers do not deal explicitly with testing models, but their relevance should be obvious. We chose these studies because they are recent, and deal with one of the best-studied fish in the world. We also examine the plausibility of one model that was well tested.

Guay et al. (2000) used a 2-D hydraulic model and observations of depth, velocity, and substrate at points occupied and not occupied by fish to develop a numerical habitat model for juvenile (parr) Atlantic salmon. They used logistic regression to develop a resource selection function they called the Habitat Probabilistic Index (HPI), using stepwise backward regression to get the simplest statistically significant model. For logistic regression, humped relations between the habitat index and a habitat variable are achieved by including the variable with a positive constant, and the variable squared with a negative constant, as done here for depth. They also developed a PHABSIM-style Habitat Suitability Index (HIS), using the standard variables of depth, velocity, and substrate, but using a method to weight the value from the suitability curve for each variable.[1] The biological models are:

$$\text{HPI} = 1 / \left(1 + e^{-\lambda}\right)$$
$$\lambda = -3.067 + 8.461D + 2.86V + 0.093S - 6.203D^2$$
$$\text{HSI} = I_D^{0.30} \cdot I_V^{0.38} \cdot I_S^{0.32}$$

where D (cm), V (m^{-s}), and S (m) are water depth, vertically averaged velocity and median substrate size (D_{50}), and I_D, I_V and I_S are the relevant values from the suitability curves.

The models assign a habitat value to each cell or tile in the study area, and the density of fish in the tiles can be plotted over the estimated habitat values of the tile. Because relatively few of the cells were occupied and the fish are scattered, no tile is likely to support two fish, so the density indicates the proportion of the tiles with that index value that are occupied. Guay et al. tested both the HPI and the HSI against an independent data set from the same stream (Figure 7.6). This plots the observed density of Atlantic salmon parr over the habitat values calculated by the biological models, with a line fit to the plotted data using polynomial regression.

This way of testing the method has several good features. First, the test used independent data. Second, the test is clear and easy to understand, and highly informative. The optimal result for a resource selection function would be a straight line slanting up from the origin, showing that the probability that the tile will be occupied is approximately proportional to the index value. A line fit to the results for the HPI is concave upward, but not strongly so, and reaches zero near the origin of the graph, showing that the HPI is approximately a resource selection function, as intended.

Third, the new index (HPI) is tested against another model, a standard PHABSIM-style index (HSI), and performs better. The line fit to HSI densities is concave downwards, so clearly it performs poorly as a resource selection function, although originally it was assumed to be one (Bovee and Cochnauer 1977). Alternatively, the HSI could be interpreted as a suitability index sensu Fretwell (1972). Then, because relatively few tiles were occupied and the fish were scattered, the "ideal free distribution" could apply, such that the fish would select the tiles with the highest suitability. In this case, the fit line should turn steeply upward at higher values of the index. Clearly, the curve fit to the HSI does not. In either case, managing a stream to provide more habitat with HSI values of 0.9 or 1.0 would seem counterproductive.

Fourth, Guay et al. (2000) also reported the fit of their hydraulic model to measured water velocity, even though it was poor. Too often, this aspect of model performance is ignored, even though it is obviously important. Why the model nevertheless functioned well is discussed below.

There are, however, problem with the r^2 values shown in the figure, and their interpretation in the paper. The r^2 values describe the scatter around the lines fit by polynomial regression, which does matter, but not more than the scatter around lines drawn by eye, and the shape of the curve is more important, given the purpose of the models. To see this, suppose that all the points in Figure 7.6a fell exactly on the line, so that r^2 equaled 1. The density

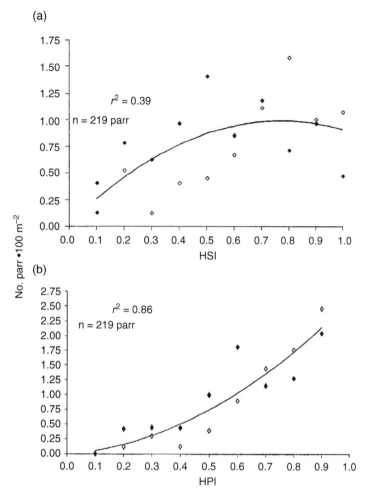

Figure 7.6 The relation between fish density and (a) the Habitat Suitability Index, and (b) the Habitat Probability Index, at two rates of discharge. Open diamonds 2.2 m s⁻¹ filled diamonds 3.2 m s⁻¹. Lines fit by polynomial regression. Source: Guay et al. (2000). © 2008 Canadian Science Publishing or its licensors. Reproduced with permission.

of parr would still decrease at the highest values of the index, which is contrary to any sensible meaning of the index.

Guay et al. (2000) also showed the preference curves for the PHABSIM-style model, and the curve for depth (Figure 7.7) may help explain the poor performance of the model; biologically, it fails the plausibility test. Unfortunately, in our experience, implausible preference or suitability criteria are not uncommon in PHABSIM studies, perhaps because they are developed from too few observations.

The plausibility of models with more than one variable, such as the HPI model, is harder to examine graphically, but it can and should be done. As a first step to examining the HPI model, a plot of HPI over λ (Figure 7.8) shows that essentially all the action lies with λ in the interval −4 to 4, and most of it between −2 and 2.

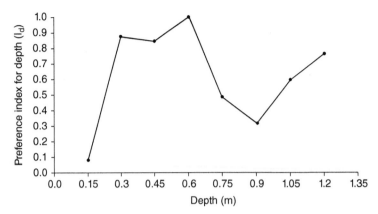

Figure 7.7 The suitability curve for depth used in Guay et al. (2000). Source: Guay et al. (2000). © 2008 Canadian Science Publishing or its licensors. Reproduced with permission.

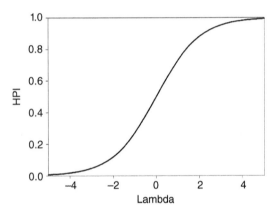

Figure 7.8 HPI as a function of λ (lambda). Source: John Williams.

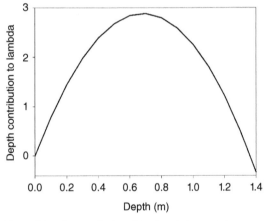

Figure 7.9 The contribution of depth to λ. Source: John Williams.

In the model, λ increases linearly with sediment size and water velocity, but is quadratic in depth (see Eq. 7.1 above), so that λ first increases and then decreases with depth, with a peak at $D = 0.68\,\mathrm{m}$ (Figure 7.9). HPI would be maximal in habitat with the highest water velocity, coarsest substrate, and a depth of 0.68 m. Eighty percent of the fish were seen with substrate between 3 and 6 cm, and the fitted constant for substrate is small, so evidently substrate size is a less important variable for the HPI than velocity or depth. For the range of the data,

with maximum water velocity of $1.2\,\mathrm{m\,s^{-2}}$ and D_{50} of 16 cm, λ will peak at 4.71, with the intercept at -3.067, and contributions from depth of 2.86, substrate of 1.49, and velocity of 3.43. Given the small size of these fish, 10 cm or less, it seems implausible that the optimal velocity is $1.2\,\mathrm{m\,s^{-2}}$, even if the fish are actually holding at a lower velocity near the substrate.

Holding substrate constant at the modal value, plots of HPI over depth at different values of velocity show that curves of HPI are humped, as would be expected, and as velocity increases the hump becomes higher and broader (Figure 7.10). Eighty per cent of the parr occupied depths and velocities indicated by the shaded region in Figure 7.10, showing that there were few fish in areas with velocity higher than 0.6 m s⁻¹, or depths greater than 0.72 m, again raising questions about the plausibility of the HPI values for high depths and velocities.

Contours of HPI can be plotted directly with two of the variables as axes (Figure 7.11). At $S = 0.4$, HPI is not highly sensitive to water velocity. This may help to explain why Guay et al. (2000) got good results with HPI despite the apparently poor accuracy of the velocity predictions on a cell-by-cell basis ($r^2 = 0.17$; Guay et al. 2001). It could also be informative to plot the original data on similar axes, which might identify values of the physical variables where the index would be more or less reliable. For example, based on Figure 3 in Guay et al. (2000), it seems that there are few data for large values of S, so that predictions of the HPI in areas with large substrate may be less reliable. Most fish were in areas with HPI less than 0.8, because only a small part of the study area had

higher HPI values, and none had values higher than 0.9 (Figure 6b in Guay et al. (2000)). It might also be useful to plot deviations from model predictions in a similar way. The point is that simply fitting a model to data is not enough; it should be examined carefully before it is used.

Subsequently, Guay et al. (2003) applied both models to the adjacent Escoumins River, to test the transferability of the two models. The flow at the Escoumins River study site is somewhat deeper and faster than in the Sainte-Marguerite study sites, and the substrate is coarser. The transferability is important for both practical and scientific reasons; studies will be less expensive if models developed elsewhere can be used, and good transferability is evidence that the response of the fish to the variables in the model do not depend strongly on the geographical context. This is sometimes called the "generality" of the model. The HPI model performed fairly well, but the HSI model did not (Figure 7.12)

Although not emphasized in the two papers just discussed, both mention that fish sampling was done between 10 : 00 and 16 : 00 on days when cloud cover

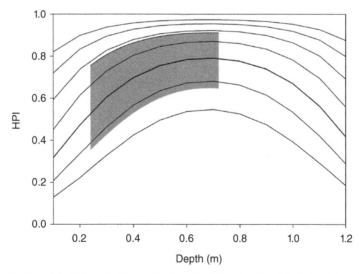

Figure 7.10   HPI as a function of depth for velocities ranging from 0 to 1.2 m s⁻¹, with substrate D_{50} of 4 cm. The shaded area indicates the depths and velocities occupied by 80 % of the 308 juvenile Atlantic salmon observed (D, 0.24–0.72; V, 0.15–0.75). The total range of depth, velocity and substrate utilized was 0.12–1.2 m, 0.05–1.2 m s⁻¹, and 1.2–9 cm. Source: John Williams.

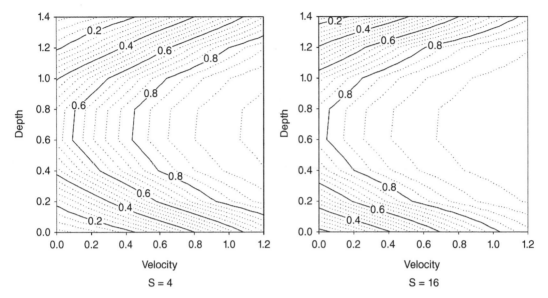

Figure 7.11 Contours of HPI with substrate size (D_{50}) of 4 cm (left), and 16 cm (right). Source: John Williams.

was less than 25%. Because of evidence in the literature that Atlantic salmon parr may take cover in the stream bed during cloudy weather, perhaps confirmed by their field experience, Girard et al. (2003) investigated the effect of cloud cover on the number of parr observed in the Escoumins River: many more parr were observed under clear skies.

Girard et al. (2003) also examined the effect of cloud cover on the performance of the HPI models developed with data collected under clear (low), intermediate, or cloudy (high) conditions, and found that the model performed poorly in cloudy conditions. The three models are:

$$HPI = 1/(1 + e^{-\lambda}), \text{and}:$$

$$\lambda_L = -2.19 + 5.73D + 3.03V + 0.52S - 6.29D^2$$
$$(\text{low cloud cover})$$

$$\lambda_I = -3.06 + 9.88D - 3.31V - 4.13S - 9.68D^2$$
$$(\text{intermediate cloud cover})$$

$$\lambda_H = -11.25 + 27.65D + 4.21V + 61.34S - 32.43D^2$$
$$(\text{high cloud cover})$$

The parameters for the cloudy sky model are clearly different, and this carries through to the results (Figure 7.13). The relationships between fish density and HPI from models developed for clear and intermediate conditions are similar for both clear and cloudy conditions, although much weaker for cloudy conditions. However, the HPI model developed for cloudy conditions fails to predict fish density in cloudy conditions, and the relationship with fish under clear conditions is marginal.

Girard et al. (2003) asserted that "HPI models developed under low to intermediate cloud cover could adequately predict habitat quality and fish distributions under any cloud cover," citing the high values of r² in Figure 7.13(a) to (e). However, this misinterprets the r² values, as discussed above. The shapes and slopes of the curves in (a) and (b) are roughly similar, but the fit line in (b) is nearly flat. Moreover, the fit lines in (a) and (c) do not approximate a resource selection function as well as the fit line for the Sainte-Marguerite River model (Figure 7.6b); instead, they resemble the results of a good suitability model, sensu Fretwell. These points

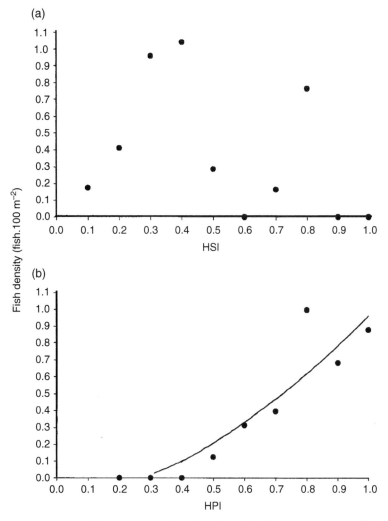

Figure 7.12   The relationship between fish densities and values of (a) HSI and (b) HPI. Source: Guay et al. (2003). © 2008 Canadian Science Publishing or its licensors. Reproduced with permission.

aside, however, testing the model under different conditions is clearly good practice,

We have found no more papers on the HPI model after 2003, and perhaps the reasons why. Perhaps because of evidence from the literature that Atlantic salmon parr may be more active at night than during the day even when water temperature is high, Imre and Boisclair (2005) investigated this in the Sainte-Marguerite River, and found that parr were indeed

more active at night. A follow-up study (Bedard et al. 2005) showed a spatial and temporal pattern to the observations, and the authors concluded their brief paper by noting that "These findings suggest that a substantial part of the overall population can move in the slower flowing and near shore areas at night. Future sampling strategies for habitat use studies and population assessments should take into consideration this heterogeneity in the night-time

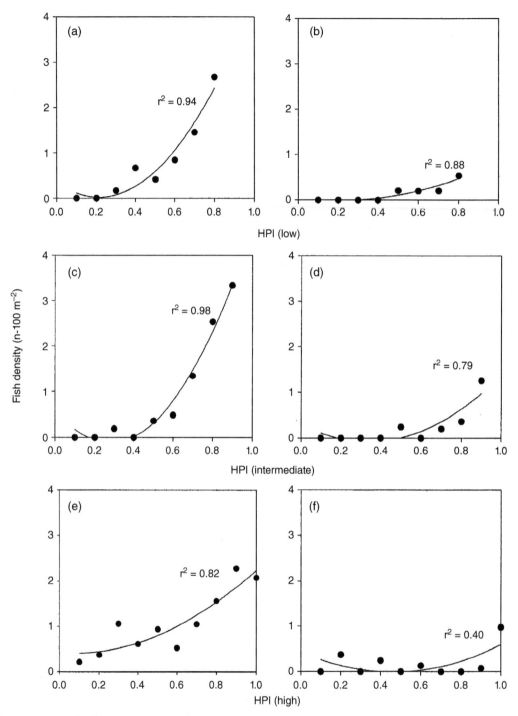

Figure 7.13 Observed fish density under clear (a, c, e) and cloudy (b, d, f) conditions plotted over HPI from models developed from observations under from low (a, b), intermediate (c, d), and high cloud conditions; thus, (a) and (b) show the results of the model developed for clear conditions with data from clear and cloudy conditions, etc. Source: Girard et al. (2003). © 2008 Canadian Science Publishing or its licensors. Reproduced with permission.

spatial distribution of Atlantic salmon parr." In other words, the mid-day observations used to develop and test the HPI were biased samples. The consequences of this bias for model results are not known. However, there is good reason to think that similar complications occur with other fishes as well (Reebs 2002; Railsback et al. 2005).

For her Master's thesis, Bouchard (2005) investigated temporal variability in counts of parr over four visits in reaches of 50–200 m, and found that the coefficient of variation decreased from about 50 % at 50 m to 30 % at 200 m. Subsequently, Bouchard and Boisclair (2008) explored habitat selection by Atlantic salmon parr in the Sainte-Marguerite River at coarser spatial scales. Working at night, divers counted parr in 32 reaches each 200 m long, with fish density recorded every 10 m. These 10 m counts were aggregated in "analytical units" (AUs) 50, 100, and 200 m long. For each AU, they developed 15 "local" variables such as measures depth, velocity, and presence of wood, ten "lateral" variables such as overhanging vegetation or tributary confluence, and eight "longitudinal" variables such as distance from the nearest tributary or from the river mouth, from either data recorded during the dives, or from maps.

Then, using a stepwise selection method involving statistical significance, and each type of variable, they tried to fit models to the AUs. They were unable to select any models for the 50 m AUs, or using the lateral variables, but did select models for the 100 and 200 m AUs, using boulders, the presence of woody debris, and, for the 100 m AU, the percentage of "smooth water surface" for local variables, with a better fit at 200 m. The selection of boulders for these models is similar to the results of Turgeon and Rodríguez (2005), discussed above regarding model selection, who found also that habitat selection of parr in the water column selected depended on whether the fish were active or resting.

The selected longitudinal models for both 100 and 200 m AUs used only one variable, the distance upstream to a "sedimentary link," but this model did not explain as much of the variance in the data as the models using local data. The sedimentary links are reaches in which the substrate becomes finer going downstream, and are bounded by locations where coarse sediments enter the river, such as tributary mouths or locations of active bank erosion into glacial deposits (Coulombe-Pontbriand and Lapointe 2004), and so are strongly related to the presence of boulders.

Returning to the point at the beginning of this discussion, we see that although Guay et al. (2000) did a good job of testing the fit of their HPI model to their data, closer examination and subsequent investigations revealed various problems: the model is implausible for part of the range of the data, the model was developed and tested on biased samples (because of the time of day and clear skies), and the model did not consider a habitat variable subsequently recognized as important (boulders). There are even more complications. Habitat selection can vary among individuals, and over time for individuals (Roy et al. 2013). How much juvenile Atlantic salmon forage during the day is also affected by whether they will migrate the following spring or remain in the stream for another year (Metcalfe et al. 1998), that is, on their juvenile life history patterns, which are affected by both genes and the environment, and vary within and among populations. And, the morphology of the parr is affected by whether they settle in pools or riffles (Paéz et al. 2008; see also Senay et al. 2015 regarding other fishes), so that parr can grow to become better suited for different microhabitat conditions, which likely affects their habitat preferences. What should we make of this? We agree with Michael Lapointe (2012), a co-author on several of the HPI papers, who concluded that a more comprehensive, multi-scale view of habitat selection is needed.

7.12 Conclusions

The following are general conclusions that can be drawn from this chapter:

- Models are best used in EFA to help people think, not to provide answers.

- Don't use a model unless you have a reason to do so.
- Few people use modeling with the ambition of skewing results to benefit special interests or of deliberately applying flawed methods. Nevertheless, many modelers end up doing so, and are then tempted to rationalize what they have done. This is a human thing to do, and there is little point to blaming people for being human. However, we should anticipate and try to counter this behavior with systematic analyses of how a conclusion was reached.
- Turning conceptual models into numerical models entails many choices about just what to model and how to model it. Even for established methods such as PHABSIM, such choices include the shape of suitability curves to be used and the method by which they will be combined. Models may therefore be biased if these choices are strongly influenced by the data to which the model will be applied.
- Sampling is an important and generally neglected part of EFA. Except for hydrological models, data used to apply EFA models to specific situations are collected by some kind of sampling, raising the questions what, where, and how to sample. It is important to be clear about objectives of the sampling.
- All models are simplifications that deal only with selected aspects of the world. Thus, all EFA models should be tested to increase or decrease confidence in the model for a given purpose but not to yield a binary verdict such as valid or not valid.
- Experiments can test whether a change in a flow regime leads to the biological response predicted by an EFM and so are the most powerful tests of EFMs. Doing them well is difficult and requires multiple years.
- EFMs such as PHABSIM and MesoHABSIM are built on species distribution models. As a result, modelers should be concerned about goodness of fit, generality, transferability, imperfect detection of individual fish, and how plausible the model assumptions or results are in terms of current knowledge. The utility of an EPM models for its intended purpose should be considered, not just its predictive power; a model predicting that fish will be in water will be accurate, but not useful.
- Testing hydraulic models against independent data is important but seldom done. Generally, these models are calibrated in terms of water surface elevation, but are used to estimate flow velocity and depth averaged over small areas defined by the model grid. Accuracy will be affected by approximations in the model, by errors or approximations in the input data describing the shape of the channel, including errors in measurements of discharge in the calibration data, and by the calibration process itself. It should be standard practice to present scatter plots and other graphical depictions of observed versus simulated values, and comparisons should also be summarized with statistics.
- Expert opinion is ubiquitous in EFA. When models are used, there is always judgment involved in which models are selected and how they are implemented. Hydrologically based methods are really a class of expert opinion methods in that expert judgment goes into picking some flow that is calculated from flow records for gaged streams or from regional regression equations for ungaged streams. One virtue of Bayesian Network models is that they can use both expert opinion and modeling results or data in a transparent and quantified way.

Note

1 This avoids the often criticized assumption that the three variables are equally important, although the weighting is fairly equal in this case.

CHAPTER 8

Dams and channel morphology

G. Mathias Kondolf[1], Remi Loire[2], Hervé Piégay[3] and Jean-Réné Malavoi[4]

[1] University of California Berkeley and Collegium - Institute for Advanced Study, University of Lyon, Lyon, France
[2] Electricité de France and University of Lyon, Lyon, France
[3] CNRS UMR Environnement Ville Société, University of Lyon, Lyon, France
[4] Electricité de France, Paris, France

Summary

Most EFAs are done for rivers below dams which trap sediment and change the flow regime downstream. Dams disrupt the flow of sediment and the capacity of the stream to convey sediment, with broadly predictable consequences for morphology and composition of the channel below the dam and on the biota that uses the channel and adjacent areas. When dams trap sediment, the reservoirs they impound lose capacity over time, and may eventually fill up, and downstream river reaches and coasts are deprived of the sediment supply needed for their maintenance. Dams modify the interaction of flow and aquatic and riparian vegetation below them, often leading to vegetation encroachment and reduction in channel capacity. By changing the flow regime, dams can affect deposition of fine sediment and associated nutrients on floodplains below the dam. Finally, most dams also interfere with the transport of wood, again with broadly predictable consequences for the structure and habitat value of the channel below the dam and on the aquatic biota. EFAs need to take into account problems caused by dams, and include ways to analyze and manage them, including restoring flow and sediment regimes downstream of dams.

8.1 Introduction

Most EFAs take place in rivers below dams, which typically store water by design, and store sediment as an unintended consequence. (Wang and Kondolf 2013). Dam-induced changes in flow regime are typically accompanied by reductions in the river's sediment load as reservoirs trap sediment, creating conditions of sediment starvation directly below the dam, which can cause dramatic changes to the aquatic and riparian biota. Reservoirs trap 100% of the river's bedload (the coarse sediment that moves along the channel bed by rolling, sliding, and bouncing, consisting of gravel and sand), and a percentage of the suspended load (sand, silt, and mud held aloft in the water column by turbulence), which varies as the ratio of the reservoir storage capacity to the mean annual inflow of water. Storing water and sediment results in changes in flow and sediment load downstream of dams, almost always producing downstream changes in alluvial channel form (e.g. incision, narrowing) and bed material composition (e.g. clogging, armoring).

Reservoirs with large storage (relative to flow in the river) reduce high flows, reducing the dynamism of the river channel downstream. Gravel beds formerly mobilized every year or two may go for years without being moved, allowing riparian vegetation to establish in the active channel, and fine sediments to accumulate within the gravel (clogging). The reduced flow may not transport sediment delivered to the river below the dam by tributaries, promoting channel aggradation and potentially increasing flooding risk. Without frequent mobilization of the bed, the active channel typically narrows by encroachment of woody riparian vegetation and deposition of sediment (Williams and Wolman 1984), but not universally. If peak flows are only slightly reduced but bedload is fully starved by the reservoir, a channel can degrade and become progressively armored and clogged. If both are significantly reduced, the channel may not degrade but undergo a significant narrowing due to riparian vegetation encroachment. The complexity of alluvial channel forms depends upon the availability of coarse sediment (sand and gravel) that compose bars and riffles. When the supply of coarse sediment is reduced because of trapping by upstream dams, alluvial channel form typically simplifies, as bars and riffles are eroded away without being replaced by deposition of sediments from upstream.

Dams also trap large wood, and historically such wood has mostly been collected and "disposed" of (e.g. by burning), because it was viewed as creating management problems, such as debris jams at downstream bridges and blocking passage of migratory fishes (Piégay and Landon 1997). In many river systems, large wood is critically important in creating channel complexity (Gregory et al. 2003), and where wood is denied to the river system below dams, downstream channel complexity can decline as a consequence.

To mitigate dam-induced impacts, controlled high-flow releases designed to mimic the action of natural floods are increasingly required in licenses for dams and as part of programs to restore river function. These deliberate, high-flow releases thus constitute one component of environmental flow requirements for maintenance of aquatic and riparian habitat, and reflect an evolution of environmental flow requirements from simple minimum flows to include periodic high flows to mimic flood effects on channels or on ecological processes (Kondolf and Wilcock 1996; Yarnell et al. 2015). Various names have been applied to these high-flow releases (e.g. flushing flows, channel maintenance flows), but Loire et al. (2019) propose using morphogenic flows as better reflecting the explicit geomorphic objectives for the flow releases (Table 8.1). Such morphogenic flow releases are usually for ecological purposes, but they can be also implemented for other objectives such as risk management (e.g. maintaining channel capacity to prevent flooding risk or bank erosion, preventing reservoir siltation). To restore habitat lost to sediment starvation and lack of wood downstream of dams, restoring flows only may not achieve the objectives. Sometimes it is better to introduce coarse sediment and wood, as done in many rivers mechanically or by promoting bank erosion.

There is a large consensus in the scientific literature about the necessity to release morphogenic flows (e.g. Konrad et al. 2011; Montgomery and Bolton 2003; Mürle et al. 2003; Poff et al. 1997; Robinson and Uehlinger 2003). The last two decades have seen morphogenic flows prescribed to restore river channels by mimicking effects of natural floods (Kondolf 1998; Konrad et al. 2011; Rivaes et al. 2015; Wu and Chou 2004). Importantly, some empirical tests of implementing such flows have also been conducted (reported below).

While some studies have emphasized interactions of water flows with sediment (Wohl et al. 2015), such as sediment supplied from downstream tributaries (e.g. Grams et al. 2013), most published work has focused on flow releases needed to accomplish geomorphic goals. Even if a post-dam flow regime were to mimic precisely the pre-dam flow regime, the river system would be severely altered by the loss of its sediment load (Kondolf and Wilcock 1996; Wohl et al. 2015). Thus, increasingly, partial restoration of sediment load is prescribed along with morphogenic flows (Kondolf 1997; Tena et al. 2012). Coordinating morphogenic flows with sediment augmentations has seldom been reported, but is likely to become more common (e.g. Kantoush and Sumi 2011).

Table 8.1 Potential objectives of morphogenic flows.

	Ecological or management objectives	Specific objectives	Flow requirements
Flushing flows	Restore/enhance riffle habitat	Remove surficial fine sediment	Mobilize sand
		Remove interstitial fine sediment	Mobilize gravel
		Maintain gravel "looseness"	Mobilize gravel
Channel-maintenance flows	Restore/enhance pool habitat	Scour accumulated fine sediments	Net transport of sand out of pools
	Maintain active channel width and topographic diversity	Prevent encroachment of vegetation by uprooting seedlings or drowning terrestrial plants	Mobilize gravel throughout cross-section
			Inundate floodplains
	Create/build floodplain habitats	Produce vertical accretion on floodplain	Produce muddy overbank flow
	Create diverse multi-age riparian habitat	Induce channel migration and thereby create diverse geomorphic surfaces	Erode banks, deposit point bars, overbank deposits

Source: adapted from Kondolf and Wilcock (1996).

Loire et al. (2019) analyzed publications on morphogenic flows through 2016. The literature review focused on factors relevant to setting the dimensions of morphogenic flows and results from the experience implementing such flows. Of over 103 articles and reports identified as relevant to the topic, about a dozen were guides published to inform dam managers on how to design such flow releases. The objective most commonly cited in the published papers was to improve conditions for spawning fish. The case studies reported were mostly in the USA, Australia, New Zealand, South Africa, Switzerland and Spain. In the past two decades, experimental flow releases have been undertaken and monitored, the best-documented being those on the Colorado River in the Grand Canyon (US), the Spöl (Switzerland), and the Ebro (Spain).

In the following sections we clarify the context of channel changes downstream of dams and approaches to mitigating negative effects, which include morphogenic flows, addition of coarse sediment in downstream reaches starved of sediment, and input of large wood. We then discuss steps involved in designing morphogenic flow releases. Morphogenic flow releases cannot be successfully applied everywhere, and even where they can, they improve ecological function only through their interactions with sediment supply and large wood.

8.2 Diagnosing the problem and setting objectives

The first step is to clearly assess past or anticipated dam effects on peak flow and sediment supply and transport, based on an analysis of drivers (flow and bedload transport modifications) and the downstream channel responses (physical, biological and chemical). Actions taken are first based on a fundamental distinction between rivers where flows are intended to maintain pre-dam conditions following construction of a new dam and on rivers in which the channel has already adjusted to a post-dam flow and sediment regime. For example, a formerly wide, gravel-bedded channel that has narrowed would require a lower flow for bed mobilization than would be the case for the original wide channel because depth would be greater for a given flow. However, the bed may have coarsened, requiring a higher flow (shear stress) to mobilize.

For determining morphogenic flows below an existing dam, it is necessary to assess whether the bed is still adjusting to new conditions or has already adjusted (Arnaud et al. 2017; Rollet et al. 2014). Different reaches along the same river may be in different states of adjustment as a function of their location with respect to tributary sediment sources, the size, location and operational regimes of the dam, and the type of river channel (alluvial, bedrock-controlled, etc.). For rivers in western North America, Schmidt and Wilcock (2008) estimated sediment supply and sediment transport capacity by river reach, thereby determining reaches in sediment deficit or surplus. Multiple distinct reaches were evident, in some cases switching from deficit to surplus with distance downstream of dams as tributaries entered the mainstem (as illustrated on the Green–Colorado River system). Quantifying the sediment balance by reach can clarify the problems and provide insights into how morphogenic flows and sediment management can help to mitigate the identified impacts. Most published studies concern dams that have been long-established, and channels that have already undergone at least partial adjustment to the post-dam flow and sediment regime.

8.3 Managing sediment load

8.3.1 Existing dams

To partially restore sediment load in a regulated stream, coarse sediments are most commonly added to downstream channels by mechanical means. These coarser fractions preferentially deposit in deltas at the upstream end of reservoirs. In some cases, sediment has been mechanically removed from reservoir deltas and placed in the downstream channel to augment sediment supply (Jones and Stokes 2003; Kantoush and Sumi 2010). Even though this solution replaces the downstream sediment supply with the same sediments transported by the river, it is rarely done because of logistical and legal impediments to dredging the deltas and transporting the sediments around the reservoir and dam, and the cost of such

operations. Where sediment has been mechanically added to the channel downstream (discussed further below), the sediment (usually gravel or gravel-sand mixtures) has mostly been derived from other sources, such as terrace gravels, floodplain gravel pits, or in some cases, gravel mines on tributary streams.

Adding gravel to river channels below dams is commonly termed gravel augmentation or gravel replenishment. It has been widely undertaken in North America and Europe, in the vast majority of cases to restore spawning habitat for fish, especially salmon or trout. In northern California, over 200 000 m³ was added to the Sacramento River, 30 000 m³ to the Trinity River, and over 45 000 m³ to Clear Creek, between 1976 and 2013. On the Trinity River, the first such projects were undertaken in 1976 to create artificial riffles, with lines of boulders across the stream to hold the gravel in place. The river's transport capacity was greatly reduced by Trinity Dam, so the placed gravels did not immediately wash out, as occurred with similarly designed projects on the Merced River (Kondolf et al. 1996). By the early 1990s, releases of morphogenic flows were coordinated with gravel augmentation (Wilcock et al. 1996). Planners have measured the transport of gravels downstream of Trinity Dam by morphogenic flows (and natural floods spilling over the dam) and sought to compensate for these gravel losses from the reach with gravel additions. Thus, the restoration project evolved to have the explicit goal of building of bars and complex channel habitat through additions of coarse sediment and release of flows to transport and redeposit the sediment in natural channel forms (Gaeuman 2014); resulting ecological benefits were documented, such as processing particulate organic matter, inducing hyporheic exchange, and creating thermal complexity (Ock et al. 2015).

Similarly, on the Uda River below the Murou Dam, Japan, sediment replenishment has been undertaken to restore sediment transport and channel complexity since 2006. In the first five years of the restoration program, natural flows spilling from the dam were sufficient to transport the added sediment in the first year, but in the subsequent four years, morphogenic flows were released to achieve desired sediment mobility (Kantoush and Sumi 2011). Increasingly, sediments are added to reaches below dams in Japan to support development of gravel bars and other complex channel features (Ock et al. 2013).

As summarized by Ock et al. (2013, p. 54), such efforts to restore sediment loads and fluvial processes below dams require systematic planning that accounts for specific objectives and local restrictions of the river basin, river and reservoir characteristics, and coordinating "flushing flows (magnitude, frequency, and timing), determining quantity (amount added) and quality (grain size and source materials) of coarse sediment, and selecting an effective implementation technique for adding and transporting sediment...." Dams vary widely in their settings (flow, sediment load, presence of tributaries downstream, channel slope, etc.), in their size relative to the river flow, and in their design and operation (size and location of outlets, reservoir geometry, etc.). To assess dam-induced disruptions to a pre-dam sediment balance, a sediment budget (Reid and Dunne 2016) can provide a framework within which to analyze information on the transport capacity of the river (with and without morphogenic flows) and the quantity and caliber of sediment supplied from tributaries and other downstream sources, as a basis for specifying morphogenic flows and, if needed, supplying sediment to downstream reaches. Programs of coupled gravel additions and morphogenic flows are expensive and consequently not widespread, but prescribing a morphogenic flow alone without accounting for sediment supply will usually not achieve ecological goals envisioned for the flows.

8.3.2 Proposed dams
Designing dams to pass sediment
Mechanically adding sediment downstream of dams is expensive. It is more efficient to employ gravity to deliver sediment to the channel downstream of dams by passing sediment through or around dams, for which a range of techniques can be used (Morris and Fan 1998; Wang and Hu 2009; Annandale 2013; Kondolf et al. 2014a; Annandale et al. 2016). The

summary below is drawn primarily from Kondolf et al. (2014a). For smaller dams, the most sustainable approach (where feasible) is to pass the sediment load around or through the dam. Water can be diverted to an off-channel reservoir only during lower flows, when water is relatively sediment free, while allowing sediment-laden floodwaters to pass by in the main river. A sediment bypass can divert part of the incoming sediment-laden waters into a tunnel around the reservoir, so they never enter the reservoir at all, but rejoin the river below the dam. Sediment can also be sluiced by maintaining sufficient velocities through the reservoir to let it pass through without allowing it to deposit. Alternatively, the reservoir can be drawn down to scour and re-suspend sediment in the reservoir, and transport it downstream. This involves complete emptying of the reservoir through low-level gates. Density current venting involves opening dam outlets when turbidity currents pass through the reservoir so that they can remain intact and exit the reservoir via the outlets, carrying most of their sediment with them. Sluicing, flushing, and density current venting pass sediments in suspension, which tend to be the finer fractions of the sediment load, but can include significant sand. Sluicing and flushing work best on reservoirs that are narrow, have steep channel gradients, and have storage that is small relative to the river flow. As shown by Sumi (2008), flushing has been effective for reservoirs that impound less than 4 % of the mean annual inflow (Figure 8.1). Large reservoirs with year-to-year carry-over storage are poor candidates for such sediment pass-through approaches, and it is generally most efficient to take sediment management into account at the outset of the design and operation of dams.

Minimizing sediment trapping through strategic dam planning
Strategic dam planning at the river basin scale is an often overlooked opportunity to minimize impacts of dams on sediment supply. Such planning should involve recognizing the cumulative effects on sediment supply of multiple dams in a river network and

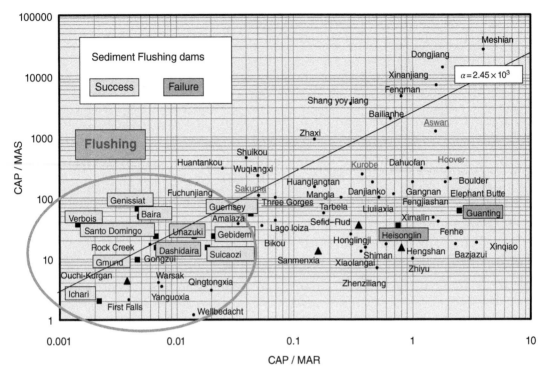

Figure 8.1 Plot of flushing projects from diverse environments showing that successful cases are characterized by impoundment ratios of 0.04 or less. That is, reservoir storage capacity (CAP) divided by mean annual inflow to the reservoir (MAS) should be less than 0.04. Source: Kondolf et al. (2014a), used by permission of AGU/John Wiley & Sons.

consequent geomorphic impacts (e.g. Kondolf et al. 2014b). New dams should be located to minimize disruption of sediment supply (Schmitt et al. 2018). Such dams should also be designed to maximize their ability to pass sediment around or through reservoirs (discussed above). Overall, there is large, but so far mostly missed, potential to develop and manage dams more sustainably for both reservoirs and rivers (Wild et al. 2016).

Throughout the developing world there is an explosion of dam building, motivated largely by a push for hydroelectricity, with an anticipated doubling of global hydroelectric capacity within the next two decades (Zarfl et al. 2015). As demonstrated for the major downstream tributary of the Mekong River (the Sre Pok, Se San and Se Kong system), strategic dam planning could have resulted in a dam

portfolio producing 68 % of the basin's hydroelectric power potential while trapping only 21 % of its sand load. The current portfolio is the result of project-by-project construction of dams, without a strategic trade-off analysis or planning (Figure 8.2). As a result, the dams produce 51 % of the basin's hydroelectric capacity while trapping 91 % of its sand load, mostly because of early construction of downstream dams in the Sre Pok and Se San basins (Figure 8.2) (Schmitt et al. 2018). In an effort to preserve remaining connectivity of sediment sources in the basin, the Natural Heritage Institute (a US-based NGO) and the National University of Laos developed a plan (adopted by the Laotian government) to site new hydropower dams in the Se Kong River basin only upstream of existing dams. The plan follows a strategic analysis for planned and built dams to minimize

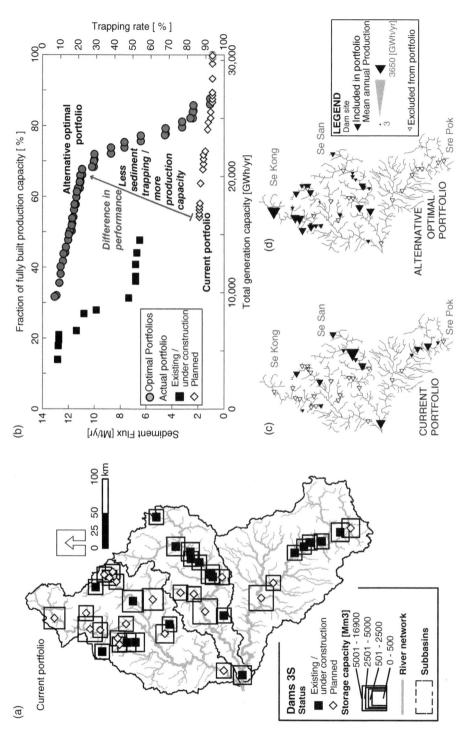

Figure 8.2 Power generation and sediment trapping from dam building in the Sre Pok, Se San, and Se Kong rivers (the "3S basin"), the largest downstream tributary to the Mekong River (drainage area 80 000 km²). (a) The current 3S dam portfolio includes 21 dams built or under construction, and 21 more at various planning stages. (b) Increased power-generation capacity and cumulative sediment trapping with construction of the current dam portfolio (black squares for existing dams, white diamonds for planned dams) and alternative portfolios with an optimal trade-off between sediment trapping and power production (grey circles). The arrow indicates a dam portfolio with higher power production but lower sediment trapping compared to the current portfolio (see arrow). Optimal portfolios were identified based on analysis of 17 000 alternative dam portfolios (not shown). The optimal portfolio compares favorably to the currently planned development because of a different spatial configuration of dams in the network. (c) The current dam portfolio includes dams downstream in the Sre Pok and Se San. (d) The alternative, optimal portfolio relies more on dams in the headwaters and on lower sediment-yield portions of the basin. The optimal portfolio greatly reduces environmental impacts and reservoir sedimentation, and also produces higher economic benefits. Source: Adapted from Schmitt and Kondolf (2017).

additional sediment trapping in the basin (NHI and National University of Laos 2018).

Unfortunately, most dams have been (and continue to be) built on an individual, project-by-project basis, without analysis of cumulative effects of multiple dams on a river network, much less strategic planning to minimize impacts. In these cases, maintaining habitat downstream of dams could involve a combination of morphogenic flows, sediment augmentation, and adding large wood. However, before undertaking these actions, a simple sediment budget and assessment of geomorphic processes and habitat conditions should be conducted to inform the restoration actions. The sediment budget should compare downstream sediment supply with the energy available to transport it, to ascertain if the reach has a sediment deficit or surplus, and to what degree. Likewise, assessing post-dam channel adjustments and their implications for aquatic habitat will inform potential options for restoration.

8.3.3   Obsolete dams

While some dams are deliberately built to trap sediment to prevent its delivery to downstream reservoirs or urban areas, this is a questionable strategy at best, because once these sediment-control dams fill, they become problems in and of themselves (Wang and Kondolf 2013). Most dams are designed to trap water and the accumulation of sediment is an undesirable side effect. While some dams have persisted as usable features for centuries (Kondolf and Farahani 2018), most have shorter lives as their finite storage capacities fill with sediment (Kondolf et al. 2014a). Besides filling with sediment, dams can become obsolete through changes in technology, such as the replacement of the mechanical energy provided by water mills with coal-fired steam engines (Tvedt 2015), through the deterioration of their structural elements, such as decay of concrete through alkali–aggregate reaction (Fournier and Bérubé 2000), or through new environmental requirements that make the dam uneconomical. Four hydroelectric dams on the Klamath River (Oregon and California, USA) are a case in point. The Klamath River historically supported large runs of salmon, upon which Indian tribes depended. The four hydroelectric dams blocked access to headwater reaches of the mainstem. These dams, in combination with dams on the Trinity River (the Klamath's principal tributary) and with habitat degradation on other formerly important tributaries, caused severe decreases in salmon populations. When the dams recently came up for license renewal, the costs of retrofitting them with fish passage structures (of dubious effectiveness) and other environmental costs led the utility that owned the dams to decide that dam removal was preferable to potential costs of keeping the dams operating, given their relatively small power production (Leslie 2017).

Removal of obsolete dams has become an important trend in restoration of rivers, to restore passage for migrating fish, as well as for restoring longitudinal continuity for navigation and the transport of sediment and nutrients (O'Connor et al. 2015). After an initial period of unnaturally high sediment load resulting from erosion of stored reservoir deposits, dam removal can renaturalize flow and sediment regimes, as well-documented after the removal of Arase Dam on the Kuma River (Japan) in 2015 (Kobayashi et al. 2016) and Elwha and Giles Canyon Dams on the Elwha River (Washington state, USA), as described below.

One of the principal technical challenges to dam removal is managing sediment accumulated behind dams, which is greatest for large dams behind which large volumes of sediment are stored and available for release in a large pulse if the dam is removed. The largest dam removal to date is that of Giles Canyon and Elwha Dams on the Elwha River (Magirl et al. 2015). The Elwha River dams provided electrical power to a pulp and paper mill at below-market rates, but when the hydroelectric projects came up for relicensing, there was strong pressure by the Elwha Klallam Tribe, and eventually by federal and state environmental agencies, to remove the dam to restore salmon runs. Fish passage measures were extremely expensive and unlikely to be effective for most salmon species, and in any event would not have fully restored potential salmon runs. Before the US Federal Energy Regulatory Commission made a final decision, the US Congress passed the "Elwha River Ecosystem and Fisheries Restoration Act" (Public Law 102-495), which called for full restoration and provided for the US Government to buy the dams and provide replacement power from the Bonneville Power Authority (Gowan et al. 2006). With removal of the two dams in 2011–2013, about 7 million m³ of sediment was flushed from the former reservoir sites in relatively modest high flows, producing aggradation and increased channel complexity in the reaches downstream. Salmon immediately began to ascend the river to spawn (East et al. 2015; Magirl et al. 2015).

In the Elwha River case, many factors aligned to make dam removal feasible and extremely beneficial ecologically. The dam owner was willing to be rid of the dams, the federal government endorsed the solution and provided funds to implement it, and importantly, the downstream lands affected by channel aggradation and increased flooding were primarily tribal lands, and the Tribe was strongly supportive of dam removal. The upstream habitats opened by dam removal have been used by five species of Pacific salmon and steelhead trout; they remained in excellent condition by virtue of their inclusion within Olympic National Park, so the potential ecological benefit was large. In many cases, dam removal is more contested and complex institutionally. In the Coast and Transverse Ranges of California, at least four small water-supply reservoirs completely filled with sediment by the late twentieth century, reflecting the high sediment yields of these geomorphic provinces and small capacity of these dams. All four dams have densely settled river corridors downstream, with extremely high-value real estate, so the potential liability for increased flooding from downstream aggradation is a key issue.

- San Clemente Dam on the Carmel River, was considered seismically unsafe due to foundation issues and was removed in 2015 at a cost of over $83 million USD. Its sediment was mostly stabilized in place, and the river was rerouted past the stored sediment through a tributary stream valley (CalAm (California-American Water Company) 2017).
- Matilija Dam on the Ventura River suffers from safety problems due to deterioration of its concrete from alkali–aggregate reaction and being full of sediment; it is planned for eventual removal, starting with creation of large outlets that will permit sediment to be flushed out during floods, at a cost estimated at $111 million USD (Johnson 2017).
- Searsville Dam on San Francisquito Creek is to be managed in place with either a large outlet retrofit into the dam or a sediment bypass around the sediment-filled structure (Krieger 2015).
- For Ringe Dam on Malibu Creek, removal of the dam and truck transport of some of its stored sediment directly to the sediment-starved coast has been identified as the preferred alternative to restore stream functions (USACE 2017).

In all these cases, managing the accumulated sediments and reducing their potential for raising flood levels downstream have emerged as key issues to be addressed.

Dams eventually become obsolete, and removing or otherwise managing obsolete dams is expensive. When dams are built, however, obsolescence seems far enough in the future that the associated costs typically have been ignored. This is a mistake. If time to obsolescence and cost of removal had been considered in the original cost –benefit analyses, it is likely a number of dams never would have been built. There is no reason why such costs should not be considered as part of contemporary dam projects, however.

The vast majority of removals have been of small dams, many former mill ponds (e.g. Doyle et al. 2005). While such dams may have little economic value today in terms of power production, they may have cultural significance for local populations, and proposals to remove them (to improve fish passage and sediment transport continuity) often generate controversy (e.g. Barraud 2017). As compelling a topic as it is, dam removal is unlikely to be a viable option in the vast majority of cases, at least in the short run (<100 years). We must manage with the dams in place as best we can. Thus, looking for solutions to improve longitudinal continuity with dams in place remains an imperative.

8.4 Specifying morphogenic flows

Like environmental flows in general, morphogenic flows are well-suited to adaptive management, discussed elsewhere in this book. It is difficult to predict channel responses to a given flow release because of each river's unique geology and history, including past adjustments to the channel and its bed, as well as continuing adjustments by managers, such as adding coarse sediment. Thus, morphogenic flows are best approached initially as a series of experiments, whose results need to be well monitored and assessed to inform future releases. Once flows producing the desired geomorphic and ecological response are identified, the intensity of monitoring can be reduced and morphogenic flows can be integrated into routine management of the reservoir and regulated river, along with sediment releases or downstream augmentation.

8.4.1 Three common approaches to specifying morphogenic flows

Morphogenic flows have commonly been based on one of three underlying approaches: self-adjusted channel methods, sediment entrainment methods and direct calibration methods (Kondolf and Wilcock 1996). Self-adjusted channel methods assume that the channel is in equilibrium with the prevailing flow regime and sediment load, and that all channels achieve this state of adjustment with a discharge whose frequency is similar from place to place. The concepts of bankfull, dominant, or channel-forming discharge are most relevant to the overall, mean behavior of rivers. They may apply poorly to any particular channel, because the channel may not be in equilibrium, and generalizations of effective discharge from the literature may not apply to the

river (e.g. Nolan et al. 1987). In any case, plots of channel dimensions as a function of flow or drainage area typically show wide scatter, even within a given region (Leopold and Maddock 1953), a reminder of the hazards of over-generalization. Specifying a statistic from the flow record as the morphogenic flow implicitly assumes a self-adjusted channel. This is clearly the easiest approach in terms of data requirements and probably the most widely used, but it is not justified in most cases, because the settings in which morphogenic flows are most often specified – downstream of dams – are by definition unlikely to be in equilibrium. This simple fact argues strongly for approaches that base morphogenic flows on specific functions, such as mobilization of bed sediments.

Sediment entrainment methods do not rely on assumptions of equilibrium, but may be applied to the existing channel based on local properties (channel size and slope, bed material) without calibration. Sediment entrainment methods (without calibration) can be used to provide initial estimates of the flows needed for specific functions such as bed mobilization, but such sediment transport functions are notoriously imprecise. This is discussed in the excellent review of sediment transport concepts and approaches by Wilcock et al. (2009).

Direct calibration methods require field observations of bed movement, sediment transport, or changes in fine sediment content of gravel beds, through the use of tracers (Hassan and Roy 2016), computation of effective discharge from flow records and sediment rating curves, or other evidence. These can be observations undertaken during experimental releases (e.g. Wilcock et al. 1996) or by taking advantage of naturally occurring floods to derive empirical data on bed mobility (Cullis et al. 2015; Zurwerra et al. 2016). In most cases the cost of forgone power is sufficiently high to justify an adaptive management program of monitoring bed mobility (and/or other desired flow effects) to better specify the flow required. Ideally, flows should be sufficient to achieve their intended purpose, but need not consume more water than needed.

Statistical analysis can provide a useful basis for specifying morphogenic flows. Using a structured Bayesian approach described in other chapters, Webb et al. (2015a) modeled the likely effects of 50 days of inundation in reducing the extent of terrestrial vegetation that had invaded rivers affected by flow diversions after a 10-year drought, with initial states of terrestrial vegetation coverage estimated from opinions of experts. The modeling indicated significant reductions in terrestrial vegetation extent in all but one of seven rivers studied, providing results that can guide management decisions and allocations of water for environmental restoration.

8.4.2 Clear objectives needed

A clear statement of objectives is essential for effectively specifying morphogenic flows, not just to estimate flows that needed to meet the objectives, but also, later on, to assess whether the objectives were met. Objectives that are broadly stated may be difficult to translate into definable flows, and thus may be of little practical use, regardless of their importance. Each objective should be associated with hypotheses regarding the particular physical processes to be activated and the changes in the stream channel that they are expected to create, as well as estimates of the range of flows needed. Then, costs, constraints, and tradeoffs associated with that objective can be addressed, the compatibility among different objectives can be evaluated, and flows can be implemented adaptively. For example, to specify flows for "channel maintenance" requires knowledge of specific channel processes such as bed and bar mobilization before flows can be specified, unless it is assumed that a flow with a given return period can serve to achieve a suite of functions (as in a self-adjusted channel approach) (Kondolf and Wilcock 1996). Morphogenic flows are often specified based on five basic parameters: magnitude, frequency, duration, seasonality and, increasingly, the form of the hydrograph (Pruitt and McKay 2013). The specified flow commonly represents a compromise among factors representing the intensity of the mobilization of the channel bed, the volume of water

needed to realize a desired morphologic effect, and the effects of the flows on the purposes of the dams.

8.4.3 Magnitude

The magnitude of a flow is the most obvious variable to specify. Under the self-adjusted channel paradigm, the flow magnitude may be specified based on pre-dam flows of a given recurrence, such as the flow that occurred every 1.5 or 2 years under pre-dam conditions, based on assumptions about "bankfull discharge." With the assumption of equilibrium, messy details are avoided and a further assumption can be made that replicating a given flow from the pre-dam hydrograph will serve to restore/maintain the channel and its sediment quality. This approach has been widely used in practice, for example in hydroelectric license renewal applications before the US Federal Energy Regulatory Commission. There has been little empirical evidence to support the approach, and there are strong reasons to believe it does not work in many cases because of ongoing channel adjustments, which would render assumptions of equilibrium invalid.

Under a sediment entrainment approach, a morphogenic flow is predicted to mobilize the bed, bars, or other features, depending on the objective (Wilcock et al. 1996; Gaeuman 2012; Batalla et al. 2014). The entrainment flow depends on the channel form which, as noted above, may have changed post-dam, commonly by narrowing. It also depends on the grain size of the surface to be mobilized; usually this is the active channel bed, but it could also be bar tops and other surfaces. The presence of a surface armor (or pavement) implies a high flow is needed to initiate motion and a lower flow is needed to maintain transport of the subsurface grains once exposed (Beschta et al. 1981; Tonkin and Death 2014). More than one flow can be required to achieve a given objective. For example, in the Colorado River below Glen Canyon Dam, a high flow was prescribed to erode sediments from tributary deltas, followed by a moderate flow to build sand deposits that would provide beaches for human use when exposed at low flows (Collier et al. 1997). Flows specified for one objective may be insufficient or too high for another objective, as illustrated by the conflict between flows needed to mobilize the active channel bed vs flows needed to mobilize bar tops to prevent establishment of woody riparian species; the latter flow could result in such a high rate of transport in the active channel that much of its gravel would be lost to downstream

transport (Figure 8.3) (Kondolf and Wilcock 1996). Direct observation of sediment mobilization can test models used in sediment entrainment methods, as was implemented in the Trinity River, California (Wilcock et al. 1996).

The magnitude of flow that can be released from a dam is often constrained by outlet capacity. For

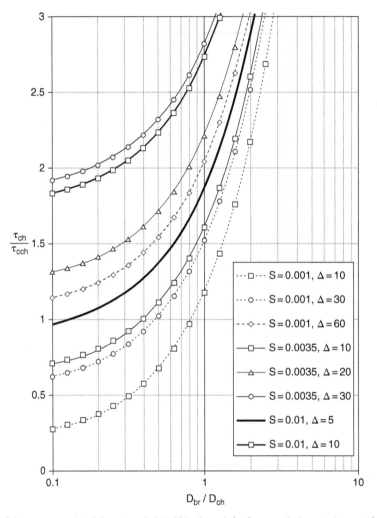

Figure 8.3 Ratio of shear stress to critical shear stress (tch/tcch) in channels for flows producing entrainment of sediment on the top of an adjacent bar. This ratio is plotted as a function of the ratio of grain size for the bar and channel (**D**br/**D**ch) for channel slope, **S**, of 0.001, 0.0035, 0.01 and various values of dimensionless bar height, **D**; tch/tcch increases directly with all of **D**br/**D**ch, **S**, and **D**. It generally takes values that do not allow selective transport of sand in the channel but results in wholesale transport of channel gravels. Source: Kondolf and Wilcock (1996), used by permission of AGU/John Wiley & Sons.

example, restoration of Clear Creek, California, has been severely constrained by difficulties in achieving high-flow releases from Whiskeytown Reservoir because the dam was built with only small outlet pipes, incapable of releasing flows required to mobilize the downstream channel. The reservoir receives and passes on water as part of an inter-basin transfer, so the reservoir operators have unusual control over water levels, and high-flow releases are made by bringing reservoir levels up to the point that water spills through the "glory-hole" spillway (Pittman 2013). Nevertheless, an engineering analysis that explored options for retrofitting the dam with larger outlet work concluded that the risk of destabilizing the dam was too high.

8.4.4 Duration

The duration of the morphogenic flow needed depends upon the objectives and the condition of the river. To flush accumulated sediments from a reach, key variables include grain size of the accumulated sediments, slope of the reach, channel form, and length of the reach. Fine sediments generally travel in suspension at the same velocity as the water, so once they are flushed from the bed, they move out of a reach at the speed of the wave (Mürle et al. 2003; Zurwerra et al. 2016). However, coarser sediments moving as bedload travel at a much slower rate and thus the time required to remove them from a reach will be greater (Wilcock et al. 1996; Dollar 2000; Zurwerra et al. 2016). As noted by Collier et al. (1997), the duration needed to flush sand depends in large measure on the magnitude of the flow: they observed a doubling of flow in the Colorado River yielded an eightfold increase in transport rate. The exponential relationship of sediment transport to flow is well-established. It implies that releasing a higher flow for a shorter period is generally more efficient (in terms of water used) than releasing a moderate flow for a longer period to remove a given volume of sediment from a reach.

Given that the cost of water released for a morphogenic flow can be very high in terms of lost power revenue or other uses, duration is a key variable.

Often the magnitude is easier to specify (e.g. critical discharge to mobilize a gravel bed) so duration becomes the variable with greatest uncertainty and largest implications for cost.

Because lost revenues from foregone hydropower generation and deliveries to other water users can be so costly, relying on sediment transport from morphogenic flows alone may not be an economical approach to removing large quantities of tributary-derived fine sediment accumulated in a mainstem. For example, the Trinity River mainstem by the 1980s had accumulated large quantities of coarse sand delivered by tributaries, partly from accelerated erosion from extensive clear-cut logging. As a result, early restoration efforts included repeated sand removal from pools by heavy equipment, which excavated the bed to depths that would trap sand delivered from upstream (Wilcock et al. 1996).

8.4.5 The hydrograph

Relatively little research has been published evaluating methods to set the shape of the hydrograph for morphogenic flows, including the rising limb, the maximum flow itself, and the falling limb. Commonly the rate of increase is constrained for ecological and safety reasons (e.g. to avoid drownings). The rate of decrease is constrained to minimize ecological effects, such as stranding of fish on the floodplain and geomorphic impacts such as bank collapse from positive pore pressure of still-saturated banks. Moreover, as noted above, the ability to release flows in a given hydrograph may be limited by outlet capacity and operational constraints (Wilcock et al. 1996; Dollar 2000; Rivaes et al. 2015; Wilcox and Shafroth 2013).

The rate of flow increase and decrease influences processes of erosion and sedimentation (Petts et al. 1985; Watts et al. 2009). Appropriately sized high-flow releases can be more effective in transporting bed sediments because of their flashiness (Batalla and Vericat 2009). However, because they are typically for a short time only, the total volume transported is limited (Tena et al. 2012).

Most published hydrographs for morphogenic flows specify only the peak flow. Depending on the objectives of the flow, multiple peaks may be prescribed, as was done on the Ebro River where a double peak was tested. The first peak was designed to detach algae from the bed gravels and the second, larger peak to flush the detached algae out of the reach (Batalla and Vericat 2009). However, macrophytes in the active channel increased bed roughness and resulted in higher river stages for a given flow, with the result that some low-lying villages were threatened with flooding. Informed by model simulations, Batalla et al. (2014) designed a new hydrograph with the objective of having a lower peak flow, which would still be effective in transporting and removing macrophytes. They used the concept of a "chasing wave" introduced by Hicks and Goring (2009) to design hydrographs with three peaks, each of increasing magnitude and duration, which successfully detached and transported algae from the bed (Batalla et al. 2014) (Figure 8.4). By the time the peaks reached the gaging station 12 km downstream, they had merged into a single (somewhat

attenuated) peak (Figure 8.4). On the Trinity River, California, gravel injections at the Lowden Ranch restoration site were coordinated with morphogenic flows released from Lewiston Dam upstream and resulting transport rates were measured (Gaeuman 2014) (Figure 8.5).

It has proved difficult to predict the volume of sediments transported (Eder et al. 2014) and whether the maximum transport occurs before, during, or after the peak flow (Beschta et al. 1981; Petts et al. 1985; Jakob et al. 2003; Tena et al. 2012). Differences among river reaches arise from multiple factors influencing erosion and transport, such as vegetation within the active channel, fine sediment matrix and soil moisture.

8.4.6 Seasonality

Because ecological restoration is the primary driver for creating morphogenic flows, seasonal timing is commonly dictated (or constrained) by the seasonality of biological processes. For example, on the Bill Williams River (Arizona), a flow release was timed to correspond to the timing of release of native willow

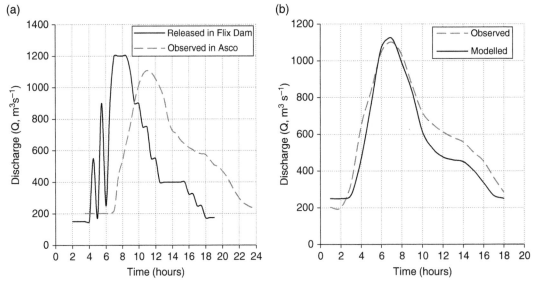

Figure 8.4 Hydrographs for morphogenic flow releases from Flix Dam on the Ebro River, Spain. (a) Hydrograph of release from dam on November 19, 2013 (solid line) with hydrograph of flow recorded at the Asco gauging station 12 km downstream (lighter dashed line). (b) Hydrograph at Asco (lighter dashed line) compared to model results (solid line). Source: Figure 9, Batalla et al. (2014), used by permission of University of Rioja, Logroño, Spain.

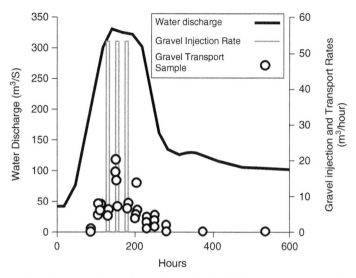

Figure 8.5 Coordination of gravel augmentation at the Lowden Ranch rehabilitation site, Trinity River, with morphogenic flow releases in 2011, showing flow release hydrograph (solid line), magnitude and timing of gravel injection during the release (shaded rectangles), and magnitude and timing of measured gravel transport rates (open circles). Source: Gaeuman (2014), used by permission of John Wiley & Sons, Publishers.

seeds, so the flow would distribute them as it did under pre-dam conditions (Wilcox and Shafroth 2013). Because life histories of native species are typically keyed to the pattern of high and low flows that naturally occur on a river, the seasonal requirements of species can normally be met by imitating the natural flow regime. The timing of incubation of salmon eggs can be another important factor influencing the timing of morphogenic flows. During the period of incubation (typically 6–10 weeks in the winter or spring), embryos are vulnerable to scour. While there was and is clearly some such loss of incubating embryos in natural systems, avoiding such losses is often cited as a concern in designing morphogenic flows on salmon-bearing rivers (May et al. 2009).

Coordination with seasonal high flows on tributaries is another important consideration. The extent and depth of fine sediments to be removed depends in large measure on the hydrologic conditions in the preceding months and the delivery of sediment from the tributaries (Tena et al. 2012; Batalla et al.

2014). In some cases, flows that can be released from dam outlets are insufficient to accomplish initially defined objectives, so artificial releases can be timed to "piggyback" onto natural high flows entering from tributaries to reach the needed flow magnitudes (May et al. 2009). Because tributaries are likely to be contributing not only flow but also sediment load (more so than releases from the reservoir), the coordination with tributary inflow may be more complicated than simply adding flows together. For example, on the Trinity River, deposition of suspended sands from overbank flows built up natural levees along the river channel. These levees served to further concentrate flow within the main channel, such that even when flows were high enough to spread out on the adjacent valley bottom, the natural levees would confine the river to the main channel. To avoid exacerbating this problem, the morphogenic flow releases would best be timed to avoid coinciding with high flows on tributaries (Kondolf and Wilcock 1996; Wilcock et al. 1996). A contrary example is the Colorado River below Glenn Canyon Dam, where a goal was to deposit suspended sand on marginal sand bars to build the beaches, and where high-flow releases are now timed to follow immediately upon high flows on the Paria River, the major tributary just downstream of the dam, which is now the largest sediment contributor in this part of the river (Grams et al. 2013).

8.4.7 Recurrence

The frequency of occurrence needed for morphogenic flows over a multi-year period depends on the functions for which the flow is released. In many cases, natural recurrence intervals and flow patterns are commonly imitated (Poff et al. 1997), implicitly assuming a self-adjusted channel. Of course the artificial flows have reduced magnitude and duration compared to historical hydrographs, to leave water in the reservoir for human use. As discussed below, to prevent encroachment of terrestrial woody vegetation, frequent scouring flows (every one to two years) are required to prevent establishment; releasing a five-year flow every five years would not achieve the same result (Kondolf and Wilcock 1996). Thus, it can make a big difference how a given "block" of available water is used for morphogenic flows. If artificial flow releases are too low to mobilize the bed

or perform other functions, even if made frequently or for long duration, they are a waste of water (at least for the morphogenic purposes intended). If the available water is used for higher flows but these flows are spaced too far apart in time, they may be ineffective by virtue of changes in channel conditions that occur in the long intervals between the flows. Scheurer and Molinari (2003) found that releases of 40 m3 s⁻¹ were most effective on the River Spöl, and closely tracked the annual snowmelt peak flows in the river pre-dam. Such flows have been deliberately released at least annually since the morphogenic flows began in 2000 (Figure 8.6) (Robinson 2012).

In other cases, the recurrence of morphogenic flows may be set in response to other events occurring in the river system, rather than upon a fixed periodicity. Natural floods still occur (though modified by upstream dams), and these can be more powerful than the artificial flows released, as documented on the Ebro River (Batalla et al. 2014). On the Jogne and Grande Eau rivers in Switzerland, one geomorphically effective flow per year is judged sufficient to maintain channel processes. Consequently, if a high flow occurs naturally, its effectiveness is assessed by a team of experts based on flow records and their

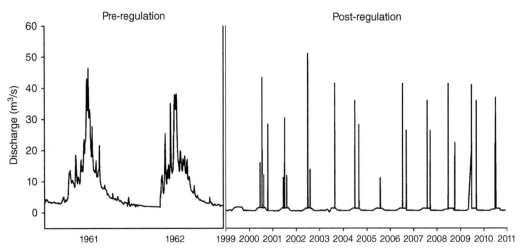

Figure 8.6 Hydrographs for the River Spöl for water years 1961–1962, reflecting the pre-dam flow regime, and 2000–2011, showing the morphogenic releases made at least annually (with the exception of 2005) and generally achieving 40 m³ s⁻¹. Source: Robinson (2012), used by permission of John Wiley & Sons.

field assessment of channel changes resulting from the flow, to determine whether an artificial flow is needed in addition to the natural flood (Zurwerra et al. 2016). As noted above, morphogenic flows on the Colorado River below Glen Canyon Dam are timed in response to flash floods on a downstream tributary and the consequent input of sediment to the mainstem (Melis 2011, Grams et al. 2013), and similarly Mürle et al. (2003) based the recurrence of high-flow releases on the Spöl upon the intensity of sediment supply from its tributaries. As another complication, a given flow release hydrograph on a river may produce very different results depending on condition of the channel prior to the flow. The latter is strongly influenced by how recently high flows had occurred and how large they were (Eder et al. 2014), and on variations in sediment supply from downstream tributaries.

8.5 Flows for managing vegetation in channels

Managing terrestrial and aquatic vegetation within river channels is a common motivation for mor-phogenic flows. As noted above, in the absence of frequent scouring floods, terrestrial woody species can establish within the active channel, increasing hydraulic roughness, narrowing the channel cross-section, both by virtue of the trees themselves and by the sediment they trap, a phenomenon often referred to as *vegetation encroachment* (Figure 8.7) (Kondolf and Wilcox 1996).

To prevent vegetation encroachment requires that the bed be mobilized frequently to scour seed-lings before they can establish. The resistance to scour offered by willow seedlings is non-linear: in many rivers, seedlings less than two years old can be scoured by a flood flow with a return period of two years, but seedlings that are five years old cannot be removed by a five-year flood flow. Thus, simply re-instating the pre-dam high-flow regime on a channel will not remove vegetation already-established (Cooper and Andersen 2012; Wilcox and

Shafroth 2013), and a high-flow regime designed to prevent vegetation encroachment usually needs to involve frequent scouring flows, every one to two years (Kondolf and Wilcock 1996), ideally timed to account for the seasonality of seedling dispersal and establishment. High flows can destroy vegetation not only by scouring the substrate in which they are rooted, but also by burial in sediment or prolonged inundation (Hughes and Rood 2003; Jourdain 2017).

In many rivers, steady clear-water releases from reservoirs (usually with high dissolved nutrient loads) make ideal conditions for algae growth on the bed. Without high flows to disrupt the bed or coarse-sediment-laden flows to scour bed gravels, algae can persist, with increasing accumulations of dead algae, which can lead to oxygen depletion, foul odors, and bacteriological pollution. The principal motivation for morphogenic flows on many rivers, such as the Ebro River (Batalla et al. 2014), has been removal of accumulated algae, a by-product of organic pollution.

8.6 Constraints

Morphogenic flows must be considered in the con-text of ecological or management constraints, some of which have been noted above, others we introduce in this section. It is worth bearing in mind that even if morphogenic flows could be provided without constraints, the result is likely to be channel adjust-ments such as bed incision and coarsening, rather than rebuilding of lost habitats, unless there is a supply of sediment to balance the erosive/transport energy of the flow.

8.6.1 Minimizing cost of foregone power production and other uses of water

The potential benefits of morphogenic flows must be balanced against the costs of foregone hydropower production and, in cases where the water is not captured for use downstream, foregone municipal and agricultural uses. Thus, the cost of water used in a morphogenic flow is a constraint. The value of

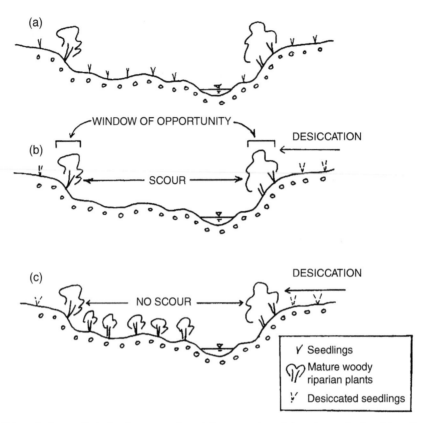

Figure 8.7 Schematic diagram illustrating the process of vegetation encroachment downstream of a dam: (a) seedling distribution following annual flood recession, (b) the "window of opportunity" for establishing riparian vegetation between the zone of scour and zone of desiccation in an unregulated channel, and (c) and encroachment of vegetation into the channel after reducing flood peaks by an upstream reservoir and eliminating scour. Source: Kondolf et al. (1996), used by permission of AGU/John Wiley & Sons.

this environmental water depends not only on the amount needed but also on timing of the release. Water released during the wet season is more likely to be replaced in the reservoir by new water flowing in from upstream than releases made during the dry season (Kondolf and Wilcock 1996).

8.6.2 Preserving spawning gravels

The flows that mobilize gravels not only allow fine sediment and seedlings to be flushed from the gravel bed, but can also transport gravels downstream, exacerbating problems of gravel starvation downstream of dams. As detailed by Kondolf and Wilcock (1996), this constraint puts a premium on specifying

flows capable of transporting sand while minimizing gravel transport, and also means that some sort of coarse sediment supply downstream of the dam is needed: either natural supply from unregulated tributaries and/or mechanical additions of sediment. There can be an inherent conflict between mobilization of gravel bar surfaces to prevent vegetation encroachment and preserving spawning gravels. Flows needed to scour seedlings of woody riparian species (especially on bars elevated above the river bed) are typically far greater than flows required to simply mobilize the river bed and flush fine sediment from the gravels. Such vegetation-scouring flows can result in substantial sediment transport and consequent loss of spawning gravels to downstream transport. In this context, the downstream transport by morphogenic flows would need to be balanced by increased supply of coarse sediment.

8.6.3 Preventing flooding and bank erosion

With older dams, human encroachment on the flood-plain can constrain morphogenic flows. Downstream of large reservoirs that control large floods, land-use regulatory authorities commonly re-zone former floodplain areas as no longer flood-prone and there-fore suitable for development. On the Trinity River, the magnitude of the so-called 100-year flood (the standard widely used to regulate land use on flood-plains in the US) was reduced by Trinity Dam to a small fraction of its pre-dam flow. As a result, only a small area of the bottomland is now considered to be within the 100-year floodplain. In some cases even parts of the former active channel are now considered outside the 100-year floodplain. Avoiding impacts to houses that have been constructed on formerly active in-channel bars, and to small bridges built over the post-dam narrowed channel, have been significant challenges to implementing morphogenic flows on the Trinity River. Similarly, bank erosion by morpho-genic flows can generate complaints from bankside residents, many of whom (as on the Trinity) occupy sites that were formerly in the active channel.

8.7 Conclusions

The concept of EFAs has evolved and expanded from minimum flows required to fill a static channel with sufficient water (and suitable velocities) for fish, to include flows that trigger geomorphic processes that maintain channel form and sediment quality, usually to benefit fish and other aquatic life. Such morphogenic flow releases are increas-ingly required from dams, such as US hydropower dams facing license renewals before the Federal Energy Regulatory Commission. Unfortunately, these releases are often based on simple statistics such as 200 % of the mean flow (based on the Ten-nant Method described in Chapter 6), and may not account for actual conditions in the channel, nor for sediment fluxes. Even though morphogenic flows are designed to interact with bed sediments, ques-tions of sediment supply are often not addressed in prescriptions for such flows. Moreover, the very situation that calls for morphogenic flows in some cases can be (largely) avoided through implemen-tation of sustainable sediment management prac-tices, such as passing sediment around or through reservoirs, and through strategic planning of dam placement within a river network. As with environ-mental flows generally, uncertainty about the flows needed for morphogenic purposes makes adaptive management the best approach.

CHAPTER 9

Improving the use of existing evidence and expert opinion in environmental flow assessments

Summary

Environmental flow assessments have long taken advantage of advanced hydrologic and hydraulic modelling to compute flow regimes and hydraulic habitat. In contrast, the models that translate these predictions to ecological responses are very simple, often implicit, and usually based heavily on expert opinion. We propose an approach to improve the use of existing knowledge in the ecological predictions of environmental flow assessments. It centers on using the existing literature to construct an evidence-based conceptual model, quantifying the links in that conceptual model using formal expert elicitation, and then updating the links with empirical data using Bayesian methods. This allows us to make evidence-based predictions of ecological responses during the environmental flow assessment process. The method also is consistent with principles of adaptive management: as we learn through monitoring and evaluation, model predictions can be updated with empirical data.

9.1 Introduction

As discussed earlier (Chapter 6), there are many environmental flow assessment methods. Many of the commonly used methods today make considerable use of hydrological and hydraulic modeling methods to describe flow environments and hydraulic habitat. For example, the FLOWS method in the Australian state of Victoria (DEPI 2013) uses one-dimensional modeling with US Army Corps of Engineers' HEC-RAS modeling suite (Brunner 2010) to translate discharge volumes to estimates of available hydraulic habitat. Other methods employ more sophisticated hydraulic modeling approaches to develop estimates of habitat under different flow conditions (e.g. Meso-HABSIM; Parasiewicz 2007).

In contrast, most methods are far less rigorous when attempting to translate forecasts of hydraulic habitat provision into actual ecological responses. There is an overreliance on the assumption of the field-of-dreams hypothesis (Palmer et al. 1997) – that the provision of hydraulic habitat alone will be sufficient to elicit the ecological response desired. Where predictions go beyond an assumption of a linear relationship between habitat provision and ecological response, these are often largely informed by expert opinion (Stewardson and Webb 2010). While expert opinion is a valuable resource, and can be used when empirical data are lacking (Martin et al. 2005), the informal and ad-hoc methods often used to elicit expert-based predictions are prone to

Environmental Flow Assessment: Methods and Applications, First Edition. John G. Williams, Peter B. Moyle, J. Angus Webb and G. Mathias Kondolf.
© 2019 John Wiley & Sons Ltd. Published 2019 by John Wiley & Sons Ltd.

bias and overconfidence (de Little et al. 2018). Countering the trend of using expert opinion to inform models of ecological response to flow alteration, the ecological limits of hydraulic alteration (ELOHA) framework (Poff et al. 2010) has been successful in increasing the number of empirical flow–response relationships being used in environmental flow assessments (Arthington et al. 2012; McManamay et al. 2013; Rolls and Arthington 2014).

In earlier chapters of this book, we have criticized some of the commonly used approaches to environmental flow assessment. In this chapter, we propose one potential method that could improve the quality of flow–ecology response modeling in these assessments. Rather than concentrate entirely on either expert opinion or empirical data, the proposed method makes use of both via Bayesian methods. In addition, it makes more explicit use of the body of knowledge in the published literature than is normally the case. The result, we believe, is a method that makes best use simultaneously of all the information available to inform environmental flow assessment. In addition, it provides realistic estimates of the uncertainty of ecological response predictions, allows more complex model structures than the bivariate flow–response relationships currently being used, and is fully compatible with adaptive management and learning, thereby allowing empirical data to play a greater and greater role in predictions as monitoring data accumulate.

The proposal concentrates on the prediction of ecological responses, as we believe this has received insufficient attention in the past. However, we acknowledge that other phases of the environmental flows assessment process are also in need of novel approaches that improve the rigor and evidence base of all aspects of the assessment.

9.2   Overview of proposed method

The proposed method is based heavily on that of Webb et al. (2015a). That method centered on using the literature to develop an evidence-based conceptual model of ecological responses to changing flow regimes. Linkages in the model were initially quantified using expert knowledge via a formal expert elicitation process, with this quantification providing prior probability values for a hierarchical Bayesian analysis of monitoring data. The final output was an assessment of the success or otherwise of an environmental flows program, and an ability to make detailed predictions of ecological responses under different flow regimes. The method was thus primarily designed for post-hoc assessment of environmental flows as part of a robust monitoring and evaluation framework.

Here, we vary the method somewhat in order to make it more useful for predictions a priori of ecological responses as part of an environmental flow assessment process. Specifically, we loosen the requirement for the hierarchical Bayesian model-based analysis of extensive empirical data. This change is motivated primarily by the observation that, for most environmental flow assessments, empirical data, while perhaps not entirely lacking, are scarce. The hierarchical modeling phase of the Webb et al. (2015a) method presupposes the existence of substantial data to help fit the model.

The proposed method thus has the following steps (Figure 9.1).

1   Systematic assessment of the literature to develop the structure of the flow–response model,
2   Translation of the conceptual model into a Bayesian Belief Network (BBN) structure,
3   Quantification of the relationships in the BBN using formal expert elicitation to reduce effects of bias and overconfidence,
4   Updating the relationships using empirical data. If extensive data are available, this should be done using hierarchical Bayesian methods; if data are scarce, the updating can be done within the BBN.

The following section provides more detailed background on the approaches used in each phase, including the theoretical justification for their use. This is then followed by an illustration of the method through a case study examining golden perch (*Macquaria ambigua*) a flow-cued iconic native fish species in southeastern Australia.

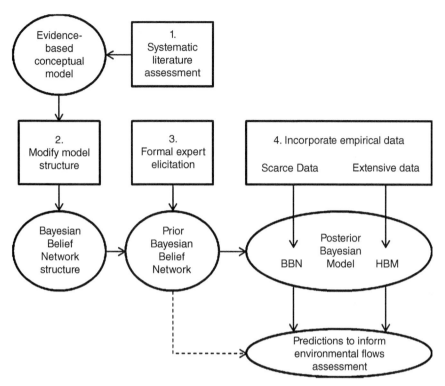

Figure 9.1 Workflow for the proposed method for modeling ecological responses to environmental flows. Boxes are actions, with numbers corresponding to the step numbers described below. Circles/ovals are outputs from an action, with some of these outputs necessary inputs to the next phase. Abbreviations: BBN – Bayesian Belief Network; HBM – Hierarchical Bayesian Model. Source: Angus Webb.

9.3 Basic principles and background to steps

9.3.1 Literature as a basis of an evidence-based conceptual model

A quantified ecological response model must be based upon a conceptual model. Conceptual models are often created with little consideration of the state of knowledge as captured in the scientific literature. The majority of literature reviews undertaken in ecology and environmental science are narrative reviews that are unstructured, difficult to repeat, and usually do not attempt to test any sort of hypotheses (Roberts et al. 2006). Systematic reviews, which are common in several fields, most notably the health sciences, undertake formal assessments

of hypotheses, using evidence from the literature as data, and testing hypotheses using tightly formulated methods (Khan et al. 2003). Full systematic reviews are time consuming and costly undertakings, but are acknowledged as the gold-standard approach to evidence synthesis (CEE 2013). Rapid evidence synthesis (Webb 2017) is an emerging approach to evidence synthesis that attempts to use systematic methods to assemble evidence from the literature, but more quickly and with less expense (Box 9.1). Systematic synthesis of evidence from the literature can be used to test the evidence for or against hypothesized cause–effect linkages in a proposed conceptual model. Linkages should only be retained if supported by the evidence or if the evidence is equivocal.

Box 9.1   Comparison of Different Types of Evidence Assessment from the Literature

An increasing number of methods are being employed in different fields to assemble, summarize, and synthesize evidence from the literature (Cook et al. 2017). The different methods apply different levels of rigor to the phases of evidence assessment, and hence have different levels of confidence in the conclusions. In the table below, level of confidence in the conclusions increases from left to right. These differing levels of rigor also imply differing amounts of time, money, and need for expertise to complete an evidence synthesis – factors that can be of overriding concerns in time- and budget-limited applications such as environmental flow assessments.

Type of evidence synthesis	Narrative review	Rapid evidence synthesis	Systematic review
Purpose	Provide a qualitative review of the literature on a topic	Provide a rapid evaluation of evidence to test a hypothesis	Provide a transparent repeatable evaluation of the evidence for a hypothesis

Source: Adapted from Cook et al. (2017).

9.3.2 Translate the conceptual model into the structure of a Bayesian belief network

For the next phases of the framework, the conceptual model needs to be translated into a BBN. BBNs are graphical models that depict causal relationships through nodes (state variables) and arcs (causal linkages among variables). They can use both empirical and expert-based data. BBNs are widely used in natural resource sciences (McCann et al. 2006), and there are several examples where they have been used to predict ecological responses to changing flows (Shenton et al. 2011, 2014; Chan et al. 2012). The conceptual model developed in Step 1 may be in a structure that translates immediately to a BBN. It is also possible that the conceptual model may need to be simplified to some extent to create a BBN for which the causal linkages can be quantified.

9.3.3 Quantify causal relationships in the BBN using formal expert elicitation

The strength of BBNs to be able to employ expert-derived data is also a potential weakness if those data do not truly reflect the knowledge of the experts. Unstructured approaches to expert elicitation are widely used, including in the quantification of causal relationships in BBNs (e.g. Shenton et al. 2011, 2014; Chan et al. 2012), but these approaches have been shown to be negatively affected by biases, expert overconfidence, and groupthink (Kuhnert et al. 2010; Speirs-Bridge et al. 2010; de Little et al. 2018). Several methods for formal expert elicitation have been developed that reduce these impacts. For the most part, these revolve around helping experts to better specify their own uncertainty (Speirs-Bridge et al. 2010), and making sure that different experts provide independent opinions, unaffected by group-level deliberations (McBride et al. 2012; Wintle et al. 2013). These methods can be used to populate the conditional probability tables that quantify the causal relationships among nodes of the BBN, thereby providing a numerical ecological response model.

9.3.4 Update causal relationships using empirical data

Paradoxically, BBNs do not have to be Bayesian, and indeed many are not. Bayesian modeling provides a mathematical framework for updating a belief (e.g. in a parameter value) in the face of new data (Webb et al. 2010b). The majority of published BBNs would more rightly be considered as prior models, as the quantified relationships derive from conditional probability tables populated once, often by experts. In the case that there truly are no empirical data to update the expert-based relationships, the prior model could be used to make predictions to inform the environmental flow assessment (dotted pathway in Figure 9.1). This is the least preferred option, however. Conversely, if extensive data are available, then these can be combined with the prior model within a hierarchical Bayesian analysis to produce a posterior model. These are code-based models that are purpose-written to fit the data set and knowledge available (Box 9.2).

However, it is often the case that only a small data set will be available to support the environmental flow

Box 9.2 Bayesian Belief Networks vs. Hierarchical Bayesian Models

Bayesian methods have increased in use in natural-resource management applications in recent years (McCarthy 2007). Two Bayesian methods are proposed in this chapter, and it is worth providing some more detail on their similarities and differences to help readers to navigate these ideas.

As described above, Bayesian Belief Networks (BBNs) are graphical models constructed from *nodes* that depict model state variables, and *arcs* that link these nodes via conditional probability relationships. Chief among the advantages of BBNs is that they are simple to create and update, and can incorporate mixed types of data (e.g. empirical, expert-derived, model outputs). Continuously distributed variables (e.g. abundance of a fish species) are *discretized* into a number of node states based on thresholds or qualitative descriptions (e.g. low, medium, high). This discretization has advantages for populating the conditional probability relationships with expert knowledge (see de Little et al. 2018), but more importantly allows rapid updating of the probability distributions of *child* nodes (the outcome) based on simple point-and-click manipulations of the states of *parent* nodes (the driving variables). Their simplicity of use and construction, along with this ability to rapidly update model predictions, makes BBNs an ideal method for participatory model building and communicating otherwise complex predictions to stakeholders. As noted above, BBNs do not have to be Bayesian; a prior model can be used without having to update with empirical data. Pearl (2000) provides a comprehensive introduction to the theory and use of BBNs.

Hierarchical Bayesian models allow the representation of complex ecological processes via flexible formulas, considering both continuous and categorical variables, and allowing different types of relationships (e.g. the shape of a curve) between model variables. This flexibility has seen an explosion in the use of hierarchical Bayesian models in ecology because of their ability to fit the messy data and poor experimental designs that often characterize ecological research (Clark 2005). Hierarchical models exploit the similarity of different *sampling units* (e.g. sites) to reach stronger conclusions for each unit than is possible by treating them independently. Coupled with their ability to incorporate prior knowledge to strengthen predictions (McCarthy and Masters 2005), hierarchical Bayesian models will almost always outperform "standard" statistical approaches. The tradeoff is that models must be directly coded into software, and this requires considerable expertise. Once coded, the models use *Markov chain Monte Carlo* simulation (Andrieu et al. 2003) to estimate posterior distributions for model parameters, and can also make predictions for outcomes under different scenarios. However, unlike BBNs, these scenarios need to be included in code and cannot be rapidly updated. For a comprehensive treatment of Bayesian inference methods, including hierarchical models, see Gelman et al. (2004) and Gelman and Hill (2007).

assessment. In this case, the updating of the prior model may more effectively be done within the BBN itself. Bayesian Network software has the ability to update prior relationships using empirical data through such approaches as the expectation maximization (EM) algorithm (Dempster et al. 1977). In that case, the expert-derived prior conditional probability tables are updated to a posterior model with data. The influence of the expert knowledge on model output is not lost, but is reduced by the incorporation of empirical data. The more data available, the greater the reduction in the influence of expert opinion on model outputs.

9.4 Case study: golden perch (*Macquaria ambigua*) in the regulated Goulburn River, southeastern Australia

Golden perch is an iconic species of inland waterways of south-eastern Australia. Its natural range extends well into the north of the country, but has contracted under river regulation (Zampatti and Leigh 2013). Adults commonly reach 40 cm in length and can live for more than 20 years. Golden perch are a frequent target of environmental flow management in southeastern Australia because of their iconic status, but also because the adults are known to spawn only when there is a high flow event late in the austral spring (November–December). Here, we illustrate the method described above with a case study for golden perch from the Goulburn River, Victoria, where the species is one focus of the environmental flows program (GBCMA 2016). However, we note however that these results have not yet been used in any environmental flow assessment for this species. The results are drawn from Smith (2013) and reported in greater detail therein.

9.4.1 Evidence-based conceptual model of golden perch responses to flow variation

We used the Eco Evidence method for rapid evidence synthesis (Norris et al. 2012; Webb et al. 2015b) to develop an evidence-based conceptual model. Eco Evidence uses systematic methods to search and synthesize evidence from the literature, testing the evidence base for and against hypothesized linkages in a conceptual model. The synthesis relies on a modified form of "vote-counting" – counting the number of papers in favor of a hypothesis versus the number against. There are two refinements in Eco Evidence. First, each "evidence item" – the finding from a study – is assigned a weight from 1 to 10 based upon the strength of the experimental design and consequent susceptibility to confounding. Second, evidence weights are summed and assessed against a threshold score (20 points) to reach one of four conclusions: support for hypothesis, support for alternate hypothesis, inconsistent evidence, and insufficient evidence.

A systematic search of five literature databases and evaluation of the references returned yielded 46 evidence items across eight sub-hypotheses encapsulated in the preliminary conceptual model of golden perch response to flow variation (Smith 2013). These hypotheses are the solid lines shown in Figure 9.2, with the dashed lines remaining as untested or assumed linkages. Four of the hypotheses returned a finding of "support for hypothesis" and the other four "insufficient evidence" (Table 9.1). With none of the hypotheses refuted by the evidence assessment, they were all carried forward to the next stage of the analysis as potential pathways in a BBN model.

9.4.2 Bayesian belief network structure of the golden perch model

The conceptual model was translated into the structure of a BBN using Netica® software (Figure 9.3). There were several changes in the BBN model structure compared to the conceptual model. First, effects of flow and non-flow drivers were linked directly to adult and young-of-year abundance, rather than via nodes specifying the provision of habitat. Habitat provision is still implicit in the model, but including it explicitly within the BBN would have complicated the expert elicitation process. Second, although we did not test effects of water quality as part of the

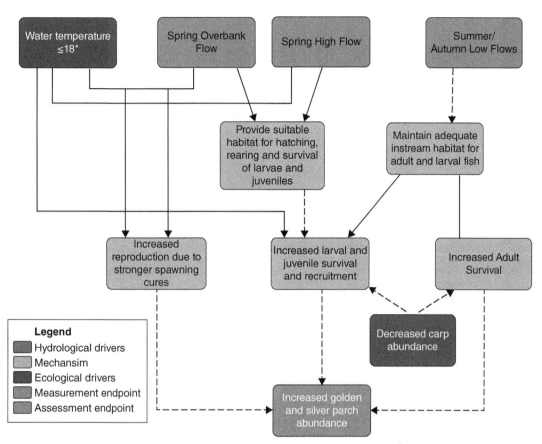

Figure 9.2 Evidence-based conceptual model of the processes driving golden perch populations in Australian lowland rivers. Arrows are testable hypotheses, with solid arrows being those prioritized for testing and quantification in the study. Source: Reprinted from Smith (2013) with permission.

evidence synthesis step, the effect of stream water salinity and dissolved oxygen levels were included as new nodes. This inclusion runs counter to our own statements above about ensuring that conceptual models are based on evidence. However, experts we consulted with regards to structuring the BBN considered these an important inclusion and it was important to maintain their support moving forward to the expert elicitation step. Finally, we had not directly tested the evidence for or against the effects of carp presence on golden perch even though it was included in the conceptual model (this decision was based on available resources). However, again at the

urging of independent experts, we included this in the BBN model for expert elicitation.

9.4.3 Expert-based quantification of effects of flow and non-flow drivers on golden perch

We employed formal expert elicitation to populate the conditional probability tables in the BBN model. The framework (de Little et al. 2018) is based on the four-point elicitation method of Speirs-Bridge et al. (2010), and more generally, the Investigate, Discuss, Estimate, Aggregate (IDEA) framework for knowledge generation (Hanea et al. 2017). Experts

Table 9.1 Results of the Eco Evidence rapid evidence assessment for eight hypotheses from the preliminary conceptual model.

Cause–Effect hypothesis tested	Number of evidence items located	Evidence points		Conclusion
		Supporting hypothesis	Refuting hypothesis	
Increased spring high flows will cause an increase in reproduction	8	26	5	Support for hypothesis
Increased spring overbank flows will cause an increase in reproduction	6	7	13	Insufficient evidence
Water temperatures ≥18 °C will cause an increase in reproduction	7	9	11	Insufficient evidence
Increased spring high flows will cause an increase in recruitment	5	22	7	Support for hypothesis
Increased spring overbank flows will cause an increase in recruitment	4	8	5	Insufficient evidence
Water temperatures ≥18 °C will cause an increase in juvenile survival	4	13	0	Insufficient evidence
Decreased salinity will cause an increase in fish survival	6	24	0	Support for hypothesis
Increased dissolved oxygen will cause an increase in fish survival	6	20	4	Support for hypothesis

Evidence points supporting or refuting the hypothesis are points summed across all evidence items either supporting or refuting that hypothesis as calculated using the default weights of Norris et al. (2012).

are asked to provide their lowest estimate of a parameter given a set of driving conditions, their highest estimate, their best estimate, and their confidence that the true value lies between the lowest and highest estimates. This form of questioning has been shown to reduce the problem of expert overconfidence. Groupthink and the effects of personalities and reputations are managed by having each expert in the workshop initially answer questions completely independently. Following this first round of estimates, the estimates – translated into probability distributions – are showed to the group for discussion and possible revision (although the initial estimates are not discarded). This step helps to reduce outlying estimates in the group-level ensemble estimate, but does not overly inflate the confidence of estimates. Being presented with too many questions can lead to "expert fatigue" (Burgman et al. 2011). We used the interpolation methods of (Cain 2001) to reduce the number of questions that needed to be asked.

Here, five experts participated in the expert workshop. A portion of the resulting conditional probability table for abundance of golden perch young of year is shown in Table 9.2 to illustrate results. There was also an equivalent table for adult abundance.

For abundance of golden perch young-of-year, the elicitation suggested that overbank spring flows was the most important variable, but with the presence of water temperatures over 18 °C also being quite important. For adult abundance, the water-quality variables were most important, with summer–autumn low flows playing little role. Moving from the best to the worst combinations of driving variable states saw the probability of high number of golden perch young-of-year fall from 98 % to 23 %. Adult abundances were less sensitive, with the change from best to worst conditions leading to a reduction in the probability of high numbers of adults from 17 % to 7 %.

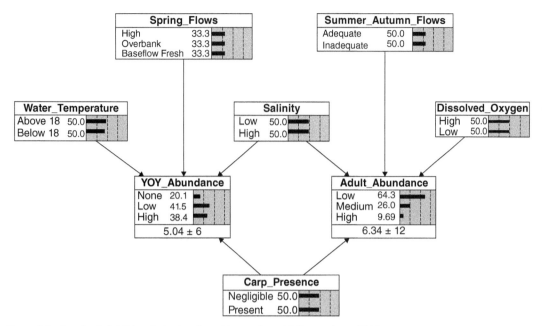

Figure 9.3 Bayesian Belief Network structure for golden perch model. Source: Angus Webb.

Table 9.2 First eight rows of the conditional probability table for golden perch young-of-year abundance as populated by expert elicitation, illustrating all results for "High" spring flows.

Elicited / Interpolated	Driving variables				Young-of-year abundance (% prob. of class)		
	Spring flows	Water temp	Carp presence	Salinity	None (0)	Low (1–15)	High (16–25)
E	High	≥18°C	Negligible	Low	0.9	31.4	67.7
E	High	≥18°C	Negligible	High	20.9	40.3	38.8
E	High	≥18°C	Significant	Low	7.8	40.6	51.6
I	High	≥18°C	Significant	High	21.1	45.8	33.1
E	High	<18°C	Negligible	Low	20.3	46.4	33.3
I	High	<18°C	Negligible	High	24	49.3	26.7
I	High	<18°C	Significant	Low	21	49.4	29.6
I	High	<18°C	Significant	High	23.5	51.1	25.4

Overall, 7 of the 24 scenarios were elicited from experts (E) with the other probabilities able to be interpolated (I). For each scenario, probabilities sum to 100 % across the three abundance classes (None, Low, High).

9.4.4   Inclusion of monitoring data to update the golden perch BBN

The case study reported here was one where few empirical data were available to update the prior model. Hence we used probability updating within the BBN. We updated the conditional probability tables using fish monitoring data from the Victorian Environmental Flows Monitoring and Assessment

Program (Webb et al. 2010a, 2014), coupled with discharge and water quality data from the state government monitoring network. Mismatches between the sites and timings of the two sets of data meant that water quality data were "missing" for a sizable proportion of the fish data. Moreover, the rarity of particularly young golden perch in Victorian rivers meant that the abundance data were dominated by zeros. Once abundances were converted to the categorical states shown in Figure 9.3, the data were used to update the conditional probability tables of the prior network using the EM algorithm, which is built into the Netica software.

Against expectations, the incorporation of the monitoring data did not improve performance of the network. In fact, the Bayesian Network for young-of-year abundance failed to update from the expert-derived prior model altogether, probably at least partly driven by the very low occurrences of golden perch young-of-year. For adults, following incorporation of the monitoring data, the low sensitivity of golden perch abundance to driving conditions noted above was exacerbated (the probability of high abundances ranged from 0.6% to 7.7%). More importantly, however, the direction of effect was reversed; the highest posterior probability of high adult abundance was predicted what we would consider as poor conditions of inadequate summer–autumn flows and high salinity.

Examination of the data revealed that a large number of the scenarios possible in the BBN were not experienced within the data set. High and overbank spring flows were very rare, as were spring water temperatures <18 °C and high salinities. Also, the very low numbers of fish obtained in the sampling data meant that when fish were present, it was inevitably at 'low' abundances according to the thresholds set in the BBN, even when the environmental conditions could be considered to be favorable. The abundance thresholds were set as part of the expert elicitation process, in consultation with fish biologists working in a different system (southern New South Wales). Expectations of abundance were higher for those biologists than the numbers observed in the Goulburn River data. The combination of the prior model

predicting higher abundances than the data, coupled with the uneven coverage of the data set across combinations of driving conditions with "favorable" conditions being more common, led to the unexpected posterior result (Smith 2013).

9.5 Discussion

The reliance of environmental flow assessments on the use of expert opinion to translate modeled predictions of hydraulic or hydrologic habitat to forecasts of ecological response has been largely justified on the basis of there being too few empirical data available to create robust ecological response models (Stewardson and Webb 2010). However, the informal ways in which expert opinion has been used in environmental flow assessments leaves these studies more at risk of bias and overconfidence in their predictions. Moreover, while knowledge from the literature is implicitly used in expert-based assessments (as it forms part of the experts' knowledge base), it has rarely been explicitly considered. Finally, the increasing number of case studies using the empirical data to create flow response relationships within ELOHA assessments (e.g. Arthington et al. 2012; McManamay et al. 2013; Rolls and Arthington 2014) demonstrates that relevant data do exist and are becoming more common.

The approach proposed here attempts to simultaneously maximize the value gained from evidence in the literature, knowledge of experts, and empirical data. Amid legislative and ethical requirements to use "best available science" to inform environmental management decisions in contested decision spaces (Ryder et al. 2010), the framework we have outlined here offers a potential pathway forward for environmental flow assessments.

9.5.1 Improved use of knowledge from the literature

In the framework described in this chapter, the conceptual model structure is informed by rigorous literature analysis. We only include linkages in the model if they are consistent with our current understanding as presented in the literature, or if they are hypothesized, but have not been tested. Many environmental flow assessments begin with a literature review, but the extent to which this informs the subsequent predictions of ecological response is usually unclear. We argue that this approach fails to make use of the considerable knowledge contained within the scientific literature, and consequently falls some way short of "evidence-based practice" as best practiced in medical research and patient management (Stevens and Milne 1997).

The Collaboration for Environmental Evidence (www.environmentalevidence.org) has adapted the medical model for systematic review of the literature for applications in environmental science and management (CEE 2013). Environmental flow assessments should be aiming for this level of rigor when making recommendations regarding scarce and contested water resources.

Full systematic review using the CEE (2013) methods should be regarded as best practice, but we acknowledge that the expense and expertise required to undertake a full systematic review is considerable. As an alternative, methods for rapid evidence synthesis (Webb 2017) represent a considerable increase in the rigor of evidence assessment compared to narrative reviews, and can be completed in a fraction of the time and cost of a full systematic review. The Eco Evidence method for rapid evidence synthesis, used in our example here, has been successfully applied to a number of questions regarding effects of flow regime alteration on aquatic animals, plants, and ecosystem processes (Greet et al. 2011; Webb et al. 2012, 2013; Miller et al. 2013), and so is a good candidate for use within the framework proposed here.

9.5.2 Improving the basis of Bayesian networks for environmental flows

The evidence-based conceptual model is converted into a Bayesian Network model for which parameterization is undertaken using the best available information. While BBNs have become quite common in natural-resource management applications, including environmental flows research, far

too often there is insufficient rigor in the processes used to develop network structure and populate the conditional probability tables that specify the relationships. The outcome is a quantitative model that looks good and is able to furnish predictions of ecological responses, but for which the predictions themselves are unreliable. In many ways, this is a worse outcome than having no model at all. It's also possible that a high-quality model may be produced, but the justification for structural choices and parameterizations are not sufficiently documented. In this case, stakeholders cannot assess the quality of the model, which may lead to mistrust. The process outlined here aims to address these shortcomings. There are several reasons why we believe the Bayesian Network approach is a sound basis for predictive modeling in environmental flows assessment.

Perhaps most importantly, modeled relationships include uncertainties. Where we have little knowledge of a precise relationship between two or more variables, the model will not claim otherwise. In the prior (expert-based) model described above, the probability of observing high numbers of golden perch young of year only shifts by 20 % when changing from low spring flows to overbank flow events (a major change in environmental drivers). This reflects the knowledge of the experts that the process is far from deterministic, but also their level of uncertainty in how large the effect would be. Over time, we would hope to use monitoring data to reduce the level of uncertainty in the prediction. However, the stochastic nature of the relationship will always remain because of irreducible uncertainty in the relationship (Lowe et al. 2017).

The node–link structure of BBNs allows a nuanced model structure where driving variables can interact in meaningful ways to affect the outcome. Even in the simple model presented here, such interactions occur. For example, the prior model predicts that if water temperatures are below 18 °C, then an improvement in spring flow conditions from base flow fresh to overbank flows only improves the probability of high young-of-year abundance by 8 %. However, when temperatures are over 18 °C, then corresponding probability change is more than 30 %. A simple bivariate flow relationship, such as those used in many ELOHA assessments (e.g. Arthington et al. 2012) cannot capture such an interaction. Where several flow–response curves are combined to try to create more realistic ecological response models (e.g. Bryan et al. 2013), these interactions are simplified to averages, geometric averages, or taking a minimum across several curves, none of which captures the interactions very well.

With regards to combining data and knowledge, BBNs, and indeed all Bayesian methods, allow progressive overwriting of expert opinion with accumulating empirical data through the application of Bayes' rule. This is different to the sudden replacement of expert opinion when a data set becomes available; such a replacement would rely on extensive data being available to completely replace the prior modeled relationships. Bayes' rule offers a halfway solution, retaining the influence of expert opinion but modifying it with empirical data. In the example here, the inclusion of empirical data provided a sobering reality check. The fact that data had only been collected over a portion of the range of conditions covered by the model, coupled with the fact that observed numbers of fish were always low, led to a skewing of the conditional probability tables. Here, even with Bayes' rule operating, it appears necessary to collect data over a wider range of conditions before attempting to update the model. Data collection has continued in the Goulburn River in recent years (Webb et al. 2016, 2017a), and there may now be sufficient information on golden perch abundance to more sensibly update the conditional probability tables.

Such an updating is consistent with the "inner loop" (small adjustments) of an adaptive management cycle (Horne et al. 2018), with minor updates to model parameterization being used to make updated predictions of responses in the next cycle. However, over time, the accumulation of data and consequent improved knowledge may suggest that a change in model structure is required. This is more consistent with adaptive learning through the "outer loop" of an

adaptive management program (Webb et al. 2017d). The ability for BBNs to incorporate new data as they become available makes them ideal for use within an adaptive management program (Horne et al. 2018).

The contrary result returned when updating the prior model with the small amount of monitoring data available provides a cautionary lesson that having some data is not always better than having none. Here, the incorporation of a data set dominated by zeroes and with "experience" of only a portion of the conditions we wished to model, actually detracted from a set of coherent expert-based predictions. This further demonstrates the value of well-elicited expert opinions. Without the prior model, we would have no way of making robust predictions of golden perch performance under combinations of different flow and non-flow conditions.

9.5.3   Hierarchical Bayesian methods as best practice

Although the BBN model can be used to generate predictions capable of informing an environmental flow assessment, if extensive data are available assessors should endeavor to use hierarchical Bayesian methods to make the predictions. Like the use of systematic review rather than rapid evidence synthesis for developing the conceptual model, the hierarchical Bayesian approach can be regarded as the best practice.

Hierarchical Bayesian models are likely to be less affected by uncertainty in modeled relationships than BBNs, especially when the data are analyzed on top of expert-based prior probability distributions. The hierarchical model structure allows the assessor to bring together data from different sampling units (e.g. rivers) to inform predictions at several scales (Webb et al. 2015a). By linking the estimation of model parameters together through the hierarchical structure, predictive uncertainty is reduced – a process known as borrowing strength (Gelman et al. 2004). More precise predictions will lead to better informed environmental flow recommendations. For a detailed example of applying the hierarchical Bayesian modeling step described in this framework, including the assessment of how incorporation of prior information provides more precise predictions that a 'data-only' approach to the analysis, see Webb et al. (2018).

9.5.4 Piggy-backing on existing knowledge

The description of the framework as provided above implicitly assumes that the environmental flows assessment is starting from scratch, and that no similar syntheses or models exist for the species of interest. This is often not the case. Webb et al. (2017b) outline a process for developing numerical models for environmental flow assessments, considering wider possibilities than the Bayesian methods described here. As part of this, they provide a workflow for assessing and employing existing models to improve efficiency (Table 9.3).

A similar philosophy should be applied to the different steps outlined in this chapter. For example, a systematic literature assessment may exist for the endpoint of interest, or for a functionally related endpoint. In this case, a briefer check of new literature may be sufficient to objectively, but still rigorously, assess the state of knowledge. More likely, conceptual or numerical models may exist for the species of interest, or a closely related species, at

Table 9.3 Taking advantage of existing models in environmental flow assessments.

Status of existing models	Work required for new flow assessment
No relevant model	Full model development required
Model for functionally related objective	Assess the model structure, adjust as necessary and re-parameterize
Model for same objective, but different location	Re-parameterize model
Model for same objective, and at same location	Run model for flow scenarios

Where models already exist, these should be used to reduce the amount of effort for new modeling.
Source: Compiled from Figure 14.6 in Webb et al. (2017b), with modifications.

similar locations to where the assessment is being done. It is often the case that environmental flows assessments are done independently for different locations, but the same species may be the priority ecological endpoints. In this case, models from existing assessments could be used to (i) set the hypotheses for testing in the systematic literature assessment, (ii) inform the structure of a resulting BBN, or (iii) be used in addition to (or possibly instead of) the expert elicitation process to generate informed prior probability distributions for a Bayesian model. The general message is that the process outlined in this chapter should not be used to the exclusion of substantial work in similar locations or for similar species. Our aim is to take full advantage of existing knowledge and data, and other models and assessments are a potential source for both.

9.5.5 Resourcing improved practice

Improved practice in environmental flow assessment will not come for free. Rapid evidence synthesis using Eco Evidence can be completed for a cost in the order of $10 000 per endpoint (e.g. fish species) (Webb et al. 2017c), and expert elicitation using the full de Little et al. (2018) method is likely to cost around the same amount to recruit and pay experts, pay transport costs, host workshops, etc. (J.A. Webb unpublished data). Moving from rapid evidence synthesis to full Systematic Review, and moving from empirically-updated BBNs to hierarchical Bayesian analyses will increase the expense further. However, much of the criticism leveled at environmental flow assessment methods in this book has centered on their failure to follow scientific method, and short cuts taken on rigorous methods. It's difficult to see how practice can be improved significantly without at least some extra expense.

On top of the monetary costs lies the consideration of the extra time an environmental flow assessment may take by employing these approaches compared to more traditional and less formal methods. However, some of these tasks (e.g. evidence synthesis from the literature and the organization of expert workshops to quantify BBNs) can take place simultaneously. Although we have presented the steps here as a linear progression, this is not entirely necessary for their implementation.

9.5.6 Accessibility of methods

We have presented two pathways to improved predictions of ecological responses for environmental flows assessment, both of which greatly improve the use of all available evidence. While we recommend the use of hierarchical Bayesian models to make the final predictions when sufficient data are available, we acknowledge that these methods require considerable expertise to implement. The single greatest advantage of the BBN-based pathway is its accessibility. Compared to hierarchical Bayesian methods, BBNs do not require high-level expertise. Basic creation and operation of Bayesian Networks can be learned in a 2–3-day short course, and these are common and affordable.

Similarly, while full systematic review methods require considerable training and expertise in specific methods such as quantitative meta-analysis (Gurevitch and Hedges 2001), rapid evidence synthesis using an approach such as Eco Evidence is more accessible. It can be undertaken by anybody with experience of reading scientific papers, and while domain expertise is an advantage, it is not necessary for undertaking the literature assessment. Finally, the expert elicitation methods employed in the case study can similarly be undertaken by anybody reasonably well versed in using spreadsheets and with the patience to facilitate all-day workshops.

In recent work, Webb et al. (2017b) reviewed the increasing range of analytical methods for modeling ecological responses as part of environmental flow assessments. They noted the emergence of specialized modeling methods such as hierarchical Bayesian models, machine-learning approaches, and functional linear models. While all these methods could add considerable rigor to an environmental flow assessment, they require substantial expertise to implement to the point where an "ecological modeling specialist" would become a necessary member of an environmental flows assessment panel (Webb et al. 2017b). While such an approach is highly desirable, and should be regarded as best practice, the less onerous approaches described here also increases the rigor of environmental flow assessments well beyond the level of many current assessments, but without the need for such expertise.

9.6 Summary

In this chapter, we have outlined an approach for improving the evidence basis and rigor of predictions of ecological responses to restored flow regimes within environmental flow assessments. It is certainly not the only way to improve the rigor of environmental flow assessments, and indeed the ELOHA method (Poff et al. 2010) has already made considerable strides towards improving the use of empirical evidence in the prediction of ecological responses. The appeal of the methods presented here lies in their ability to operate, regardless of whether empirical data are scarce or plentiful, and also to create more nuanced ecological model structures than the bivariate flow–response relationships that typify ELOHA assessments. Incorporating these methods into environmental flows assessments will increase costs, but not prohibitively so. More importantly, these approaches would move environmental flows assessments to a level of rigor where they could rightly claim to be based on best-available-science and to be consistent with best practice in evidence-based environmental management.

CHAPTER 10

Summary conclusions and recommendations

In this book, we have treated environmental flow assessment (EFA) mainly as a problem in applied ecology, although we recognize that it has important legal, economic, social and moral aspects as well (Horne et al. 2018). We noted some of these aspects in Chapter 1 and elaborated on a few in later chapters. We have also emphasized that EFA is a human activity, and so subject to human behavior.

In Chapter 1, we recalled the claim that three of us made over two decades ago, that "…currently no scientifically defensible method exists for defining the instream flows needed to protect particular species of fish or aquatic ecosystems" (Castleberry et al. 1996, p. 20). We went on to recommend adaptive management as the appropriate management response. In subsequent chapters in this book, we have tried to describe the advances that have been made with methods for EFA, the reasons why both the claim and the recommendation still hold true, and present what we think is currently the best analytical approach for adaptive management. Here, we recapitulate the main points, offer conclusions and recommendations, make brief comments on particular methods, and close with a checklist for EFA.

10.1 Conclusions and recommendations

10.1.1 Confront uncertainty and manage adaptively

Because of the dynamism and complexity of riverine ecosystems, predictions of the ecological effects of changes in flow regimes are uncertain at best. Therefore, confront uncertainty and manage adaptively

Environmental flow assessment is about predicting the uncertain ecological consequences of changes in the flow regime of a stream. In Chapter 1, we listed some of the reasons why predictions of future conditions in a river or stream will be uncertain, and subsequent chapters elaborate on these and discuss other reasons. For example, besides water, streams also carry sediment and other inorganic matter, wood and other organic matter, organisms, chemicals and heat. All of these affect stream ecosystems, so EFA needs to take account of more than water and fish. Because our understanding is limited, managing a regulated or modified stream is inevitably an experiment, whether we want it to be or not; the challenge is to make it an informative experiment. Adaptive management, discussed in Chapters 6 and 9, attempts to do just that, as well as inform future changes in the flow regime if appropriate.

Environmental Flow Assessment: Methods and Applications, First Edition. John G. Williams, Peter B. Moyle, J. Angus Webb and G. Mathias Kondolf.
© 2019 John Wiley & Sons Ltd. Published 2019 by John Wiley & Sons Ltd.

Water managers and regulators face a dilemma. On the one hand, they need to make decisions about projects or policies that will change flow regimes, and, on the other, they have uncertain information about the environmental consequences of their decisions. For several decades, adaptive management has been recommended as the best response to this situation for management of streams, for example by Poff et al. (1997), and for ecosystem management generally (DeFries and Nagendra 2017). Adaptive management recognizes that assessments are uncertain and allows management to be modified in light of new information. We recommend an "evidence-based" approach (Webb et al. 2015a, b; Chapters 6 and 9). Generally, the approach involves applying best practices to each step in the process of developing and applying Bayesian models, which seem a natural choice for adaptive management.

Given that management is experimental, monitoring is necessary to get the experimental results. Although some questions may be answered quickly, most will not, so long-term monitoring is a critical part of adaptive management. It is also a critical resource for improved scientific understanding of the effects of water developments on rivers, so stable funding for monitoring should be an integral part of water management.

10.1.2 Methods for EFA

There have been major advances in models that are useful for assessing biological aspects of EFA, particularly hierarchical Bayesian statistical models, Bayesian network models, dynamic energy budget models, state-dependent life history models, dynamic occupancy models, and individual-based models. There have also been major advances in hydraulic and hydrological models. Despite these advances, precise predictions of the overall ecological responses of streams to significant changes in flow regimes remain beyond the reach of science, unless the changes in flow regime are extreme (Chapter 1). More progress has been made on predicting the responses of particular components of ecosystems, but even here, predictions are uncertain,

and best given as interval estimates or probability distributions, unless the change crosses some known threshold, such as the temperature tolerance of an important species.

The most important advance for EFA, in our view, has been the use of Bayesian methods, both Bayesian network models and hierarchical Bayesian models (Chapters 5, 6 and 9). These methods have the great virtues of dealing explicitly and quantitatively with uncertainty, and of incorporating prior knowledge into the analysis in a formal and transparent way. They also encourage a more scientific kind of thinking about EFA that is appropriate for adaptive management. Bayesian methods encourage people to start with some understanding of the river system in question, seek evidence that makes them more or less confident of their understanding, and then repeat the process, with appropriate adjustments to their understanding. We think this is how science should work.

Hierarchical Bayesian modeling is well suited for complex problems, including problems with variables that cannot be measured directly and have to be estimated from data on other variables. It is also well suited for dealing with multiple streams at once, and so for approaches such as ELOHA that deal with groups of streams. Unfortunately, using hierarchical Bayesian models properly, or even understanding how they work, requires more statistical expertise than is common among workers in EFA (Webb et al. 2017b). This is a major disadvantage.

Bayesian network models are easier to use and to understand, and so seem a better choice for most assessments until more people with good statistical skill become involved in EFA. However, Bayesian network models also have serious limitations, such as no feedback loops, and it would be wrong to restrict attention to conceptual models that lack such loops. When the necessary expertise and sufficient data are available, hierarchical Bayesian models probably should be used; otherwise, Bayesian network should be the default choice for most assessments, although other methods may be needed as well. Bayesian network models can also be used to quantify conceptual models and to develop prior distributions for hierarchical Bayesian models from expert elicitation (Chapters 6 and 9).

Non-Bayesian methods still have a roll in EFA. Assessments are likely to require various less complex statistical analyses that can be done more easily with standard (frequentist) methods than with Bayesian methods. Managers are more likely to be familiar with the frequentist methods, and so more comfortable with their results.

Simulation models can also have important roles in EFA (Chapter 5). Simulation models incorporate descriptions of what we think are the most important aspects of the situations at hand, and so calculate the consequences of how we think the world works. Simulations can give good predictions for physical systems that are well understood, as with hydraulics, but these should be tested with independent data. Simulations are much less useful for predicting what will happen with biological systems. If a simulation model has many adjustable parameters, it is often possible to choose values that give a good fit to observations, but fail badly with predictions.

For early planning, or when funding is severely limited, simple but precautionary hydrological methods seem appropriate (Chapter 5). Such methods may also be adequate where stresses on stream ecosystems are mild and flow data are scarce. However, these hydrological methods are not a substitute for at least basic understanding of the stream ecosystems in question.

Reviewing the relevant literature should be part of any serious assessment. Systematic assessments of the literature are more informative and transparent than traditional narrative reviews (Chapters 5 and 9). Full systematic reviews should be used for important questions if resources are available; otherwise, use rapid evidence assessments

Major assessments will entail collecting new data. Using statistically proper sampling plans for collecting data, and appropriate methods to analyze the data, will make the assessment more accurate and defensible, and allow for calculating and reporting interval estimates, rather than just point estimates

(Chapter 5). The data will also be of more general scientific value. Practitioners who claim that they can identify "representative reaches" should not be believed.

Hydraulic rating methods for EFA seem unreliable and are best avoided, especially when channel cross-sections are estimated from biased (non-random) samples. Hydraulic geometry relations estimate the central tendencies of channel width and depth, but ignore the variances (Chapter 5). Although using points of rapid change in a curve of wetted width over discharge to set flow standards has intuitive appeal, good evidence that habitat value is proportional to wetted width is lacking (Chapter 6).

Habitat association models assume that occupancy is a good indicator of habitat quality. Reasons that this assumption can fail are well known, as discussed in Chapters 5 and 6. As shown by examples in Chapter 7, habitat selection is affected by many environmental variables, especially at fine spatial scales, and fish select habitat at multiple spatial scales simultaneously. Accordingly, habitat association models such that work only at the microhabitat scale, such as PHABSIM, should be avoided.

Drift-foraging or net energy intake models can help clarify thinking, but in the absence of good information on drift they should not be used to quantify habitat as a function of discharge. Even with good information on the drift and good hydraulic modeling, the model results apply to only one aspect of habitat, even for drift-feeders. Models based on the influence of water velocity on capture efficiency can add another dimension to the output of hydraulic models, but again, apply to only one aspect of habitat. On the other hand, Individual-based models that incorporate net energy intake models, such as InSALMO, allow for exploring a broader range of ideas about how fishes use habitats (Chapter 6).

Models should be used in EFA to aid thinking about the situation at hand. It follows that using models that you do not understand is dangerous. In earlier chapters this point was made explicitly about hierarchical Bayesian models, but it applies as well to other kinds; the risk of model abuse is too great. It is much easier to get a computer model to run than it is to get it to run properly. Models entail assumptions that may not be obvious, and violating them can produce dubious results. If you want to use a model that you do not understand well, hire someone who does.

10.1.3 Recommendations on monitoring

Our recommendation for adaptive management implies the need for a monitoring program, so we offer the following recommendations on monitoring, adapted from Williams (2006). In adaptive management, management actions are treated as experiments, testing ideas about how species or ecosystems will respond to the actions. In this context, monitoring is one part of getting the experimental results. The other part is analyzing the monitoring data to assess the ideas.

Monitoring programs traditionally measure attributes of populations, especially abundance, and attributes of habitats such as water temperature, discharge, etc. These are necessary, but attention should also be given to measuring attributes of individuals, such as growth rates, and indices of physiological condition such as lipid content. Detecting changes in populations is difficult, but changes are more easily detected in organismal variables (Osenberg et al. 1994; Fordie et al. 2014). Organismal variables typically have better statistical properties, and also provide evidence on the mechanisms by which populations may be affected by changes in flow regimes.

In most cases, monitoring should be linked to specific hypotheses or management activities. Basic biological monitoring is often justified, especially where lack of data inhibits the development of useful hypotheses (Power et al. 2001), but there is a danger that monitoring for "status and trends" can become rote, producing data that are never thoughtfully analyzed or critically evaluated for utility. This danger can best be avoided by making data analyses that address specific questions, as well as exploratory analyses, basic parts of the monitoring program. This will clarify also which aspects of the monitoring program are producing useful data, and which

are not. Monitoring produces better and more useful data if the person responsible for analyzing the data is closely associated with the data collection. This gives the analyst a better understanding of the strengths and weaknesses of the data, and promotes more careful work by the field crews.

Sometimes monitoring can involve a census, as when all fish are counted as they pass a weir. More often, monitoring involves sampling in time or space. In such cases, more information can be obtained if the monitoring program uses a probabilistic (random) design (Chapter 5). Such a sampling design justifies inferences about the parts of the system that are not sampled; a design using deliberately selected samples does not.

To be useful, monitoring data must be converted into information. As a first step, summary statistics should be supplemented with appropriate graphical summaries, such as scatter plots, box plots, and cumulative frequency distributions. Different graphical summaries emphasize different aspects of the data, so using several is often appropriate.

Models are another way to get information from the monitoring data. In an important sense, modeling is simply a formal way of thinking about the data. Bayesian methods generally provide a more useful guide to management of natural resources than do the more traditional methods, and are easily updated as new monitoring data becomes available. With frequentist methods, more attention should be given to developing estimates of effect size and confidence intervals, and much less to tests of statistical significance (Stewart-Oaten 1996; Steidl et al. 1997; Johnson 1999, Chapter 5).

Long-term monitoring is a critical part of adaptive management and a critical resource for improved scientific understanding of the effects of water developments on rivers, so stable funding for monitoring should be an integral part of water management. Many municipalities have "percent for art" ordinances, which impose a fee on large capital improvement projects, usually 0.5–2% of cost. This pays for public art, which typically is incorporated into the project.

Regulators should do something similar with water projects, based either on cost or amount of water diverted, to fund monitoring.

Analyses of simulated data that include realistic amounts of uncertainty can help determine whether a proposed monitoring program will provide useful answers to the questions it is designed to address (Ludwig and Walters 1985; Williams 1991). Monitoring can be expensive, and such simulations reduce the risk of expensive failures.

Most sampling methods introduce bias. For example, if the target organisms are not equally susceptible to the sampling gear, the samples will be biased. Unfortunately, this is the usual case. The efficiency of sampling gear often depends on the size of the fish, as is generally recognized. Or, it may be impossible to use the sampling gear in water that is too shallow, too deep, flowing under brush and trees, etc. Sometimes, biases can be more subtle. For example, it is common to sample juvenile salmonids with traps facing upstream, in order to catch fish moving downstream, but this practice may give a distorted picture of fish behavior if the fish are not migrating, or migrating very slowly.

Extra sampling during unusual events or situations, such as the aftermath of floods, can sometimes produce particularly valuable information. By their nature, unusual events are hard to plan for, but funding for contingency sampling can be made part of the monitoring program. The utility of contingency sampling probably depends upon close supervision of the field program.

10.1.4 Recommendations for assessments

The circumstances of EFAs vary too much for any list of specific steps to make sense, but it seems useful to offer some general guidance on the assessment process, similar to that for modeling in Chapter 7.

Writing down a good description of your stream or region, the available data, and the broad questions that you are trying to answer, is almost always the place to start. Our experience is that writing clarifies thinking, and often changes it. And, if you consult with others, as you should, having a good description of your situation is essential, unless you want to pay them to figure it out for themselves. Sooner or later, you will need this description for some kind of report. You can expect this description to change as the EFA progresses and as you learn more about the system, but modifying a description that you already have is much less difficult than starting from scratch.

Questions adapted from Healey (1998), provide a good framework for this task. What kind of a stream ecosystem do you have? What kind of a stream ecosystem did you have? What kind of a steam ecosystem do you want? Within this framework, EFA should try to answer two additional questions: What kind of a stream ecosystem can you have, and how can you get it? As a practical matter, you also need to consider the question: What kind of an EFA can you have? Budgets and data that are available will constrain your options.

Along with writing descriptions and questions, developing conceptual models of various aspects of your system should start early. Like any model, they should be tested and refined or rejected as appropriate. If the conceptual models do not involve feedback loops, they can and probably should be turned into Bayesian Network models.

Clarifying the financial and human resources available or potentially available for the assessment should be another early step, so that resources can be reasonably allocated over different aspects of the assessment, matching the assessment to the situation. All else equal, however, there will be a trade-off between the resources available and the uncertainty in the results.

Select appropriate methods for the assessment. We recommend the evidence-based approach described in Chapters 6 and 9 if appropriate resources and data are available. If they are not, consider simple but conservative methods applied in the context of adaptive management.

Methods or models that purport to give clear guidance to managers, such as curves of point estimates of habitat value over discharge, are almost certainly untrustworthy, and should be avoided.

Avoid methods that you or people working with you do not understand. In earlier chapters this point was made explicitly about hierarchical Bayesian models, but it applies equally to other kinds; the risk of model abuse is too great.

The results of the assessment will almost certainly be more uncertain that those paying for it would like, so it will be tempting to understate that uncertainty when the assessment is written up and presented. This temptation should be resisted as much as possible.

Assessments are usually done in the context of disputes over water, and are more likely to be acceptable to the parties to the dispute if the parties are consulted during the assessment process, and the process is open and transparent. Technical material should be presented in enough detail to satisfy experts, but most of it should go into appendices. Assessments should inform decisions, and can do this only if decision makers and other readers can understand them. Language that is hard to understand, because it is too technical, too ornate, or simply too poorly written, subverts this purpose.

10.2 A checklist for EFA

We conclude with a checklist for EFA that sums up much of what we have advocated in this book. The list is not exhaustive, and not all of these points will be applicable to all assessments, but we hope it will be helpful.

1 Initial phases
   - Has a description been written of the project setting, the resources at risk, and the issues to be addressed in the assessment?
   - Have affected parties been notified?
   - Will affected parties have opportunities to provide input?
   - Has the conceptual framework for the assessment been described?
   - Have initial conceptual models for the effects of the project been developed?
   - Have the financial and other resources available for the assessment been determined and described?

2 Literature review
   - Has the literature relevant to the major issues of the assessment been reviewed?
   - Did the review find conflicting findings or opinions on the major issues?
   - If so, are these described?
   - Was the review structured?
   - Did the literature review examine evidence for or against links in the proposed conceptual models?
   - Were conceptual models modified based on the literature review?
   - Did the review support any choices of focal species?
   - Did the review support any choices of methods?

3 Expert elicitation
   - Were independent experts consulted during the assessment?
   - Did they make recommendations regarding methods?
   - Were the experts asked to describe the reasoning underlying their recommendations?
   - Was expert option obtained using a structured approach?
   - Were the conceptual models modified in light of the expert opinion?

4 Sampling
   - Was sampling done as part of the assessment?
   - Have the reasons for the sampling program and the questions to be addressed been described?
   - Are the sampling universe and the sampling frame described?
   - Is part of the sampling universe unavailable for logistical reasons? If so, is this described?
   - Are the sampling sites selected with a probability-based sampling design?
   - Is sampling of adequate duration and frequency?
   - Are sampling methods sufficiently accurate to meet the objectives?

- Have potential biases in sampling methods been described and assessed?

5 Data Management
- Are data collection techniques and procedures described?
- Are data properly checked for errors?
- Are data securely archived?
- Are archived data readily available to others?

6 Analysis and Interpretation
- Are any numerical models used based on the conceptual models developed from the literature or expert opinion?
- Are the assumptions of the models described? Are they plausible?
- Are alternative models, or forms of the model, also used?
- If informative priors are used for Bayesian models, are they clearly described and justified?
- Are model results tested against alternative models, or variations of the model?
- Are model selection criteria used to select a model or average over models?
- Are model results tested with independent data?
- Are alternative hypotheses and analytical methods properly considered?
- If statistical significance was used to assess where associations are meaningful, are the methods for data collection and analysis specified in advance?
- Is uncertainty properly described and taken into account?
- Are the analytical techniques and models used appropriate and properly applied?
- Do the analyses identify and assess benefits as well as risks from management alternatives?

- Can the analyses identify opportunities for additional learning?
- Will findings from the assessment be reported in the professional literature?

7 Wrapping Up
- Were there unexpected findings?
- Did these change the course of the assessment?
- Did the assessment use different methods from those originally proposed?
- Does the assessment recommend adaptive management?
- Does the assessment recommend additional monitoring?
- If so, is a detailed monitoring plan provided?
- Does it articulate hypotheses to be addressed?
- Is the proposed monitoring of adequate duration for its purposes?
- Will the physiological condition of organisms be monitored as well as numbers?
- Has the monitoring program been simulated?
- Will adequate funding be provided for the monitoring?

8 Writing and Revisions
- Is there a detailed description of the site and setting?
- Has critical information been presented graphically?
- Has the writing been edited for clarity?
- Is technical jargon kept to a minimum?
- Are there plain language summaries of technical information?
- Are technical details put in appendices?
- Does the assessment clearly describe its own shortcomings and limitations?

Literature cited

Acreman, M.C. and Dunbar, M.J. (2004). Defining environmental river flow requirements – a review. *Hydrology and Earth System Sciences* 8: 861–876.

Ahmadi-Nedushan, B., St-Hilaire, A., Bérubé, M. et al. (2006). A review of statistical methods for the evaluation of aquatic habitat suitability for instream flow assessment. *River Research and Applications* 22: 500–523.

Al-Chokhachy, R., Wenger, S.J., Isaak, D.J., and Kershner, J.L. (2013). Characterizing the thermal suitability of instream habitat for salmonids: a cautionary example from the Rocky Mountains. *Transactions of the American Fisheries Society* 142: 793–801.

Anderson, K.E., Harrison, L.R., Nisbet, R.M., and Kolpas, A. (2013). Modeling the influence of flow on invertebrate drift across spatial scales using a 2D hydraulic model and a 1D population model. *Ecological Modeling* 265: 207–220.

Anderson, K.E., Paul, A.J., McCauley, E. et al. (2006). Instream flow needs in streams and rivers: the importance of understanding ecological dynamics. *Frontiers in Ecology and the Environment* 4: 309–318.

Andrieu, C., de Freitas, N., Doucet, A., and Jordan, M.I. (2003). An introduction to MCMC for machine learning. *Machine Learning* 50: 5–43.

Annandale, G.W. (2013). *Quenching the Thirst: Sustainable Water Supply and Climate Change*. North Charleston, SC: CreateSpace.

Annandale, G.W., Morris, G.L., and Karki, P. (2016). *Extending the Life of Reservoirs: Sustainable Sediment Management for Dams and Run-of-River Hydropower*. Washington, DC: The World Bank Group.

Annear, T., Chrisholm, I., Beecher, H.A. et al. (2004). *Instream Flows for Riverine Resource Stewardship*. Cheyenne, WY: Instream Flow Council.

Araki, H., Cooper, B., and Blouin, M.S. (2007). Genetic effects of captive breeding cause a rapid, cumulative fitness decline in the wild. *Science* 318: 100–103.

Armanini, D. G., Denmartini, D., Idígoras, C. A., Kunnansaari, T., Monk, W. A., St-Hilaire, A. and Curray, R. A. 2015. Environmental flow guidelines for resource development in New Brunswick. *Energy Institute Technical Report #2015-01*.

Armstrong, J.D. (2010). Variation in habitat quality for drift-feeding Atlantic salmon and brown trout in relation to local water velocity and river discharge. In: *Salmonid Fisheries: Freshwater Habitat Management*, (ed. P. Kemp), 1–27. Oxford: Wiley-Blackwell.

Armstrong, J.D. and Nislow, K.H. (2012). Modeling approaches for relating effects of change in river flow to populations of Atlantic salmon and brown trout. *Fisheries Management and Ecology* 19: 527–536.

Arnaud, F., Piégay, H., Béal, D. et al. (2017). Monitoring gravel augmentation in a large regulated river and implications for process-based restoration. *Earth Surface Processes and Landforms* https://doi.org/10.1002/esp.4161.

Arthington, A.H. (2012). *Environmental Flows: Saving Rivers in the Third Millennium*. Berkeley, CA: University of California Press.

Arthington, A. H., Batan, E., Brown, C. A., Dugan, P., Halls, A. S., King, J. M. and Ninte-Vera, C. V. 2007. *Water Requirements of Floodplain Rivers and Fishes: Existing Decision Support Tools and Pathways for*

Environmental Flow Assessment: Methods and Applications, First Edition. John G. Williams, Peter B. Moyle, J. Angus Webb and G. Mathias Kondolf.
© 2019 John Wiley & Sons Ltd. Published 2019 by John Wiley & Sons Ltd.

Development. Comprehensive Assessment of Water Management in Agriculture Research Report 17. Colombo, Sri Lanka: International Water Management Institute.

Arthington, A.H., Conrick, D.L., and Bycroft, B.M. (1992b). *Environmental Study of Barker–Barambah Creek. Scientific Report: Water Quality, Ecology and Water Allocation Strategy.* Brisbane: Centre for Catchment and In-stream Research (Griffith University) and Water Resources Commission (Dept. of Primary Industries).

Arthington, A. H., King, J. M., O'Keefe, J. H., Bunn, S. E., Day, J. A. et al. 1992a. Development of an holistic approach for assessing environmental flow requirements of riverine ecosystems. *Proceedings of an International Seminar and Workshop on Water Allocation for the Environment.* Armidale, NSW.

Arthington, A.H., MacKay, S.J., James, C.S. et al. (2012). *Ecological limits of hydraulic alteration: a test of the ELOHA framework in south-east Queensland,* Waterlines Report Series No. 75. Canberra, Australia: Australian National Water Commission.

Arthington, A.H., Naiman, R.J., McClain, M.E., and Nilsson, C. (2010). Preserving the biodiversity and ecological services of rivers: New challenges and research opportunities. *Freshwater Biology* 55 (1): 1–16.

Ayllón, D., Almodóvar, A., Nicola, G.G., and Elvira, B. (2012). The influence of variable habitat suitability criteria on PHABSIM habitat index results. *River Research and Applications* 28: 1179–1188.

Bachman, R.A. (1984). Foraging behavior of free-ranging wild and hatchery brown trout in a stream. *Transactions of the American Fisheries Society* 113: 1–32.

Bain, M.B. and Boltz, J.M. (1989). *Regulated streamflow and warmwater stream fish: a general hypothesis and research agenda,* Biological Report 89-18. Washington, DC: U.S. Dept. of the Interior, Fish and Wildlife Service.

Baker, E.A. and Coon, T.G. (1997). Development and evaluation of alternative habitat suitability criteria for brook trout. *Transactions of the American Fisheries Society* 126: 65–76.

Baltz, D.M., Moyle, P.B., and Knight, N.J. (1982). Competitive interactions between benthic stream fishes, riffle sculpin, *Cottus gulosus,* and speckled dace, *Rhinichthys osculus. Canadian Journal of Fisheries and Aquatic Sciences* 39: 1502–1511.

Barraud, R. (2017). Removing mill weirs in France: the structure and dynamics of an environmental controversy. *Water Alternatives* 10 (3): 796–818.

Batalla, R.J. and Vericat, D. (2009). Hydrological and sediment transport dynamics of flushing flows: implications for management in large Mediterranean rivers. *River Research and Applications* 25: 297–314.

Batalla, R.J., Gomez, C.M., and Kondolf, G.M. (2004). River impoundment and changes in flow regime, Ebro River basin, northeastern Spain. *Journal of Hydrology* 290: 117–136.

Batalla, R., Vericat, D., and Tena, A. (2014). The fluvial geomorphology of the lower Ebro (2002–2013): bridging gaps between management and research. *Cuadernos de Investigación* 40 (1): 29–51.

Bayer, J.M. and Schei, J.L. (2009). Remote Sensing Applications for Aquatic Resource Monitoring. In: *Pacific Northwest Aquatic Monitoring Partnership Special Publication.* Cook, WA: Pacific Northwest Aquatic Monitoring Partnership.

Beakes, M.P., Moore, J.W., Retford, N. et al. (2014). Evaluating statistical approaches to quantifying juvenile Chinook salmon habitat in a regulated California river. *River Research and Applications* 30: 180–191.

Beale, C.M. and Lennon, J.J. (2012). Incorporating uncertainty in predictive species distribution modelling. *Philosophical Transactions of the Royal Society Series B* 367: 247–258.

Bedard, M.-E., Imre, I.J., and Boisclair, D. (2005). Nocturnal density patterns of Atlantic salmon parr in the Sainte-Marguerite River, Quebec, relative to the time of night. *Journal of Fish Biology* 66: 1483–1488.

Beebe, J.T. (1996). Fluid speed variability and the importance to managing fish habitat in rivers. *Regulated Rivers: Research and Management* 12: 63–79.

Bélanger, G. and Rodríguez, M.A. (2002). Local movement as a measurement of habitat quality in stream salmonids. *Environmental Biology of Fishes* 64: 155–164.

Behnke, R. J. 1986. The illusion of technique and fisheries management. *Proceedings, 22nd Conference.* University of Wyoming, Laramie, USA: Colorado–Wyoming Chapter American Fisheries Society.

Benjankar, R., Tonina, D., and Mckean, J. (2015). One-dimensional and two-dimensional hydrodynamic modeling derived flow properties: impacts on aquatic habitat quality assessments. *Earth Surface Processes and Landforms* 40: 340–356.

Bennett, N.D., Croke, B.F., Guariso, G. et al. (2013). Characterizing performance of environmental models. *Environmental Modelling and Software* 40: 1–20.

Berry, D. (2017). A *p*-value to die for. *Journal of the American Statistical Association* 112 (519): 895–897.

Beschta, R.L., Jackson, W.L., and Knoop, K.D. (1981). Sediment transport during a controlled reservoir release. *Journal of the American Water Resources Association* 17 (4): 635–641.

Bisson, P.A., Nielsen, J.L., and Ward, J.W. (1988). Summer production of coho salmon stocked in Mount St. Helens streams 3–6 years after the 1980 eruption. *Transactions of the American Fisheries Society* 117: 332–335.

Blocken, B. and Gualtieri, C. (2012). Ten iterative steps for model development and evaluation applied to computational fluid dynamics for environmental fluid mechanics. *Environmental Modelling and Software* 33: 1–22.

Blöschl, B., Sivapalan, M., Wagener, T. et al. (eds.) (2012). *Runoff Prediction in Ungaged Basins: Synthesis Across Processes, Places and Scales*. Cambridge: Cambridge University Press.

Bolker, B.M., Brooks, M.E., Clark, C.J. et al. (2009). Generalized linear mixed models: a practical guide for ecology and evolution. *Trends in Ecology and Evolution* 24: 127–135.

Bond, N.R., Lake, P.S., and Arthington, A.R. (2008). The impacts of drought on freshwater ecosystems: an Australian perspective. *Hydrobiologia* 60: 3–16.

Bondi, C., Yarnell, S.M., and Lind, A.J. (2014). Transferability of habitat suitability criteria for a stream breeding frog (*Rana boylii*) in the Sierra Nevada, California. *Herptological Conservation and Biology* 8: 88–103.

Booker, D.J. (2010). Predicting width in any river at any discharge. *Earth System Processes and Landforms* 35: 828–841.

Börk, K.S., Krovoza, J.F., Katz, J.V., and Moyle, P.B. (2012). The rebirth of California Fish and Game Code section 5937: water for fish. *University of California Davis Law Review* 45: 809–913.

Bottom, D.L., Simenstad, C.A., Baptista, A.M. et al. (2005). *Salmon at River's End: The Role of the Estuary in the Decline and Recovery of Columbia River Salmon*. Seattle, WA: National Marine Fisheries Service.

Bouchard, J. 2005. *Modélisation de la Qualité d'Habitat des Juveniles de Daumons Atlantiques (Salmo salar) à l'Echelle d'une Riviére*. MS thesis, Université de Montréal.

Bouchard, J. and Boisclair, D. (2008). The relative importance of local, lateral, and longitudinal variables on the development of habitat quality models for a river. *Canadian Journal of Fisheries and Aquatic Sciences* 65: 61–73.

Boulton, A.J., Findlay, S., Marmonier, P. et al. (1998). The functional significance of the hyporheic zone in streams and rivers. *Annual Review of Ecology and Systematics* 29: 59–81.

Bovee, K.D. (1982). *A Guide to Stream Habitat Analysis Using the Instream Flow Incremental Methodology, Instream Flow Information Paper No. 12 FWS/OBS - 82/86*. Washington, DC: U.S. Fish and Wildlife Service.

Bovee, K. D. 1988. Use of the instream flow incremental methodology to evaluate influences of microhabitat variability on trout populations in four Colorado streams. *Proceedings of the 68th Annual Conference, Western Association of Fish and Wildlife Agencies*. Albuquerque, NM.

Bovee, K.D. (ed.) (1996). *The Complete IFIM: A Coursebook for IF 250*. Fort Collins, CO: U.S. Geological Survey.

Bovee, K.D. and Cochnauer, T. (1977). *Development and Evaluation of Weighted Criteria, Probability-of-Use Curves for Instream Flow Assessments: Fisheries, Instream Flow Information Paper No. 3. FWS/OBS-77/63*. Fort Collins, CO: U.S. Fish and Wildlife Service.

Bovee, K.D., Lamb, B.L., Bartholow, J.M. et al. (1998). *Instream habitat analysis using the instream flow incremental methodology, Biological Resources Division Information and Technology Report*. Fort Collins, CO: U.S. Geological Survey.

Boyd, I.L. (2012). The art of ecological modeling. *Science* 337: 306–307.

Bradford, M.J. and Higgins, P.S. (2001). Habitat-, season-, and size-specific variation in diel activity patterns of juvenile Chinook salmon (*Oncorhynchus tshawytscha*) and steelhead trout (*Oncorhynchus mykiss*). *Canadian Journal of Fisheries and Aquatic Sciences* 58: 365–374.

Bradford, M.J., Higgins, P.S., Korman, J., and Sneep, J. (2011). Test of an environmental flow release in a British Columbia river: does more water mean more fish? *Freshwater Biology* 56: 2119–2134.

Breiman, L. (2001). Statistical modeling: the two cultures. *Statistical Science* 16: 191–231.

Brenden, T.O., Wang, L., Clark, R.D.J., and Seebach, P.W. (2007). Comparison between model-predicted and field-measured stream habitat features for evaluating fish assemblage–habitat relationships. *Transactions of the American Fisheries Society* 136: 580–592.

Brenden, T.O., Wang, L., and Seelbach, P.W. (2008). A river valley segment classification of Michigan streams based on fish and physical attributes. *Transactions of the American Fisheries Society* 137: 1621–1636.

Brunner, G.W. (2010). HEC-RAS (River Analysis System). In: *Proceedings of the North American Water and Environment Congress and Destructive Water*, 3782–3787. ASCE.

Bryan, B.A., Higgins, A., Overton, I.C. et al. (2013). Ecohydrological and socioeconomic integration for the operational management of environmental flows. *Ecological Applications* 23: 999–1016.

Bult, T.P., Haedrich, R.L., and Schneider, D.C. (1998). New technique describing spatial scaling and habitat selection in riverine habitats. *Regulated Rivers: Research and Management* 14: 107–118.

Burgman, M., Carr, A., Godden, L. et al. (2011). Redefining expertise and improving ecological judgment. *Conservation Letters* 4: 81–87.

Burnham, K.P. and Anderson, D.R. (1998). *Model Selection and Inference: A Practical Information-theoretic Approach*. New York: Springer.

Burnham, K.P. and Anderson, D.R. (2002). *Model Selection and Multi-model Inference: A Practical Information-theoretic Approach*, 2e. New York: Springer Science and Business Media, Inc.

Cade, B.S. and Noon, B.R. (2003). A gentle introduction to quantile regression for ecologists. *Frontiers in Ecology and the Environment* 1: 412–420.

Cain, J. (2001). *Planning Improvements in Natural Resource Management: Guidelines for Using Bayesian Networks to Support the Planning and Management Development Programmes in the Water Sector and Beyond*. Wallingford: Centre for Ecology and Hydrology.

Caissie, J., Caissie, D., and El-Jabi, N. (2015). Hydrologically based environmental flow methods applied to rivers in the Maritime Provinces (Canada). *River Research and Applications* 31: 651–662.

CalAm (California-American Water Company). 2017. San Clemente Dam Removal webiste, http://www.sanclementedamremoval.org/?page_id=1045, accessed October 2017

California Department of Fish and Wildlife (CDFW) (2014). *Draft Instream Flow Regime Recommendations, Big Sur River, Monterey County*. Sacramento, CA: Water Branch, Instream Flow Program, CDFW.

Campbell, E. A. 1998. *Influence of Streamflow and Predators on Habitat Choice by Trout*. Ph.D. thesis, University of California, Davis.

Cardenas, M.B., Wilson, J.L., and Zlotnik, V.A. (2004). Impact of heterogeneity, bed forms, and stream curvature on subchannel hyporheic exchange. *Water Resources Research* 40 (8): W08307.

Castleberry, D.T., Cech, J.J. Jr., Erman, D.C. et al. (1996). Uncertainty and instream flow standards. *Fisheries* 21: 20–21.

Cavalli, M., Tarolli, P., Marchi, L., and Fontana, G.D. (2008). The effectiveness of airborne LiDAR data in the recognition of channel-bed morphology. *Catena* 73: 249–260.

Cavallo, B., Kurth, R., Kindopp, J., Seeholtz, A. and Perrone, M. 2003. Distribution and habitat use of steelhead and other fishes in the Lower Feather River, 1999–2001. Interim Report, SP-F10, Task 3a. California Department of Water Resources, Division of Environmental Services.

CEE (2013). *Guidelines for Systematic Review and Evidence Synthesis in Environmental Management*. Version 4.2. Bangor, Wales: Collaboration for Environmental Evidence.

CGS (California Geological Survey) (2002). *California geomorphic provinces*, Note 36. Sacramento, CA: California Geologic Survey.

Chamberland, J.-M., Lanthier, G., and Boisclair, D. (2014). Comparisons between electrofishing and snorkeling surveys to describe fish assemblages in Laurentian streams. *Environmental Monitoring and Assessment* 186: 1837–1846.

Chan, T.U., Hart, B.T., Kennard, M.J. et al. (2012). Bayesian network models for environmental flow decision making in the Daly River, Northwest Territory, Australia. *River Research and Applications* 28: 283–301.

Chanton, J. and Lewis, F.G. (2002). Examination of coupling between primary and secondary production in a river-dominated estuary: Apalachicola Bay, Florida, U.S.A. *Limnology and Oceanography* 47 (3): 683–697.

Chapman, D.W. (1966). Food and space as regulators of salmonid populations in streams. *American Naturalist* 100: 345–357.

Chee, Y.-E., Webb, J. A., Cottingham, P. and Stewardson, M. J. 2009. *Victorian Environmental Flows Monitoring and Assessment program: Monitoring and Evaluation of Environmental Flow releases in the Campaspe River*. Report prepared for the North Central Catchment Management Authority and the Department of Sustainability and Environment. Canberra, Australia: Water Cooperative Research Center.

Chen, S.H. and Pollino, C.A. (2012). Good practice in Bayesian network modelling. *Environmental Modeling and Software* 37: 134–145.

Christie, M.R., Ford, M.J., and Blouin, M.S. (2014). On the reproductive success of early-generation hatchery fish in the wild. *Evolutionary Applications* 7: 883–896.

Christie, M.R., Marine, M., Fox, S.E. et al. (2016). A single generation of domestication heritably alters the expression of hundreds of genes. *Nature Communications* 7: 10676. (online).

Church, M. (2006). Bed material transport and the morphology of alluvial river channels. *Annual Review of Earth and Planetary Science* 34: 325–354.

Church, M. 2015. Channel stability: morphodynamics and the morphology of river. *In*: Rowinski, P. and Radecki-Paulik, A. (eds.), *Rivers – Physical, Fluvial and Environmental Processes*, pp. 281–322. Cham, Switzerland: Springer International. doi: https://doi.org/10.1007/978-3-319-17719-9.

Clark, J.S. (2005). Why environmental scientists are becoming Bayesians. *Ecology Letters* 8: 2–14.

Clark, J.S. (2007). *Models for Ecological Data: An Introduction*. Princeton, NJ: Princeton University Press.

Clark, J.S., Bell, D.M., Hersch, M.H. et al. (2011). Individual-scale variation, species-scale differences: inference needed to understand diversity. *Ecology Letters* 14: 1273–1287.

Cleveland, W.S. (1985). *The Elements of Graphing Data*. Monterey, CA: Wadsworth Books.

Cochran, W.G. (1977). *Sampling Techniques*. New York: John Wiley and Sons.

Cohen, J. (1992). Quantitative models in psychology: a power primer. *Psychological Bulletin* 112: 155–159.

Collier, M.P., Webb, R.H., and Andrews, E.D. (1997). Experimental flooding in Grand Canyon. *Scientific American* 276: 82–89.

Cooper, D.C. and Andersen, D.J. (2012). Novel plant communities limit the effects of a managed flood to restore riparian forests along a large regulated river. *River Research and Applications* 28: 204–215.

Cooper, S.D., Diehl, S., Kratz, K., and Sarnelle, O. (1998). Implications of scale for patterns and processes in stream ecology. *Australian Journal of Ecology* 23: 27–40.

Cook, C.N., Nichols, S.J., Webb, J.A. et al. (2017). Simplifying the selection of evidence synthesis methods to inform environmental decisions: a guide for decision makers and scientists. *Biological Conservation* 213: 135–145.

Cottingham, P., Stewardson, M. J. and Webb, J. A. 2005. Victorian Environmental Flows Monitoring and Assessment Program: Stage 1 Statewide Framework. *CRC Freshwater Ecology and CRC Catchment Hydrology Report*, Department of Sustainability and Environment.

Coulombe-Pontbriand, M. and Lapointe, M. (2004). Landscape controls on boulder-rich, winter habitat availability and their effects on Atlantic salmon (*Salmo salar*) parr abundance in two fifth-order mountain streams. *Canadian Journal of Fisheries and Aquatic Sciences* 61: 648–658.

Cressie, N., Calder, C.A., Clark, J.S. et al. (2009). Accounting for uncertainty in ecological analysis: the strengths and limitations of hierarchical statistical analysis. *Ecological Applications* 19: 553–570.

Crook, D.A., Robertson, A.I., King, A.J., and Humphries, P. (2001). The influence of spatial scale and habitat arrangement on diel patterns of use by two lowland river fishes. *Oecologea* 120: 525–533.

Crowder, D.W. and Diplas, P. (2002). Vorticity and circulation: spatial metrics for evaluating flow complexity in stream habitats. *Canadian Journal of Fisheries and Aquatic Sciences* 19: 553–570.

Cullis, J.D.S., McKnight, D.M., Spaulding, S.A., and Prairier, Y. (2015). Hydrodynamic control of benthic mats of Didymosphenia geminata at the reach scale. *Canadian Journal of Fisheries and Aquatic Sciences* 72 (6): 902–914.

Cutler, D.R., Edwards, T.C. Jr., Beard, K.H. et al. (2007). Random forests for classification in ecology. *Ecology* 88: 2738–2792.

Dauble, D., Hankin, D., Pizzimenti, J.J., and Smith, P. (2010). *The Vernalis Adaptive Management Program (VAMP): Report of the 2010 Review Panel*. Sacramento, CA: Delta Science Program.

Dauwalter, D.C., Rahel, F.J., and Gerow, K.G. (2009). Temporal variation in trout populations: implications for monitoring and trend detection. *Transactions of the American Fisheries Society* 138: 38–51.

Dauwalter, D.C., Wenger, S.J., and Gardner, P. (2014). The role of complexity in habitat use and selection by stream fishes in a Snake River Basin tributary. *Transactions of the American Fisheries Society* 143: 1177–1187.

Davies, B. R. and Day, J. A. 1989. *Physical and chemical attributes important in the biological functioning of river ecosystems*. South African National Scientific Programmes Report No. 162. Pretoria: Council for Scientific and Industrial Research.

Davidson, R.S., Letcher, B.H., and Nislow, K.H. (2010). Drivers of growth variation in juvenile Atlantic salmon (*Salmo salar*): an elasticity analysis approach. *Journal of Animal Ecology* 79: 1113–1121.

Davison, A.C. and Hinkley, D.V. (1997). *Bootstrap Methods and their Application*. Cambridge: Cambridge University Press.

De Little, S. C., Webb, J. A., Miller, K. A., Rutherford, L. D., and Stewardson, M. J. 2013. Using Bayesian hierrchical models to measure and predict the effectiveness of environmental flows and ecological responses. 20th International Congress on Modeling and Simulations, Adelaide, Australia, 1–6. December: 359–365. www.mssanz.org.au/modsim2013.

de Little, S.C., Casas-Mulet, R., Patulny, L. et al. (2018). Minimising biases in expert elicitations to inform environmental management: case studies from environmental flows in Australia. *Environmental Modelling and Software* 100: 146–158.

Deitch, M.J. and Dolman, B. (2017). Restoring summer base flow under a decentralized water management regime: constraints, opportunities, and outcomes in Mediterranean-climate California. *Water* 9 (1): 29. https://doi.org/10.3390/w9010029.

Deitch, M.J., Kondolf, G.M., and Merenlender, A.M. (2009). Surface water balance to evaluate the hydrological impacts of small instream diversions and application to the Russian River basin, California, USA. *Aquatic Sciences: Marine and Freshwater Ecosystems* 19: 274–284.

Deitch, M.J., Sapundjieff, M.J., and Feirer, S.T. (2017). Characterizing precipitation variability and trends in the world's Mediterranean-climate areas. *Water* 9: 259. https://doi.org/10.3390/w9040259.

Dempster, A.P., Laird, N.M., and Rubin, D.B. (1977). Maximum likelihood from incomplete data via the EM algorithm. *Journal of the Royal Statistical Society. Series B (Methodological)* 39: 1–38.

DeFries, R. and Nagendra, H. (2017). Ecosystem management as a wicked problem. *Science* 356: 265–270.

DEPI (2013). *FLOWS – A Method for Determining Environmental Water Requirements in Victoria*, 2e. Melbourne: Department of Environment and Primary Industries.

Deschénes, J. and Rodríguez, M.A. (2007). Hierarchical analysis of relationships between brook trout (*Salvelinus fontinalis*) density and stream habitat features. *Canadian Journal of Fisheries and Aquatic Sciences* 64: 777–785.

Deser, C., Knutti, R., Solomon, S., and Phillips, A.S. (2012). Communication of the role of natural variability in future North American climate. *Nature Climate Change* 2: 775–779.

Dettinger, M.D., Ralph, F.M., Das, T. et al. (2011). Atmospheric rivers, floods and the water resources of California. *Water* 3: 445–478. https://doi.org/10.3390/w3020445.

Dingman, S.L. (1989). Probability distribution of velocity in natural channel cross sections. *Water Resources Research* 25: 508–518.

Dodds, W.K. (2002). *Freshwater Ecology Concepts and Environmental Applications*. New York: Academic Press.

Dodds, W.K. (2009). *Laws, Theories, and Patterns in Ecology*. Berkley: University of California Press.

Downes, B.J. (2010). Back to the future: little-used tools and principles of scientific inference can help disentangle effects of multiple stressors on freshwater ecosystems. *Freshwater Biology* 55 (Suppl.1): 60–79.

Dollar, E.J.S. (2000). Fluvial geomorphology. *Progress in Physical Geography* 24: 385–406.

Doyle, M.W., Stanley, E.H., Orr, C.H. et al. (2005). Stream ecosystem response to small dam removal: lessons from the heartland. *Geomorphology* 71: 227–244.

Doyle, M.W., Stanley, E.H., Strayer, D.L., and Jacobson, R.B. (2005). Effective discharge analysis of ecological processes in streams. *Water Resources Research* 41: W11411.

Dunbar, M.J., Ibbotson, A., Gowing, I. et al. (2002). *Further Validation of PHABSIM for the Habitat Requirements of Salmonid Fish*, R&D Technical Report W6-036/TR. Bristol, England: Centre for Ecology and Hydrology.

Dunne, T. and Leopold, L.B. (1978). *Water in Environmental Planning*. San Francisco: W.H. Freemen and Sons.

Durance, I., Lepichon, C., and Omerod, S.J. (2006). Recognizing the importance of scale in the ecology and management of riverine fish. *River Research and Applications* 22: 1143–1152.

Duveiller, G., Fasbender, D., and Meroni, M. (2016). Revisiting the concept of a symmetric index of agreement for continuous datasets. *Scientific Reports* 6: 19401. (online).

East, A.E., Pess, G.R., Bountry, J.R. et al. (2015). Large-scale dam removal on the Elwha River, Washington, USA: river channeland floodplain geomorphic change. *Geomorphology* 228: 765–786.

Eder, A., Exner-Kittridge, M., Strauss, P., and Blöschl, G. (2014). Re-suspension of bed sediment in a small stream: results from two flushing experiments. *Hydrology and Earth System Sciences* 18 (3): 1043–1052.

Efron, B. and Tibshirani, R. (1991). Statistical data analysis in the computer age. *Science* 253: 390–395.

Efron, B. and Tibshirani, R. (1993). *An Introduction to the Bootstrap*. New York: Chapman and Hall.

Eicher, G.J. (1976). Flow stabilization and fish habitat. In: *Proceedings of the Symposium and Specialty Conference on Instream Flow Needs: Solutions to Technical, Legal and Social Problems in Increasing Competition for Limited Streamflow*, (ed. J.F. Osborn and C.H. Allman), 416–420. Bethesda, MD: American Fisheries Society.

Eitel, J.U.H., Höfle, B., Vierling, L.A. et al. (2016). Beyond 3-D: the new spectrum of lidar applications for earth and ecological sciences. *Remote Sensing of the Environment* 186: 372–392.

Elith, J. and Leathwick, J.R. (2009). Species distribution models: ecological explanation and prediction across space and time. *Annual Review of Ecology, Evolution and Systematics* 40: 467–497.

Elliott, J.G. and Capesius, J.P. (2009). Geomorphic changes resulting from floods in reconfigured gravel-bed river channels in Colorado, USA. In: *Management and Restoration of Fluvial Systems with Broad Historical Changes and Human Impacts*, (ed. L.A. James, S.L. Rathburn and G.R. Whitecare), 173–198. Special Paper of the Geological Society of America.

Elliott, J.M. (1994). *Quantitative Ecology and the Brown Trout*. Oxford: Oxford University Press.

Ellison, A.M. (1996). An introduction to Bayesian inference for ecological research and environmental decision-making. *Ecological Applications* 6: 1036–1046.

Enders, E.C., Roy, M.L., Ovidio, M. et al. (2009). Habitat choice by Atlantic salmon parr in relation to turbulence at a reach scale. *North American Journal of Fisheries Management* 29: 1819–1830.

EPRI (Electric Power Research Institute) (2000). *Instream Flow Assessment Methods: Guidance for Evaluating Instream Flow Needs in Hydropower Licensing*. Palo Alto, CA: EPRI.

Failing, L., Horn, G., and Higgings, P.S. (2004). Using expert judgement and stakeholder values to evaluate adaptive management options. *Ecology and Society* 9 (1): 13. (online).

Farmer, A.H., Cade, B.S., and Stauffer, D.F. (2002). Evaluation of a habitat suitability index model. In: *The Indiana Bat: Biology and Management of an Endangered Species*, (ed. A. Kunta and J. Kennedy), 172–179. Austin, TX: Bat Conservation International.

Fausch, K.D. (1984). Profitable stream positions for salmonids: relating specific growth rate to net energy gain. *Canadian Journal of Zoology* 62: 441–451.

Fausch, K.D. (2014). A historical perspective on drift foraging models for stream salmonids. *Environmental Biology of Fishes* 97: 453–464.

Fausch, K.D., Hawkes, C.L., and Parsons, M.G. (1988). *Models that Predict Standing Crop of Stream Fish from Habitat Variables: 1950-1985, General Technical Report PNW-GTR-213*. Portland, OR: U.S. Dept. of Agriculture, Forest Service, Pacific Northwest Research Station.

Fausch, K.D., Torgersen, C.E., Baxter, C.V., and Li, H.W. (2002). Landscapes to riverscapes: bridging the gap between research and conservation of stream fishes. *BioScience* 56: 483–498.

Fedak, K.M., Bernal, A., Capshaw, Z.A., and Gtoss, S. (2015). Applying the Bradford–Hill criteria in the 21st Century: how data integration has changed causal inference in molecular epidemiology. *Emerging Themes in Epidemiology* 12: 14.

Ferrar, A.A. (ed.) (1989). *Ecological Flow Requirements for South African Rivers*. Pretoria: Council for Scientific and Industrial Research.

Feyrer, F., Sommer, T., and Harrell, W. (2006). Managing floodplain inundation for native fish: production dynamics of age-0 splittail (*Pogonichthys macrolepidotus*) in California's Yolo Bypass. *Hydrobiologia* 573: 213–226.

Fidler, F., Wintle, B. and Thomason, N. 2012. Groups making wise judgements. *Advanced Research Projects Activity*, USA Office of the Director of National Intelligence.

Fonstad, M.A. and Marcus, W.A. (2010). High-resolution, basin-extent observations of fluvial forms and implications for understanding river form and process. *Earth Surface Processes and Landforms* 35: 680–698.

Fordie, F.J., Able, K.W., Galvez, F. et al. (2014). Integrating organismal and population responses of estuarine fishes in Macondo Spill research. *BioScience* 64: 778–788.

Fournier, B. and Bérubé, M.-A. (2000). Alkali–aggregate reaction in concrete: a review of basic concepts and engineering implications. *Canadian Journal of Civil Engineering* 27: 167–191.

Francis, R.I.C.C. and Shotton, R. (1997). "Risk" in fisheries management: a review. *Canadian Journal of Fisheries and Aquatic Sciences* 54: 1699–1715.

Freedman, D.A. (1983). A note on screening regression equations. *The American Statistician* 37: 152–155.

Freeman, M.C., Bowen, Z.H., and Crance, J.H. (1997). Transferability of habitat suitability criteria for fishes in warm-water streams. *North American Journal of Fisheries Management* 17: 20–31.

Fretwell, S.D. (1972). *Populations in a Seasonal Environment*. Princeton, NJ: Princeton University Press.

Frissell, C.A., Gresswell, W.J., Nawa, R.K., and Ebersole, J.L. (1997). A resource in crisis: changing the measure of salmon management. In: *Pacific Salmon and their Ecosystems* (ed. D.J. Strouder, P.A. Bisson and R.J. Naiman), 411–444. New York: Chapman and Hall.

Gaeuman, D. (2012). Mitigating downstream effects of dams. In: *Gravel-bed Rivers: Processes, Tools, Environments* (ed. M. Church, A.G. Roy and P.M. Biron), 182–189. Chichester, UK: John Wiley and Sons.

Gaeuman, D. (2014). High-flow gravel injection for constructing designed in-channel features. *River Research and Applications* 30: 685–706.

Gard, M. (2009). Comparison of spawning habitat predictions of PHABSIM and River2D models. *International Journal of River Basin Management* 7: 55–71.

Gard, M. (2010). Response to Williams (2010) on Gard (2009): comparison of spawning habitat predictions of PHABSIM and River2D models. *International Journal of River Basin Management* 8: 121–125.

Gard, M. (2014). Modeling changes in salmon habitat associated with river channel restoration and flow-induced channel alterations. *River Research and Applications* 30: 40–44.

Gard, M.F. (2005). Variability in flow habitat relationships as a function of transect number for PHABSIM modeling. *River Research and Applications* 21: 1013–1029.

Garshelis, D.L. (2000). Delusions in habitat evaluation: measuring use, selection, and importance. In: *Research Techniques in Animal Ecology*, (ed. L. Boitani and T.K. Fuller), 111–164. New York: Columbia University Press.

Gasith, A. and Resh, V.H. (1999). Streams in Mediterranean climate regions: abiotic influences and biotic responses to predictable seasonal events. *Annual Review of Ecology and Systematics* 30: 51–81.

GBCMA (2016). *Goulburn River Seasonal Watering Proposal 2016–2017*. Shepparton: Goulburn Broken Catchment Management Authority.

Gelman, A. (2015). The connection between varying treatment effects and the crisis of unreplicable research: a Bayesian perspective. *Journal of Management* 41: 632–643.

Gelman, A. and Hill, J. (2007). *Data Analysis Using Regression and Multilevel/Hierarchical Models*. Cambridge: Cambridge University Press.

Gelman, A. and Loken, E. 2013. The garden of forking paths: Why multiple comparisons can be a problem, even when there is no "fishing expedition" or "p-hacking" and the research hypothesis was posited ahead of time. Technical report. New York: Department of Statistics, Columbia University.

Gelman, A., Rubin, J.B., Stern, H.S., and Rubin, D.B. (2004). *Bayesian Data Analysis*. Boca Raton, FL: Chapman and Hall/CRC.

Ghanem, P., Steffler, P., Hicks, F., and Katopdis, C. (1996). Two-dimensional hydraulic simulations of physical habitat conditions in flowing streams. *Regulated Rivers: Research and Management* 12: 185–200.

Gibson, A.J., Bowlby, H.D., and Amiro, P.G. (2008). Are wild populations ideally distributed? Variation in density-dependent habitat use by age class in juvenile Atlantic salmon (*Salmo salar*). *Canadian Journal of Fisheries and Aquatic Sciences* 65: 1667–1680.

Gido, K.B. and Propst, D.L. (2012). Long-term dynamics of native and nonnative fishes in the San Juan River, New Mexico and Utah under a partially managed flow regime. *Transactions of the American Fisheries Society* 141: 645–659.

Giger, R. D. 1973. Streamflow requirements of salmonids. *Job Final Report, Anadromous Fish Project*. Oregon Wildlife Commission.

Girard, P., Boisclair, D., and Leclerc, M. (2003). The effect of cloud cover on the development of habitat quality indices for juvenile Atlantic salmon (*Salmo salar*). *Canadian Journal of Fisheries and Aquatic Sciences* 60: 1386–1397.

Gopal, B. (2013). Methodologies for the assessment of environmental flows. In: *Environmental Flows*, (ed. B. Gopal), 129–182. Delhi: National Institute of Ecology.

Gore, J.A. and Nestler, J.M. (1988). Instream flow studies in perspective. *Regulated Rivers: Research and Management* 2: 93–101.

Gowan, C. and Fausch, K.D. (2002). Why do foraging salmonids move during summer? *Environmental Biology of Fishes* 64: 139–153.

Gowan, C., Stephenson, K., and Shabman, L. (2006). The role of ecosystem valuation in environmental decision making: hydropower relicensing and dam removal on the Elwha River. *Ecological Economics* 56: 508–523.

Grams, P.E., Topping, D.J., Schmidt, J.C. et al. (2013). Linking morphodynamic response with sediment mass balance on the Colorado River in Marble Canyon: issues of scale, geomorphic setting, and sampling design. *Journal of Geophysical Research: Earth Surface* 118 (2): 361–381.

Grant, J.W.A. and Kramer, D.L. (1990). Territory size as a predictor of the upper limit to population density of juvenile salmonids in streams. *Canadian Journal of Fisheries and Aquatic Sciences* 47: 1724–1737.

Grantham, T.E., Mezzatesta, M., Newburn, D.A., and Merenlender, A.M. (2013). Evaluating tradeoffs between environmental flow protections and agricultural water security. *River Research and Applications* 30: 315–328.

Grantham, T.E., Newburn, D.A., McCarthy, M.A., and Merenlender, A.M. (2012). The role of streamflow and land use in limiting over summer survival of juvenile steelhead in California streams. *Transactions of the American Fisheries Society* 141: 585–598.

Greet, J., Webb, J.A., and Cousens, R.D. (2011). The importance of seasonal flow timing for riparian vegetation dynamics: a systematic review using causal criteria analysis. *Freshwater Biology* 56: 1231–1247.

Gregory, R., Failing, L., and Higgins, P.S. (2006). Adaptive management and environmental decision making: a case study application to water use planning. *Ecological Economics* 58: 434–447.

Gregory, S., Boyer, K., and Gurnell, A.M. (eds.) (2003). *The Ecology and Management of Wood in World Rivers*. Bethesda, MD: American Fisheries Society.

Grimm, V., Augusiak, J., Focks, A. et al. (2014). Towards better modeling and decision support: documenting model development, testing, and analysis using TRACE. *Ecological Modelling* 280: 129–139.

Grossman, G.D. (2014). Not all drift feeders are trout: a short review of fitness-based habitat selection models for fishes. *Environmental Biology of Fishes* 97: 465–473.

Grossman, G.D. and Moyle, P.B. (1982). Stochasticity in structural and functional characteristics of an Indiana stream fish assemblage: A test of community theory. *American Naturalist* 97: 423–454.

Grossman, G.D., Rincon, P.A., Farr, M.D., and Ratajczak, R.E.J. (2002). A new optimal foraging model predicts habitat use by drift-feeding stream minnows. *Ecology of Freshwater Fishes* 11: 2–10.

Gruebber, C.E., Nakagaw, S., Laws, R.J., and Jamieson, I.G. (2011). Multimodel inference in ecology and evolution: challenges and solutions. *Journal of Evolutionary Biology* 23: 699–711.

Guay, J.C., Boisclair, D., Leclerc, M. et al. (2001). Science on the edge of spatial scales: a reply to the comments of Williams (2001). *Canadian Journal of Fisheries and Aquatic Sciences* 58: 2108–2111.

Guay, J.C., Boisclair, D., Leclerc, M., and Lapointe, M. (2003). Assessment of the transferability of biological habitat models for Atlantic salmon parr (*Salmo salar*). *Canadian Journal of Fisheries and Aquatic Sciences* 60: 1398–1408.

Guay, J.C., Boisclair, D., Rioux, D. et al. (2000). Development and validation of numerical habitat models for juveniles of Atlantic salmon (*Salmo salar*). *Canadian Journal of Fisheries and Aquatic Sciences* 57: 2065–2075.

Guillermo, R.G. and Healey, M.C. (1999). Ideal free distribution theory as a tool to examine juvenile coho salmon (*Oncorhynchus kisutch*) habitat choice under different conditions of food, abundance and cover. *Canadian Journal of Fisheries and Aquatic Sciences* 56: 2362–2373.

Gurevitch, J. and Hedges, L.V. (2001). Meta–analysis: combining the results of independent experiments. In: *Design and Analysis of Ecological Experiments*, 2e (ed. S.M. Scheiner and J. Gurevitch), 347–369. New York: Oxford University Press.

Halls, A.S., Paxton, R.B., Hall, N. et al. (2013). *The stationary trawl (Dai) fishery of the Tonle sap–Great Lake, Cambodia*, MRC Technical Paper No. 32. Phnom Penh, Cambodia: Mekong River Commission.

Halls, A.S. and Moyle, P.B. (2018). Comment on "Designing river flows to improve food security futures.". *Science* 361 (6398): eaat1989.

Hämäläinen, R.P. (2015). Behavioral issues in environmental modeling – the missing perspective. *Environmental Modeling and Software* 73: 244–253.

Hamilton, D. A. and Seelbach, P. W. 2011. Michigan's water withdrawal assessment process and internet screening tool. *Fisheries Special Report 55*. Michigan Department of Natural Resources.

Hanea, A., McBride, M., Burgman, M. et al. (2017). Investigate discuss estimate aggregate for structured expert judgement. *International Journal of Forecasting* 33: 267–279.

Hardy, T.B. and Addley, R.C. (2001). Vertical integration of spatial and hydraulic data for improved habitat modeling using geographic information systems. In: *Hydroecology: Linking Hydrology and Aquatic Ecology* (ed. M.C. Acreman), 65–76. Proceedings of the Birmingham, United Kingdom Workshop, July 1999 IAHS Publication No. 266. Oxfordshire, UK: IAHS Press.

Hardy, T.B., Addley, R.C., and Saraeva, E. (2006). *Evaluation of Instream Flow Needs in the Lower Klamath River, Phase II, Final Report*. Logan, UT: Institute for Natural Systems Engineering, Utah State University.

Harris, G.P. and Heathwaite, A.L. (2012). Why is achieving good ecological outcomes in rivers so difficult? *Freshwater Biology* 57 (Suppl. 1): 91–107.

Harrison, L.R., Legleiter, C.J., Wydzga, M.A., and Dunne, T. (2011). Channel dynamics and habitat development in a meandering, gravel bed river. *Water Resources Research* 47: W04513.

Hart, B.T. (2016). The Australian Murray–Darling Basin Plan: challenges in its implementation (part 1). *International Journal of Water Resources Development* 32: 819–834.

Hart, B.T. and Pollino, C.A. (2009). *Bayesian Modeling for Risk-based Assessment of Environmental Water Allocation*. Canberra: National Water Commission.

Harvey, B.C. and Railsback, S.F. (2007). Estimating multi-factor cumulative watershed effects on fish populations with an individual-based model. *Fisheries* 32: 292–298.

Harvey, B.C. and Railsback, S.F. (2014). Feeding models in stream salmonid population models: is drift feeding the whole story? *Environmental Biology of Fishes* 97: 615–625.

Harvey, G.L. and Clifford, N.J. (2009). Microscale hydrodynamics and coherent flow structures in rivers: implications for the characterization of physical habitat. *River Research and Applications* 25: 160–180.

Hassan, M.A. and Roy, A.G. (2016). Coarse particle tracing in fluvial geomorphology. In: *Tools in Fluvial Geomorphology*, 2e (ed. G.M. Kondolf and H. Piégay), 306–323. Chichester, UK: John Wiley and Sons.

Hatfield, T., Lewis, A., Olson, D. and Bradford, M. J. 2003. *Development of Instream Flow Thresholds as Guidelines for Reviewing Proposed Water Uses*. Consultants' report for British Columbia Ministry of Sustainable Resource Management and British Columbia Ministry of Water, Land and Air Protection. Victoria, BC.

Hawkins, C.P., Kershner, J.L., Bisson, P.A. et al. (1993). A hierarchical approach to classifying stream habitat features. *Fisheries* 18: 3–12.

Hayes, J.W., Goodwin, E., Shearer, K.A. et al. (2016). Can weighted useable area predict flow requirements of drift-feeding salmonids? Comparison with a net rate of energy intake model incorporating drift-flow processes. *Transactions of the American Fisheries Society* 145: 589–609.

Hayes, J.W., Hughes, N.F., and Kelly, L.H. (2007). Process-based modeling of invertebrate drift transport, net energy intake and reach carrying capacity for drift-feeding salmonids. *Ecological Modelling* 207: 171–188.

Hazel, C.R. (1976). The reservation of instream flow for fish in California – a case study. In: *Proceedings of the Symposium and Specialty Conference on Instream Flow Needs* (ed. J.F. Orsborn and C.H. Allman), 33–55. Boise, ID: American Fisheries Society.

Healey, M.C. (1998). Paradigms, policies, and prognostications about the management of watershed ecosystems. In: *River Ecology and Management: Lessons from Pacific Coastal Ecosystems*, (ed. R.J. Naiman and R.E. Bilby), 642–661. New York: Springer.

Healey, M.C., Dettinger, M.D., and Norgaard, R.B. (eds.) (2008). *State of the Bay-delta Science.* Sacramento, CA: CALFED Science Program.

Hecht, B. (1994). South of the spotted owl: restoration strategies for episodic channels and riparian corridors in Central California. In: *Western Wetlands, Selected Proceedings of the 1993 Conference of the Society of Wetland Scientists* (ed. D.M. Kent and J.J. Zentner), 104–117. Davis, CA: University of California.

Heggenes, J. 1994. Physical habitat selection by brown trout (Salmo trutta) and young Atlantic salmon (S. salar) in spatially and temporally heterogeneous streams: implications for hydraulic modeling. Pages 12–30 in Proceedings of the 1st International Conference on Habitat Hydraulics. Trondheim, Norway, International Association of Hydraulic Research.

Heggenes, J. (1996). Habitat selection by brown trout (*Salmo trutta*) and young Atlantic salmon (*S. salar*) in streams: static and dynamic hydraulic modeling. *Regulated Rivers: Research and Management* 12: 155–169.

Heggenes, J. (2002). Flexible summer habitat selection by wild, allopatric brown trout in lotic environments. *Transactions of the American Fisheries Society* 131: 287–298.

Helsel, D.R. and Hirsch, R.M. (1992). *Statistical Methods in Water Resources.* New York: Elsevier.

Hering, D., Borja, A., Carstensen, J. et al. (2010). The European Water Framework Directive at the age of 10: a critical review of achievements with recommendations for the future. *Science of the Total Environment* 408: 4007–4019.

Hicks, DM and Goring, D. 2009. Modifying flushing flows in the Moawhango River to mitigate downstream effects. National Institute of Water and Atmospheric Research Ltd, Christchurch, New Zealand. NIWA client report no. CHC2009-164.

Hilborn, R. and Mangel, M. (1997). *The Ecological Detective: Confronting Models with Data.* Princeton, NJ: Princeton University Press.

Hilborn, R., Quinn, T.P., Schindler, D.E., and Rogers, D.E. (2003). Biocomplexity and fisheries sustainability. *Proceedings of the National Academy of Science* 100: 6564–6568.

Hill, J. and Grossman, G.D. (1993). An energetic model of microhabitat use for rainbow trout and rosyside dace. *Ecology* 74: 685–698.

Hill, M.T., Platts, W.S., and Beschta, R.L. (1991). Ecological and geomorphological concepts for instream and out-of-channel flow requirements. *Rivers* 3 (1): 198–210.

Hirzel, A.H. and Le Lay, G. (2008). Habitat suitability modeling and niche theory. *Journal of Applied Ecology* 45: 1372–1381.

Hoekstra, A.Y., Chapagain, A.K., Aldaya, M.M., and Mekonnen, M.M. (2011). *The Water Footprint Assessment Manual: Setting the Global Standard.* London: Earthscan.

Hogarth, W. (1753). *The Analysis of Beauty.* (reprinted 1997). New Haven, CT: Yale University Press.

Holm, C.F., Armstrong, J., and Gilvear, D.J. (2001). Investigating a major assumption of predictive instream habitat models: is water velocity preference of juvenile Atlantic salmon independent of discharge? *Journal of Fish Biology* 59: 1653–1666.

Horne, A.C., Szemis, J.M., Webb, J.A. et al. (2018). Informing environmental water management decisions: using conditional probability networks to address the information needs of planning and implementation cycles. *Environmental Management* 61: 347–357.

Hrachowitz, M., Aavenije, H.H.G., Blöschl, G. et al. (2013). A decade of predictions in ungauged basins (PUB) – a review. *Hydrological Science Journal* 58: 1–58.

Hubert, W.A. and Rahel, F.J. (1989). Relations of physical habitat to abundance of four nongame fishes in High Plains streams: a test of habitat suitability index models. *North American Journal of Fisheries Management* 9: 332–340.

Hughes, F.M.R. and Rood, S.B. (2003). Allocation of river flows for restoration of floodplain forest ecosystems: a review of approaches and their applicability in Europe. *Environmental Management* 32 (1): 12–33.

Hughes, N.F. and Dill, L.M. (1990). Position choice by drift-feeding salmonids: model and test for arctic grayling (*Thymallus arcticus*) in subarctic mountain streams, interior Alaska. *Canadian Journal of Fisheries and Aquatic Sciences* 47: 2039–2048.

Hughes, N.F., Hayes, J.W., Shearer, K.A., and Young, R.G. (2003). Testing a model of drift-feeding using three-dimensional videography of wild brown trout, *Salmo trutta*, in a New Zealand river. *Canadian Journal of Fisheries and Aquatic Sciences* 60: 1462–1476.

Imre, I.J. and Boisclair, D. (2005). Moon phase and nocturnal density of Atlantic salmon parr in the Sainte-Marguerite River, Quebec. *Journal of Fish Biology* 66: 198–207.

Imre, I.J., Grant, J.W.A., and Cunjak, R.A. (2010). Density-dependent growth of young-of-the-year Atlantic salmon (*Salmo salar*) revisited. *Ecology of Freshwater Fishes* 19: 1–6.

Ioannidis, J.P.A. (2005). Why most published research findings are false. *PloS Medicine* 2: e124.

Jakeman, A.J., Letcher, R.A., and Norton, J.P. (2006). Ten iterative steps in development and evaluation of environmental models. *Environmental Modeling and Software* 21: 602–614.

Jakob, C., Robinson, C.T., and Uehlinger, U. (2003). Longitudinal effects of experimental floods on stream benthos downstream from a large dam. *Aquatic Sciences* 65 (3): 223–231. https://doi.org/10.1007/s00027-003-0662-9.

James, L.A., Watson, D.G., and Hansen, W.F. (2007). Using LiDAR data to map gullies and headwater streams under forest canopy: South Carolina, USA. *Catena* 71: 132–144.

Jeffres, C.A., Opperman, J.J., and Moyle, P.B. (2008). Ephemeral floodplain habitats provide best growth conditions for juvenile Chinook salmon in a California river. *Environmental Biology of Fishes* 83: 449–458.

Jenkins, T.M. Jr. (1969). Social structure, position choice and microdistribution of two trout species (*Salmo trutta* and *Salmo gairdneri*) resident in mountain streams. *Animal Behavior Monographs* 2: 56–123.

Jenkins, T.M., Diehl, S., Dratz, K.W., and Cooper, S.D. (1999). Effects of population density on individual growth of brown trout in streams. *Ecology* 80: 941–956.

Jesson, R.J. (1978). *Statistical Survey Techniques*. New York: John Wiley and Sons.

Johnson, C.J., Nielsen, S., Merrill, E.H. et al. (2006). Resource selection functions based on use–availability data: theoretical motivation and evaluation methods. *Journal of Wildlife Management* 70: 347–357.

Johnson, D.H. (1999). The insignificance of statistical significance testing. *Journal of Wildlife Management* 63: 763–772.

Johnson, M. 2017. One of the largest dam removals in California history inches forward. Online blog "Water Deeply," published August 7, 2017. (https://www.newsdeeply.com/water/articles/2017/08/07/one-of-the-largest-dam-removals-in-california-history-inches-forward, accessed October 31, 2017).

Jones and Stokes Associates. 2003, Water Quality and Aquatic Resources Monitoring Program for the Ralston Afterbay Sediment Management Project – 2002 Annual Report, Report to Placer County Water Agency. Sacramento, CA: Jones and Stokes.

Jourdain, C. 2017. *Action des crues sur la dynamique sédimentaire et végétale dans un lit de rivière à galets: L'Isère en Combe de Savoie*. Thèse de doctorat en Océan, atmosphère, hydrologie. Université de Grenoble, France.

Jowett, I.G. (1982). Models of the abundance of large brown trout in New Zealand rivers. *North American Journal of Fisheries Management* 12: 417–432.

Jowett, I.G. (1998). Hydraulic geometry of New Zealand rivers and its use as a preliminary method of habitat assessment. *Regulated Rivers: Research and Management* 14: 451–466.

Jowett, I.G. (2002). *RHYBABSIM: River Hydraulic and Habitat Simulation*. Hamilton, New Zealand: National Institute of Water and Atmospheric Research.

Junk, W.J. (1997). General aspects of floodplain ecology with special reference to Amazonian floodplains. In: *The Central Amazon Foodplain: Ecology of a Pulsing System* (ed. W.J. Junk), 2–22. Berlin: Springer-Verlag.

Junk, W.J., Bayley, P.B., and Sparks, R.E. (1989). The flood pulse concept in river–floodplain systems. In: *Proceedings of the International Large River Symposium*, Canadian Special Publication of Fisheries and Aquatic Sciences 106 (ed. D.P. Dodge), 110–127. Honey Harbor, ON: Department of Fisheries and Oceans.

Kantoush, S. A. and Sumi, T. 2010. River morphology and sediment management strategies for sustainable reservoir in Japan and European Alps, *Annuals of Disaster Prevention Research Institute* No. 53B, Kyoto University.

Kantoush, S. A. and Sumi, T.. 2011. Sediment replenishing measures for revitalization of Japanese rivers below dams. *In: Proceedings of the 34thIAHR World Congress*, Brisbane, July 2011.

Katz, J.V.E., Jeffres, C., Conrad, J.L. et al. (2017). Floodplain farm fields provide novel rearing habitat for Chinook salmon. *Plos One* 12 (6): https://doi.org/10.1371/journal.pone.0177409.

Kelley, D.W. (ed.) (1965). *Ecological Studies of the Sacramento–San Joaquin Estuary Part 1: Zooplankton, Zoobenthos, and Fishes of the San Pablo and Suisun Bays, Zooplankton and Zoobenthos of the Delta, Fish Bulletin* 133. Sacramento, CA: California Department of Fish and Game.

Kelley, D. W., Cordone, A. J. and Delisle, G. 1960. A method to determine the volume of flow required by trout below dams: a proposal for investigations. *Unpublished report.* Sacramento: California Department of Fish and Game.

Kelley, R. (1989). *Battling the Inland Sea*, 395. Berkeley: University of California Press.

Kellner, K.F. and Swihart, R.K. (2014). Accounting for imperfect detection in ecology: a quantitative review. *PLoS One* 9: https://doi.org/10.1371/journal.pone.0111436.

Kemp, J.L., Harper, D.M., and Crosa, G.A. (1999). Use of "functional habitats" to link ecology with morphology and hydrology in river rehabilitation. *Aquatic Conservation: Marine and Freshwater Ecosystems* 9: 159–178.

Kendy, E., Apse, C.D., Blann, K. et al. (2012). *A Practical Guide to Environmental Flows for Policy and Planning*. Arlington, VA: The Nature Conservancy.

Kendy, E., Sanderson, J.S., Olden, J.D. et al. (2009). *Applications of the Ecological Limits of Hydrologic Alteration (ELOHA) in the United States*. Arlington, VA: The Nature Conservancy.

Kennard, M.J., MacKay, S.J., Pusey, B.J. et al. (2010a). Quantifying uncertainty in estimation of hydrologic metrics for ecohydrological studies. *River Research and Applications* 26: 137–156.

Kennard, M.J., Pusey, B.J., Olden, J.D. et al. (2010b). Classification of natural flow regimes in Australia to support environmental flow management. *Freshwater Biology* 55: 171–193.

Kerr, J.R., Manes, C., and Kemp, P.S. (2016). Assessing hydrodynamic space use of brown trout, *Salmo trutta*, in a complex flow environment: a return to

first principles. *Journal of Experimental Biology* 219: 3480–3491.

Khan, K.S., Kunz, R., Kleijnen, J., and Antes, G. (2003). Five steps to conducting a systematic review. *Journal of the Royal Society of Medicine* 96: 118–121.

Kiernan, J.D., Moyle, P.B., and Crain, P.K. (2012). Restoring native fish assemblages to a regulated California stream using the natural flow regime concept. *Ecological Applications* 22: 1472–1482.

King, J. and Brown, C. (2006). Environmental flows: striking the balance between development and resource protection. *Ecology and Society* 11: 26. (online).

King, J., Brown, C., and Sabet, H. (2003). A scenario-based holistic approach to environmental flow assessments for rivers. *River Research and Applications* 19: 619–639.

King J. M., Brown, C. A., Paxton, B. R. and February, R. J. 2004. Development of DRIFT, a scenario-based methodology for environmental flow assessments. *WRC Report 1159/1/04*. Rondebosch, South Africa.

King, J.M., Tharme, R.E., and De Villiers, M.S. (2008). *Manual for the Building Block Methodology (updated version)*, Water Research Commission Report No. TT 354/08. Cape Town, South Africa: Freshwater Research Unit, University of Cape Town.

Kinzel, P.J. (2009). Advanced tools for river science: EAARL and MD_SWMS. In: *PNMAP Special Publication: Remote Sensing Applications for Aquatic Resources Monitoring* (ed. J.M. Bayer and J.L. Schei), 17–26. Cook, WA: Pacific Northwest Aquatic Monitoring Partnership.

Knapp, R.A. and Preisler, H.K. (1999). Is it possible to predict habitat use by spawning salmonids? A test using California golden trout (*Oncorhynchus mykiss aguabonita*). *Canadian Journal of Fisheries and Aquatic Sciences* 56: 1576–1584.

Knight, R.R., Gregory, M.B., and Wales, A.K. (2008). Relating streamflow characteristics to specialized insectivores in the Tennessee River Valley: a regional approach. *Ecohydrology* 1: 394–407.

Kobayashi, S., Takemon, Y. and Sumi, T. 2016. Formation of gravel bars and changes in invertebrate habitats by sediment deposition during a dam removal on the Kuma River, Japan. *In: Proceedings of the 7th ICWRER Conference*, Kyoto, Japan, June 2016, pp 13-1–13-2.

Kondolf, G.M. (1997). Hungry water: effects of dams and gravel mining on river channels. *Environmental Management* 21 (4): 533–551.

Kondolf, G.M. (1998). Development of flushing flows for channel restoration on Rush Creek, California. *Rivers* 6 (3): 183–193.

Kondolf, G.M. (2006). River restoration and meanders. *Ecology and Society* 11: Article 42 (online).

Kondolf, G.M. and Farahani, A. (2018). Sustainably managing reservoir storage: ancient roots of a modern challenge. *Water* 10 (2): 117. https://doi.org/10.3390/w10020117.

Kondolf, G.M. and Wilcock, P.R. (1996). The flushing flow problem: defining and evaluating objectives. *Water Resources Research* 32 (8): 2589–2599.

Kondolf, G.M., Boulton, A., O'Daniel, S. et al. (2006). Process-based ecological river restoration: visualising three-dimensional connectivity and dynamic vectors to recover lost linkages. *Ecology and Society* 11 (2): 5. URL: http://www.ecologyandsociety.org/vol11/iss2/art5.

Kondolf, G. M., Gao, Y., Annandale, G. W., Morris, G. L., Jiang, E. et al. 2014a. Sustainable sediment management in reservoirs and regulated rivers: experiences from five continents. *Earth's Future* doi: https://doi.org/10.1002/eft2 2013EF000184 Online at http://onlinelibrary.wiley.com/doi/10.1002/2013EF000184/pdf

Kondolf, G.M., Larsen, E.W., and Williams, J.G. (2000). Measuring and modeling the hydraulic environment for assessing instream flows. *North American Journal of Fisheries Management* 20: 1016–1028.

Kondolf, G.M., Piégay, H., Schmitt, L., and Montgomery, D.R. (2016). Geomorphic classification of rivers and streams. In: *Tools in Fluvial Geomorphology* (ed. G.M. Kondolf and H. Piégay), 171–204. Chichester, UK: John Wiley and Sons.

Kondolf, G.M., Podolak, K., and Grantham, T.E. (2013). Restoring Mediterranean-climate rivers. *Hydrobiologia* 719: 527–545.

Kondolf, G.M., Rubin, Z.K., and Minear, J.T. (2014b). Dams on the Mekong: cumulative sediment starvation. *Water Resources Research* 50: https://doi.org/10.1002/2013WR014651.

Kondolf, G.M., Vick, J.C., and Ramirez, T.M. (1996). Salmon spawning habitat rehabilitation on the Merced River, California: an evaluation of project planning and performance. *Transactions of the American Fisheries Society* 125: 899–912.

Konrad, C. P. 2003. Effects of Urban Development on Floods. *US Geological Survey Fact Sheet FS-076-03*. Available online at https://pubs.usgs.gov/fs/fs07603/pdf/fs07603.pdf accessed October 2017.

Konrad, C. P. and Booth, D. B. 2002. *Hydrologic Trends Associated with Urban Development for Selected Streams in the Puget Sound Basin, Western Washington*. U.S. Geological Survey Water-Resources Investigations Report 02-4040.

Konrad, C.P., Brasher, A.M.D., and May, J.T. (2008). Assessing streamflow characteristics as limiting factors on benthic invertebrate assemblages in streams across the western United States. *Freshwater Biology* 53: 1983–1998.

Konrad, C.P., Olden, J.D., Lytle, D.A. et al. (2011). Large-scale flow experiments for managing river systems. *BioScience* 61 (12): 48–959.

Korman, J. and Higgins, P.S. (1997). Utility of escapement time series data for monitoring the response of salmon populations to habitat alteration. *Canadian Journal of Fisheries and Aquatic Sciences* 54: 2058–2067.

Korman, J.C., Perrin, C.J., and Lekstrum, T. (1994). *A Guide for the Selection of Standard Methods for Quantifying Sportfish Habitat Capability and Suitability in Streams and Lakes of British Columbia*. Vancouver, BC: Limnotek Research and Development.

Krieger, L. M. 2015. Stanford announces future of Searsville Dam. *San Jose Mercury News* (May 1, 2015).

Kuhnert, P.M., Martin, T.G., and Griffiths, S.P. (2010). A guide to eliciting and using expert knowledge in Bayesian ecological models. *Ecology Letters* 13: 900–914.

Lacey, R.W.J., Neary, V.S., Liao, J.C. et al. (2012). The IPOS framework: linking fish swimming performance in altered flows from laboratory experiments to rivers. *River Research and Applications* 28: 429–443.

Lahoz-Monfort, J.J., Guillera-Arroita, G., and Wintle, B.A. (2014). Imperfect detection impacts the performance of species distribution models. *Global Ecology and Biogeography* 23: 504–515.

Lamouroux, N. and Capra, H. (2002). Simple predictions of instream habitat model outputs for target fish populations. *Freshwater Biology* 47: 1543–1556.

Lamouroux, N. and Jowett, I.G. (2005). Generalized instream habitat models. *Canadian Journal of Fisheries and Aquatic Sciences* 62: 7–14.

Lamouroux, N., Mégigoux, S., Capra, H. et al. (2010). The generality of abundance–environment relationships in microhabitats: a comment on Lancaster and Downes (2009). *River Research and Applications* 26: 915–920.

Lamouroux, N. and Olivier, J.-M. (2015). Testing predictions of changes in flow abundance and community structure after flow restoration in four reaches of a large river (French Rhone). *Freshwater Biology* 60: 1118–1130.

Lamouroux, N. and Souchon, Y. (2002). Simple predictions of instream habitat model outputs for fish habitat guilds in large rivers. *Freshwater Biology* 47: 1531–1542.

Lancaster, J. and Downes, B.J. (2010a). Linking the hydraulic world of individual organisms to ecological processes: putting ecology into ecohydraulics. *River Research and Applications* 26: 385–403.

Lancaster, J. and Downes, B.J. (2010b). Ecohydraulics needs to embrace ecology and sound science, and to avoid mathematical artefacts. *River Research and Applications* 26: 921–929.

Lane, B.A., Dahlke, H.E., Pasternack, G.B., and Sandoval-Solis, S. (2017). Revealing the diversity of natural hydrologic regimes in California with relevance for environmental flows applications. *Journal of the American Water Resources Association* 53 (2): 411–430.

Lane, S.N. (1998). Hydraulic modeling in hydrology and geomorphology: a review of high resolution approaches. *Hydrological Processes* 12: 1131–1150.

Lapointe, M. (2012). River geomorphology and salmonid habitat: some examples illustrating their complex association, from redd to riverscape scales. In: *Gravel-bed Rivers: Processes, Tools, Environments* (ed. M. Church, P.M. Brion and A.G. Roy), 193–215. Oxford: Wiley-Blackwell.

Leclerc, M., Boudreault, A., Bechara, T.A., and Corfa, G. (1995). Two-dimensional hydrodynamic modeling: a neglected tool in the instream flow incremental methodology. *Transactions of the American Fisheries Society* 124: 645–662.

Lee, K. (1994). *Compass and Gyroscope: Integrating Science and Politics for the Environment*. Washington, DC: Island Press.

Legleiter, C.J., Overstreet, B.T., Glennie, C.L. et al. (2016). Evaluating the capabilities of the CASI hyperspectral imaging system and Aquarius bathymetric LiDAR for measuring channel morphology in two distinct river environments. *Earth Surface Processes and Landforms* 41: 344–363.

Leopold, L.B. and Maddock, T. Jr. (1953). *The Hydraulic geometry of stream channels and some physiographic implications*, Geological Survey Professional Paper 252. Washington, DC: U.S. Government Printing Office.

Leslie, J. 2017. Four dams in the West are coming down – a victory wrapped in a defeat for smart water policy. *Los Angeles Times* (November 2, 2017).

Lester, R. E., Pollino, C. A. and Cummings, C. R. 2011a. Improving ecological outcomes by refining decision support tools: a case study using the Murray Flow Assessment Tool and the Sustainable Rivers Audit. 19th International Congress on Modeling and Simulation, 2011. Perth, Australia, December 12–16, 2011.

Lester, R.E., Webster, I.T., Fairweather, P.G., and Young, W.J. (2011b). Linking water-resource models to ecosystem-response models to guide water-resource planning – an example from the Murray–Darling Basin, Australia. *Marine and Freshwater Research* 62: 279–289.

Levins, R. (1966). The strategy of model building in population biology. *American Scientist* 54: 421–431.

Lichatowich, J. (1998). *A Conceptual Foundation for the Management of Native Salmonids in the Deschutes River*. Portland, OR: Portland General Electric Company.

Lindley, S.T., Grimes, C.B., Mohr, M.S. et al. (2009). *What caused the Sacramento River fall Chinook stock collapse?* NOAA Technical Memorandum NMFS. Long Beach, CA: National Marine Fisheries Service Southwest Fisheries Science Center.

Linnansaari, T., Monk, W.A., Baird, D.J., and Curry, R.A. (2012). *Review of approaches and methods to assess environmental flows across Canada and internationally*, Research Document 2012/039. Ottawa: Canadian Science Advisory Secretariat, Fisheries and Oceans.

Littleworth, A.L. and Garner, E.L. (1995). *California Water*. Point Arena, CA: Solano Press.

Lobo, J.M., Jiménez-Valverde, A., and Real, R. (2008). AUC: a misleading measure of the performance of predictive distribution models. *Global Ecology and Biogeography* 17: 145–151.

Locke, A., Stalnaker, C., Zellner, S. et al. (2008). *Integrated Approaches to Riverine Resource Stewardship: Case Studies, Science, Law, People, and Policy*. Cheyenne, WY, USA: Instream Flow Council.

Loire, R., Piégay, H., Malavoi, J.R., and Kondolf, G.M. From environmental to morphogenic flows: evolving terminology, practice, and integration into management of regulated rivers. (2019, in review)

Lowe, L., Szemis, J., and Webb, J.A. (2017). Uncertainty and environmental water. In: *Water for the Environment: From Policy and Science to Implementation and Management* (ed. A.C. Horne, J.A. Webb, M.J. Stewardson, et al.), 317–344. Cambridge MA: Elsevier.

Ludwig, D., Hilborn, R., and Walters, C.J. (1993). Uncertainty, resource exploitation, and conservation: lessons from history. *Science* 260: 17 and 36.

Ludwig, D. and Walters, C.J. (1985). Are age-structured models appropriate for catch and effort data? *Canadian Journal of Fisheries and Aquatic Sciences* 42: 1066–1072.

Lusardi, R.A., Bogan, M.T., Moyle, P.B., and Dahlgren, R.A. (2016). Environment shapes invertebrate assemblage structure differences between volcanic springfed and runoff rivers in northern California. *Freshwater Science Science* 35: 1010–1022.

MacKay, S.J., Arthington, A.H., and James, C.S. (2014). Classification and comparison of natural and altered flow regimes to support an Australian trial of the Ecological Limits of Hydrologic Alteration framework. *Ecography* 7: 1485–1507.

MacKenzie, D.I., Nichols, J.D., Lachman, G.B. et al. (2002). Estimating site occupancy rates when detection probabilities are less than one. *Ecology* 83: 2248–2255.

MacKenzie, D.I., Nichols, J.D., Royle, J.A. et al. (2005). *Occupancy Estimation and Modeling: Inferring Patterns and Dynamics of Species Occurrence*. New York: Academic Press.

MacKenzie, D.I., Nichols, J.D., Seamons, M.E., and Gutierrez, R.J. (2009). Modeling species occurrence dynamics with multiple states and imperfect detection. *Ecology* 90: 823–835.

Magirl, C.S., Hilldale, R.C., Curran, C.C., and Duda, J.J. (2015). Large-scale dam removal on the Elwha River, Washington, USA: fluvial sediment load. *Geomorphology* 246: 669–686.

Mangel, M. and Satterthwaite, W.H. (2008). Combining proximate and ultimate approaches to understand life history variation in salmonids with application to fisheries, conservation, and aquaculture. *Bulletin of Marine Science* 107: 107–130.

Mangel, M., Talbot, L.M., Meffe, G.K. et al. (1996). Principles for the conservation of wild living resources. *Ecological Applications* 6: 338–362.

Manly, B.J. (2007). *Randomization, Bootstrap and Monte Carlo Methods in Biology*. Boca Raton, FL: Chapman and Hall.

Manly, B.J., Mcdonald, L.L., Thomas, D.L., and Erickson, W.P. (2002). *Resource Selection by Animals*. Boston: Kluwer Academic Publishers.

Manly, B.J. and Navarro Alberto, J.A. (2015). *Introduction to Ecological Sampling*. Boca Raton, FL: Chapman and Hall/ CRC Press.

Mann, J. L. 2006. *Instream Flow Methodologies: an Evaluation of the Tennant Method for Higher Gradient Streams in the National Forest System Lands in the Western U.S.* MS thesis, Colorado State University.

Marcot, B.G. (2012). Metrics for evaluating performance and uncertainty of Bayesian network models. *Ecological Modelling* 230: 50–62.

Marcot, B.G., Holthausent, R.S., Raphael, M.G. et al. (2001). Using Bayesian belief networks to evaluate fish and wildlife population viability under land management alternatives from an environmental impact statement. *Forest Ecology and Management* 153: 29–42.

Marcot, B.G., Steventon, J.D., Sutherland, G.D., and McCann, R.K. (2006). Guidelines for developing and updating Bayesian belief networks applied to ecological modeling and conservation. *Canadian Journal of Fisheries and Aquatic Sciences* 36: 3063–3074.

Marsili-Libeli, S., Giusi, E., and Nocita, A. (2013). A new instream flow assessment method: fuzzy habitat suitability and large-scale river modeling. *Environmental Modeling and Software* 41: 27–38.

Martin, T.G., Burgman, M.A., Fidler, F. et al. (2012). Eliciting expert knowledge in conservation science. *Conservation Biology* 26: 29–38.

Martin, T.G., Kuhnert, P.M., Mengersen, K., and Possingham, H.P. (2005). The power of expert opinion in ecological models using Bayesian methods: impact of grazing on birds. *Ecological Applications* 15: 266–280.

Mathur, D., Bason, W.H., Purdym, E.J. Jr., and Silver, C.A. (1985). *Canadian Journal of Fisheries and Aquatic Sciences* 42: 825–831.

May, C.L., Pryor, B., Lisle, T.E., and Lang, M. (2009). Coupling hydrodynamic modeling and empirical measures of bed mobility to predict the risk of scour and fill of salmon redds in a large regulated river. *Water Resources Research* 45: https://doi.org/10.1029/2007WR006498.

May, R.M. (2004). Uses and abuses of mathematics in biology. *Science* 303: 790–793.

McBride, M., Fidler, F., and Burgman, M.A. (2012). Evaluating the accuracy and calibration of expert predictions under uncertainty: predicting the outcomes of ecological research. *Diversity and Distributions* 18: 782–794.

McCann, R.K., Marcot, B.G., and Ellis, R. (2006). Bayesian belief networks: applications in ecology and natural resource management. *Canadian Journal of Forest Research* 36: 3053–3062.

McCarthy, M.A. (2007). *Bayesian Methods for Ecology*. Cambridge: Cambridge University Press.

McCarthy, M.A. and Masters, P. (2005). Profiting from prior information in Bayesian analyses of ecological data. *Journal of Applied Ecology* 42: 1012–1019.

Mckean, J.A., Isaak, D.J., and Wright, C.W. (2008). Geomorphic controls on salmon nesting patterns described by a new, narrow-beam terrestrial–aquatic lidar. *Frontiers in Ecology and the Environment* 6: 125–130.

McMahon, T.A. (1989). Understanding Australian streamflow – implications for instream ecology. In: *Proceedings of the Specialist Workshop on Instream Flow Needs and Water Uses* (ed. C. Teoh), 33-1–33-11. Canberra: Australian Water Resources Council.

McMahon, T.A. (1986). Hydrology and management of Australian streams. In: *Stream Protection: The Management of Stream for Instream Uses* (ed. I.C. Campbell), 23–44. East Caulfield, Victoria: Water Studies Centre, Chisholm Institute of Technology.

McManamay, R.A., Orth, D.J., Dolloff, C.A., and Mathews, D.C. (2013). Application of the ELOHA framework to regulated rivers in the upper Tennessee River basin: a case study. *Environmental Management* 51: 1210–1235.

Melis, T.S.E. (2011). Effects of three high-flow experiments on the Colorado River ecosystem downstream from Glen Canyon Dam, Arizona. *U.S. Geological Survey Circular* 1366 (February).

Menchen, R.S. (1978). *A Description of California Department of Fish and Game Management Program for the San Joaquin River System Salmon Resource*. Sacramento, CA: California Department of Fish and Game, Anadromous Fisheries Branch.

Metcalfe, N.B., Fraser, N.H.C., and Burns, M.D. (1998). State-dependent shifts between nocturnal and diurnal activity in salmon. *Proceedings of the Royal Society Series B* 265: 1503–1507.

Miller, J.R. and Kochel, R.C. (2010). Assessment of channel dynamics, in-stream structures and post-project channel adjustments in North Carolina and its implications to effective stream restoration. *Environmental Earth Sciences* 59: 1681–1692.

Miller, K.A., Webb, J.A., de Little, S.C., and Stewardson, M.J. (2013). Environmental flows can reduce the encroachment of terrestrial vegetation into river channels: a systematic literature review. *Environmental Management* 52: 1201–1212.

Milly, P.C.D., Betancourt, J., Falenmark, M. et al. (2008). Stationarity is dead: whither water management? *Science* 319: 573–574.

Modde, T. and Hardy, T.B. (1992). Influence of different microhabitat criteria on salmonid habitat simulation. *Rivers* 3: 37–44.

Moir, H.J., Gibbins, C.N., Soulsby, C., and Webb, J.H. (2006). Discharge and hydraulic interactions in contrasting channel morphologies and their influence on site utilization by spawning Atlantic salmon (*Salmo salar*). *Canadian Journal of Fisheries and Aquatic Sciences* 63: 2567–2585.

Monk, W.A., Peters, D.L., Curry, R.A., and Baird, D.J. (2011). Quantifying trends in indicator hydroecological variables for regime-based groups of Canadian rivers. *Hydrological Processes* 25 (19): 3086–3100.

Montgomery, D.R. and Bolton, S. (2003). Hydrogeomorphic variability and river restoration. In: *Strategies for Restoring River Ecosystems: Sources of Variability and Uncertainty in Natural and Managed Systems* (ed. R.C. Wissmar and P.A. Bisson), 39–80. Bethesda, MD: American Fisheries Society.

Morel-Seytoux, H.J. (2001). Groundwater. In: *Model Validation: Perspectives in Hydrological Science*, (ed. M.G. Anderson and P.D. Bates), 183–190. Chichester, UK: John Wiley and Sons.

Morgan, M.G. (2014). Use (and abuse) of expert elicitation in support of decision making for public policy. *Proceedings of the National Academy of Science* 111: 7176–7184.

Morhardt, J.E. (1988). Behavioral carrying capacity as a possible short term response variable. *Hydro Review* 7: 32–40.

Morris, G.L. and Fan, J. (1998). *Reservoir Sedimentation Handbook: Design and Management of Dams, Reservoirs and Watersheds for Sustainable Use.* New York: McGraw-Hill.

Mouton, A., De Baets, B., and Goethals, P. (2013). Data-driven fuzzy habitat models: impact of performance criteria and opportunities for ecohydraulics. In: *Ecohydraulics: An Integrated Approach* (ed. I. Maddock, A. Harby, P. Kemp and P. Wood), 93–105. Oxford: Wiley Blackwell.

Moyle, P.B. (2002). *Inland Fishes of California.* Berkeley: University of California Press.

Moyle, P.B. and Cech, J.J. (2003). *Fishes: An Introduction to Ichthyology.* Harlow, UK: Pearson.

Moyle, P.B. and Vondracek, B. (1985). Persistence and structure of the fish assemblage in a small California stream. *Ecology* 66: 1–13.

Moyle, P.B., Marchetti, P., Baldridge, J., and Taylor, T.L. (1998). Fish health and diversity: justifying flows for a California stream. *Fisheries* 23: 6–15.

Moyle, P. B., Williams, J. G. and Kirnan, J. D. 2011. *Improving environmental flow methods used in California FERC licensing.* CEC-201-037. Sacramento. California Energy Commission.

Mürle, U., Ortlepp, J., and Zahner, M. (2003). Effects of experimental flooding on riverine morphology, structure and riparian vegetation: the River Spöl, Swiss National Park. *Aquatic Sciences* 65 (3): 191–198.

Myers, R.A., Levin, S.A., Lande, R. et al. (2004). Hatcheries and endangered salmon. *Science* 303: 1980.

Naman, S.M., Rosenfeld, J.S., and Richardson, J.S. (2016). Causes and consequences of invertebrate drift in running waters: from individuals to populations and trophic fluxes. *Canadian Journal of Fisheries and Aquatic Sciences* 73: 1292–1305.

National Research Council (2007). *Colorado River Basin Water Management: Evaluating and Adjusting to Hydroclimatic Variability.* Washington, DC: National Academy of Sciences.

Nehring, R.B. and Anderson, R.M. (1993). Determination of population-limiting critical salmonid habitats in Colorado streams using the physical habitat simulation system. *Rivers* 4: 1–19.

Nehring, R.B. and Miller, D.D. (1987). The influence of spring discharge levels on rainbow and brown trout recruitment and survival, Black Canyon of the Gunnison River, Colorado, as determined by IFIM/PHABSIM models. *Proceedings of the Western Association of Fish and Wildlife Agencies and the Western Division of the American Fisheries Society* 67: 338–397.

Neyman, J. (1934). On the two different aspects of the representative method: the method of stratified sampling and the method of purposive selection. *Journal of the Royal Statistical Society* A97: 558–606.

NHI (Natural Heritage Institute) and National University of Laos. 2018. Sustainable hydropower master plan for the Xe Kong basin in Lao PDR. Submitted to the Lao Mininstry of Energy, January 2018.

Nicola, G.G. and Almodóvar, A. (2004). Growth pattern of stream-dwelling brown trout under contrasting thermal conditions. *Transactions of the American Fisheries Society* 133: 66–78.

Noack, M., Schneider, M., and Wieprecht, S. (2013). The habitat modeling system CASiMir: a multivariate fuzzy approach and its applications. In: *Ecohydraulics, an Integrated Approach,* (ed. I. Maddock, A. Harby, P. Kemp and P. Wood), 67–89. New York: John Wiley and Sons.

Nolan, K. M., Lisle, T. E. and Kelsey, H. M. 1987. Bankfull discharge and sediment transport in northwestern California,. In: *Erosion and Sedimentation in the Pacific Rim* (Proceedings of the Corvallis Symposium, August, 1987), pp. 439–449. IAHS Publ. no. 165.

Norris, R.H., Webb, J.A., Nichols, S.J. et al. (2012). Analyzing cause and effect in environmental assessments: using weighted evidence from the literature. *Freshwater Science* 31: 5–21.

Northcote, T.G. (ed.) (1969). *Symposium on Salmon and Trout in Streams.* Vancouver: University of British Columbia.

Nuhfer, A. J. and Baker, E. A. 2004. A long-term test of habitat change predicted by PHABSIM in relation to brook trout population dynamics during controlled flow reduction experiments. *Fisheries Research Report 2068.* Michigan Department of Natural Resources.

Nuzzo, R. (2014). Statistical errors. *Nature* 506: 150–152.

Ock, G., Sumi, T., and Takemon, Y. (2013). Sediment replenishment to downstream reaches below dams: implementation perspectives. *Hydrological Research Letters* 7 (3): 54–59.

Ock, G., Gaeuman, D., McSloy, J., and Kondolf, G.M. (2015). Ecological functions of restored gravel bars,the Trinity River, California. *Ecological Engineering* 83: 49–60.

O'Connor, J.E., Duda, D.D., and Grant, G.E. (2015). 1000 dams down and counting. *Science* 348: 496–497.

Olden, J.D., Konrad, C.P., Melis, T.S. et al. (2014). Are large-scale flow experiments informing the science and management of freshwater ecosystems? *Frontiers in Ecology and the Environment* 12: 176–185.

Olden, J.D. and Poff, N.L. (2003). Redundancy and the choice of hydrologic indices for characterizing streamflow regimes. *River Research and Applications* 19: 101–120.

Open Science Collaboration (2015). Estimating the reproducibility of psychological science. *Science* 349: 4719.

Opperman, J.J., Moyle, P.B., Larsen, E.W. et al. (2017). *Floodplains: Processes, Ecosystems, and Services in Temperate Regions*. Berkeley: University of California Press.

Oreskes, N. (1998). Evaluation (not validation) of quantitative models. *Environmental Health Perspectives* 106 (Supplement 6): 1453–1460.

Oreskes, N., Shrader-Frechette, K., and Belitz, K. (1994). Verification, validation, and confirmation of numerical models in the earth sciences. *Science* 263: 641–646.

Orth, D.J. (1987). Ecological considerations in the development and application of instream flow–habitat models. *Regulated Rivers: Research and Management* 1: 171–181.

Orth, D.J. and Maughan, O.E. (1982). Evaluation of the incremental methodology for recommending instream flows for fishes. *Transactions of the American Fisheries Society* 111: 413–445.

Osenberg, C.W., Schmidt, J.C., Holbrook, S.J. et al. (1994). Detection of environmental impacts: natural variability, effect size, and power analysis. *Ecological Applications* 4: 16–30.

Padmore, C.L. (1998). The role of physical biotopes in determining the conservation status and flow requirements of British rivers. *Aquatic Ecosystem Health and Management* 1: 25–35.

Paéz, D.J., Hedger, R., Bernatchez, L., and Dodson, J.J. (2008). The morphological plastic response to water current velocity varies with age and sexual state in juvenile Atlantic salmon, *Salmo salar*. *Freshwater Biology* 53: 1544–1554.

Palmer, M.A., Ambrose, R.F., and Poff, N.L. (1997). Ecological theory and community restoration ecology. *Restoration Ecology* 5: 291–300.

Parasiewicz, P. (2001). MesoHABSIM: a concept for application of instream flow models in river restoration planning. *Fisheries* 26: 6–13.

Parasiewicz, P. (2007). The MesoHABSIM model revisited. *River Research and Applications* 23: 893–903.

Parasiewicz, P., Rogers, J.N., Vezza, P. et al. (2013). Applications of the MesoHABSIM simulation model. In: *Ecohydraulics: An Integrated Approach* (ed. I. Maddock, A. Harby, P. Kemp and P. Wood), 109–124. New York: John Wiley and Sons.

Parasiewicz, P. and Walker, J.D. (2007). Comparison of MesoHABSIM with two microhabitat models: (PHABSIM and HARPHA). *River Research and Applications* 23: 904–923.

Patten, D.T., Harpman, D.A., Viota, M.I., and Randle, T.J. (2001). A managed flood on the Colorado River: background, objectives, design, and implementation. *Ecological Applications* 11: 635–643.

Payne, T.R. (2005). *RHABSIM: Riverine Habitat Simulation*. Arcata, CA: Thomas R. Payne and Associates.

Pearl, J. (2000). *Causality: Models, Reasoning, and Inference*. Cambridge: Cambridge University Press.

Pearse, D.E., Miller, M.R., Abadía-Cardoso, A., and Garzas, J.C. (2014). Rapid parallel evolution of standing variation in a single, genomic region is associated with life history in steelhead/rainbow trout. *Proceedings of the Royal Society Series B* 281: 0012.

Pecquerie, L., Johnson, L.R., Sebastiaan, A.L.M.K., and Nisbet, R.M. (2011). Analyzing variations in life-history traits of Pacific salmon in the context of Dynamic Energy Budget (DEB) theory. *Journal of Sea Research* 66: 424–433.

Peterson, J.T., Jackson, C.R., Shea, C.P., and Li, G. (2009). Development and evaluation of a stream channel classification for estimating fish responses to changing streamflow. *Transactions of the American Fisheries Society* 138: 1123–1137.

Peterson, J.T. and Shea, C.P. (2015). An evaluation of the relations between flow regime components, stream characteristics, species traits, and meta-demographic rates of warm-water-stream fishes: implications for aquatic resources management. *River Research and Applications* 31: 1227–1241.

Petts, G.E. (1996). Water allocation to protect river ecosystems. *Regulated Rivers: Research and Management* 12: 353–365.

Petts, G.E. and Amoros, C. (1996). The fluvial hydrosystem. In: *The Fluvial Hydrosystems* (ed. G.E. Petts and C. Amoros), 1–12. Dordrecht: Springer.

Petts, G.E., Foulger, T.R., Gilvear, D.J. et al. (1985). Wave-movement and water quality variations during a controlled release from Kielder Reservoir, North Tyne River, UK. *Journal of Hydrology* 80: 371–389.

Phillis, C.C., Moore, J.W., Buoro, M. et al. (2016). Shifting thresholds: rapid evolution of migratory life histories in steelhead/rainbow trout, *Oncorhynchus mykiss*. *Journal of Heredity* 107: 51–60.

Pickett, K.E. and Wilkinson, R.G. (2015). Income inequality and health: a causal review. *Social Science and Medicine* 128: 316–326.

Piégay, H. and Landon, N. (1997). Promoting ecological management of riparian forests on the Drome River, France. *Aquatic Conservation: Marine and Freshwater Ecosystems* 7: 287–304.

Piketty, T. (2015). Putting distribution back at the center of economics: reflections on *Capital in the Twenty-First Century*. *Journal of Economic Perspectives* 29: 67–88.

Pitlick, J., Cui, Y., and Wilcock, P. (2009). *Manual for computing bed load transport in gravel-bed streams*, General Technical Report RMRS-GTR-223. Fort Collins, CO: U.S.D.A., Forest Service, Rocky Mountain Research Station.

Pittman, S. 2013. 2010–2013 Clear Creek Geomorphic Monitoring: bedload sampling and gravel injection evaluation. Final report submitted to US Bureau of Reclamation Mid-Pacific Regional Office, Sacramento, California, December 2013.

Podolak, K. and Kondolf, G.M. (2015). The line of beauty in river designs: Hogarth's aesthetic theory on Capability Brown's eighteenth-century river design and twentieth-century river restoration design. *Landscape Research* 41: 149–167.

Poff, N.L., Allan, J.D., Bain, M.B. et al. (1997). The natural flow regime: a paradigm for river conservation and restoration. *BioScience* 147: 769–784.

Poff, N.L., Olden, J.D., Pepin, D.M., and Bledsoe, B.P. (2006). Placing global stream flow variability in geographic and geomorphic contexts. *River Research and Applications* 22: 149–166.

Poff, N.L., Richter, B.D., Arthington, A.H. et al. (2010). The ecological limits of hydraulic alteration (ELOHA); a new framework for developing regional environmental flow standards. *Freshwater Biology* 55: 147–170.

Poff, N.L. and Zimmerman, J.K.H. (2010). Ecological responses to altered flow regimes: a literature review to inform the science and management of environmental flows. *Freshwater Biology* 55: 194–205.

Pourret, O., Naim, P., and Marcot, B. (eds.) (2008). *Bayesian Networks: A Practical Guide to Applications*. New York: John Wiley and Sons.

Power, M.E. (1990). Effects of fish in river food webs. *Science* 250: 811–814.

Power, M.E., Bouma-Gregson, K., Higgins, P., and Carlson, S.M. (2015). The thirsty Eel: summer and winter flow thresholds that tilt the Eel River of Northwestern California from salmon-supporting to cyanobacterially degraded states. *Copeia* 103: 200–211.

Power, M.E., Dietrich, W.E., and Sullivan, K.O. (2001). Experimentation, observation, and inference in river and watershed investigations. In: *Issues and Perspectives in Experimental Ecology* (ed. W.J. Resetarits and J. Bernardo), 113–132. Oxford: Oxford University Press.

Power, M.E., Parker, M.S., and Dietrich, W.E. (2008). Seasonal reassembly of a river food web: floods, droughts, and impacts of fish. *Ecological Monographs* 78: 263–282.

Power, M.E., Stout, R.J., Cushing, C.E. et al. (1988). Biotic and abiotic controls in river and stream communities. *Journal of the North American Benthological Society* 7: 456–479.

Pruitt, B. A. and McKay, S. K. 2013. Integration of stream flow duration with hydraulic geometry in the southern Piedmont. *Proceedings of the 2013 Georgia Water Resources Conference*.

Pullin, A.S., Knight, T.M., and Watkinson, A.R. (2009). Linking reductionist science and holistic policy using systematic reviews: unpacking environmental policy questions to construct an evidence-based framework. *Journal of Applied Ecology* 46: 970–975.

R2 Resource Consultants 2004. *Instream flow claims for the Snake River Basin*, Volume 1, Part 1, Main Report. Redmond, WA.

Railsback, S.F. (2016). Why it is time to put PHABSIM out to pasture. *Fisheries* 41: 720–725.

Railsback, S.F., Gard, M., Harvey, B.C. et al. (2013). Contrast of degraded and restored stream habitat using an individual-based salmon model. *North American Journal of Fisheries Management* 33: 384–399.

Railsback, S.F. and Harvey, B.C. (2013). *Insalmo-FA: a Version of in SALMO for Facultative Anadromous Trout: Model Description and Initial Analysis.* Arcata, CA: Lang Railsback and Associates.

Railsback, S.F., Harvey, B.C., Hayse, J.W., and Lagory, K.E. (2005). Tests of theory for diel variation in salmonid feeding activity and habitat use. *Ecology* 86: 947–959.

Railsback, S.F., Harvey, B.C., Jackson, S.K., and Lamberson, R.H. (2009). *InSTREAM: the individual-based stream trout research and environmental assessment model. General Technical Report.* Albany, CA: USD Dept. of Agriculture, Forest Service, Pacific Southwest Research Station.

Railsback, S.F., Harvey, B.C., and White, J.L. (2014). Facultative anadromy in salmonids: linking habitat, individual life history decisions, and population-level consequences. *Canadian Journal of Fisheries and Aquatic Sciences* 71: 1270–1278.

Railsback, S.F. and Kadvany, J. (2008). Demonstration flow assessment: judgement and visual observation in instream flow studies. *Fisheries* 33: 217–227.

Railsback, S.F., Stauffer, H.B., and Harvey, B.C. (2003). What can habitat preference models tell us? Test using a virtual trout population. *Ecological Applications* 13: 1580–1594.

Raleigh, R.F., Zuckerman, L.D., and Nelson, P.C. (1986). *Habitat suitability index models and instream flow suitability curves: brown trout, Biological Report* 82(10.124). Washington, DC: U.S. Fish and Wildlife Service.

Recovery Science Review Panel (RSRP) (2000). *Report of 4–6 December 2000, meeting.* Seattle, WA: NOAA Fisheries, Northwest Fisheries Science Center.

Reebs, S.G. (2002). Plasticity of diel and circadian activity rhythms in fishes. *Reviews in Fish Biology and Fisheries* 12: 394–371.

Refsgaard, J.C., Henriksen, H.J., Brown, J.P., and Van Der Keur, P. (2006). A framework for dealing with uncertainty due to model structure error. *Advances in Water Resources* 29: 1586–1697.

Refsgaard, J.C., Van Der Sluijs, J.P., Højberg, A.L., and Vanrolleghem, P.A. (2007). Uncertainty in the environmental modeling process – a framework for guidance. *Environmental Modelling and Software* 22: 1543–1556.

Reid, L.M. and Dunne, T. (2016). Sediment budgets as an organizing framework in fluvial geomorphology. In: *Tools in Fluvial Geomorphology*, 2e (ed. G.M. Kondolf and H. Piégay), 357–379. Chichester, UK: John Wiley and Sons.

Reiman, B., Peterson, J.T., Clayton, J. et al. (2001). Evaluation of potential effects of federal land management alternatives on trends of salmonids and their habitats in the interior Columbia River basin. *Forest Ecology and Management* 153: 43–62.

Reiser, D., Wesche, T.A., and Estes, C. (1989). Status of instream flow legislation and practices in North America. *Fisheries* 14: 22–29.

Resler, R.R. (1976). Policies of the forest service in determining instream flow needs. In: *Proceedings of the Symposium and Specialty Conference on Instream Flow Needs*, vol. 2, (ed. J.F. Osborn and C.H. Allman), 626–630. Bethesda, MD: American Fisheries Society.

Resh, V.H., Brown, A.V., Covich, A.P. et al. (1988). The role of disturbance in stream ecology. *Journal of the North American Benthological Society* 7: 499–455.

Richter, B.D., Baumgartner, J.V., Braun, D.P., and Powell, J. (1998). A spatial assessment of hydrologic alteration within a river network. *Regulated Rivers: Research and Management* 14: 329–340.

Richter, B.D., Baumgartner, J.V., Powell, J., and Braun, D.P. (1996). A method for assessing hydrologic alteration within ecosystems. *Conservation Biology* 10: 1163–1174.

Richter, B.D., Baumgartner, J.V., Wigington, R., and Braun, D.P. (1997). How much water does a river need. *Freshwater Biology* 37: 231–249.

Richter, B.D., Davis, M.M., Apse, C.D., and Konrad, C. (2012). A presumptive standard for environmental flow protection. *River Research and Applications* 28: 1312–1321.

Rivaes, R., Rodriguez–Gonzalez, P.M., Albuquerque, A. et al. (2015). Reducing river regulation effects on riparian vegetation using flushing flow regimes. *Ecological Engineering* 81: 428–438.

Roberts, P.D., Stewart, G.B., and Pullin, A.S. (2006). Are review articles a reliable source of evidence to support conservation and environmental management? A comparison with medicine. *Biological Conservation* 132: 409–423.

Robertson, A.L., Lancaster, J., Belyea, L.R., and Hildrew, A.G. (1997). Hydraulic habitat and assemblage structure of stream benthic microcrustacea. *Journal of the North American Benthological Society* 16: 562–575.

Robinson, C.T. (2012). Long-term changes in community assembly, resistance and resilience following experimental floods. *Ecological Applications* 22 (7): 1949–1961.

Robinson, C.T. and Uehlinger, U. (2003). Using artificial floods for restoring river integrity. *Aquatic Sciences* 65 (3): 181–182.

Rollet, A.J., Piégay, H., Dufour, S. et al. (2014). Assessment of consequences of sediment deficit on a gravel river bed downstream of dams in restoration perspectives: application of a multicriteria, hierarchical and spatially explicit diagnosis. *River Research and Applications* 30: 939–953.

Rolls, R.J. and Arthington, A.H. (2014). How do low magnitudes of hydrologic alteration impact riverine fish populations and assemblage characteristics? *Ecologial Indicators* 39: 179–188.

Rose, K.A., Sable, S., Deangelis, D.L. et al. (2015). Proposed best modeling practices for assessing the effects of ecosystem restoration on fish. *Ecological Modeling* 300: 19–29.

Rosenfeld, J.S. (2003). Assessing the habitat requirements of stream fishes: an overview and evaluation of different approaches. *Transactions of the American Fisheries Society* 132: 953–968.

Rosenfeld, J.S., Bouwes, N., Wall, C.E., and Naman, S.M. (2014). Successes, failures, and opportunities in the practical application of drift-foraging models. *Environmental Biology of Fishes* 97: 551–574.

Rosenfeld, J.S., Leiter, T., Lindner, G., and Rothman, L. (2005). Food abundance and fish density alters habitat selection, growth, and habitat suitability curves for juvenile coho salmon (*Oncorhynchus kisutch*). *Canadian Journal of Fisheries and Aquatic Sciences* 62: 1691–1701.

Rosenfeld, J.S., Post, S.J., Robbins, G., and Hatfield, T. (2007). Hydraulic geometry as a physical template for the River Continuum application to optimal flows and longitudinal trends in salmonid habitat. *Canadian Journal of Fisheries and Aquatic Sciences* 64: 755–767.

Rosenfeld, J.S. and Ptolemy, R. (2012). Modelling available habitat versus available energy flux: do PHABSIM applications that neglect prey abundance underestimate optimal flows for juvenile salmonids? *Canadian Journal of Fisheries and Aquatic Sciences* 69: 1920–2012.

Rosenfeld, J.S. and Raeburn, E. (2009). Effects of habitat and internal prey subsidies on juvenile coho growth: implications for stream productive capacity. *Ecology of Freshwater Fishes* 18: 527–584.

Rosgen, D.L. (1994). A classification of natural rivers. *Catena* 22: 169–199.

Rothman, K.J. and Greenland, S. (2005). *Causation and causal inference in epidemiology. American Journal of Public Health* 95 (Supplement 1): S144–S150.

Rota, C.T., Fletcher, R.J.J., Evans, J.M., and Hutto, R.L. (2011). Does accounting for imperfect detection improve species distribution models? *Ecography* 34: 659–670.

Roy, M.L., Roy, A.G., Grant, J.W.A., and Bergeron, N.E. (2013). Individual variability of wild juvenile Atlantic salmon activity patterns: effects of flow stage, temperature, and habitat use. *Canadian Journal of Fisheries and Aquatic Sciences* 70: 1082–1091.

Roy, M.L., Roy, A.G., and Legendre, P. (2010). The relations between "standard" fluvial habitat variables and turbulent flow at multiple scales in morphological units of a gravel-bed river. *River Research and Applications* 26: 439–455.

Ryder, D.S., Tomlinson, M., Gawne, B., and Likens, G.E. (2010). Defining and using "best available science": a policy conundrum for the management of aquatic ecosystems. *Marine and Freshwater Research* 61: 821–828.

Sabo, J.L., Ruhi, A., Holtgrieve, G.W. et al. (2017). Designing river flows to improve food security futures in the Lower Mekong Basin. *Science* 358: eaao1053.

Satterthwaite, W.H., Beakes, M.P., Collins, E.M. et al. (2009). State-dependent life history models in a changing (and regulated) environment: steelhead in the California Central Valley. *Evolutionary Applications* 3: 221–243.

Satterthwaite, W.H., Hayes, S.A., Mezz, J.H. et al. (2012). State-dependent migration timing and use of multiple habitat types in anadromous salmonids. *Transactions of the American Fisheries Society* 141: 781–794.

Sawicz, K., Wagener, T., Sivapalan, M. et al. (2011). Catchment classification: empirical analysis of hydrologic similarity based on catchment function in the eastern USA. *Hydrology and Earth System Science* 15: 2895–2911.

Scheurer, T. and Molinari, P. (2003). Experimental floods in the River Spöl, Swiss National Park: framework, objectives and design. *Aquatic Sciences* 65 (3): 183–190.

Schmidt, J.C. and Wilcock, P.R. (2008). Metrics for assessing the downstream effects of dams. *Water Resources Research* 44: W04404. https://doi.org/10.1029/2006WR005092.

Schmidt, J.C., Andrews, E.D., Wegner, D.L., and Patten, D.T. (1999). Origins of the 1996 controlled flood in the Grand Canyon. In: *The Controlled Flood in the Grand Canyon, Geophysical Monograph 110* (ed. R.H. Webb, J.C. Schmidt, G.R. Marzolf and R.A. Valdez), 23–36. Washington, DC: American Geophysical Union.

Schmitt, R.J.P. and Kondolf, G.M. (2017). A global challenge. *International Water Power and Dam Construction* 69 (8): 42–44.

Schmitt, R.J.P. and Osenberg, C.W. (eds.) (1996). *Detecting Ecological Impacts: Concepts and Applications in Coastal Habitats*. San Diego, CA: Academic Press.

Schmitt, R.J.P., Bizzi, S., Castelletti, A., and Kondolf, G.M. (2018). Improved trade-offs of hydropower and sand connectivity by strategic dam planning in the Mekong. *Nature Sustainability* 1: 96–104.

Schnute, J.T. (2003). Designing fishery models: a personal adventure. *Natural Resource Modeling* 16: 393–413.

Scott, D. and Shirvell, C.S. (1987). A critique of the instream flow incremental methodology and observations on flow determinations in New Zealand. In: *Regulated Streams: Advances in Ecology*, (ed. J.F. Craig and J.B. Kemper), 27–44. New York: Plenum Press.

Senay, C., MacNaughton, C.J., Lanthier, G. et al. (2015). Identifying key environmental variables shaping within-river fish distribution patterns. *Aquatic Sciences* 77: 709.

Shapovalov, L. and Taft, A.C. (1954). *The life histories of the steelhead rainbow trout (Salmo gairdneri gairdneri) and Silver Salmon (Oncorhynchus kisutch)*, *Fish Bulletin No. 98*. Sacramento, CA: California Department of Fish and Game.

Shen, Y. and Diplas, P. (2008). Application of two- and three-dimensional computational fluid dynamics models to explore complex ecological stream flows. *Journal of Hydrology* 348: 195–214.

Shenton, W., Bond, N.R., Jian, D.L.Y., and Mac Nally, R. (2012). Putting the "ecology" into environmental flows: ecological dynamics and demographic modeling. *Environmental Management* 50: 1–10.

Shenton, W., Hart, B.T., and Chan, T. (2011). Bayesian network models for environmental flow decision-making: 1. Latrobe River Australia. *River Research and Applications* 27: 283–296.

Shenton, W., Hart, B.T., and Chan, T.U. (2014). A Bayesian network approach to support environmental flow restoration decisions in the Yarra River, Australia. *Stochastic Environmental Risk Assessment* 28: 57–65.

Shirvell, C.S. (1986). *Pitfalls of physical habitat simulation in the Instream Flow Incremental Methodology, Canadian Technical Report of Fisheries and Aquatic Sciences*. Prince Rupert, BC: Department of Fisheries and Oceans, Fisheries Research Branch.

Shirvell, C.S. (1994). Effect of changes in the streamflow on the microhabitat use and movements of sympatric juvenile coho salmon (*Oncorhynchus kisutch*) and Chinook salmon (*O. tshawytscha*) in a natural stream. *Canadian Journal of Fisheries and Aquatic Sciences* 112: 355–367.

Simon, A., Doyle, M., Kondolf, G.M. et al. (2007). Critical evaluation of how the Rosgen classification and associated "natural channel design" methods fail to integrate and quantify fluvial processes and channel response. *Journal of the American Water Resources Association* 43: 1117–1131.

Skinner, D. and Langford, J. (2013). Legislating for sustainable basin management: the story of Australia's Water Act (2007). *Water Policy* 15: 871–894.

Smith, A.K. (1973). Development and application of spawning velocity and depth criteria for Oregon salmonids. *Transactions of the American Fisheries Society* 102: 312–316.

Smith, D.L., Brannon, E.L., Shafii, B., and Odeh, M. (2006). Use of the average and fluctuating velocity components for estimation of volitional rainbow trout density. *Transactions of the American Fisheries Society* 135: 431–441.

Smith, G. B. 2013. *Making the best of available evidence to predict native fish responses to flow variation in Victorian rivers*. MEnv Dissertation. University of Melbourne, Melbourne.

Smith, S.M. and Prestegaard, K.L. (2005). Hydraulic performance of a morphology-based stream channel design. *Water Resources Research* 41: 1–17.

Snelder, T.H., Lamouroux, N., Leathwick, J.R. et al. (2009). Predictive mapping of the natural flow regimes of France. *Journal of Hydrology* 373 (1): 57–67.

So, N., Sy Vann, L. and Yumiko, Y. 2007. *Study of the catch and market chain of low value fish along Tonle Sap River, Cambodia and implications for management of their fisheries – A preliminary study*. Consultancy report for the World Fish Center. Inland Fisheries Research and Development Institute and World Fish Center's Greater Mekong Region. Phnom Penh, Cambodia.

Somach, S.L. (1990). The American River decision: balancing instream protection with other competing beneficial uses. *Rivers* 1: 251–263.

Sonia, T.M., Conzelman, C.P., Byrd, J.D. et al. (2013). Predicting the effects of proposed Mississippi River diversions on oyster habitat quality; application of an oyster habitat suitability index model. *Journal of Shellfish Research* 32: 629–638.

Souchon, Y., Sabaton, C., Deibel, R. et al. (2008). Detecting biological responses to flow management: missed opportunities; future directions. *River Research and Applications* 24: 506–518.

Southern Waters and Hagler Bailly. 2013. Neelum/ Kishenganga River water diversion, Environmental flow assessment in response to the partial decision of the court of arbitration of Kishenganga case technical report. Consulting report by Southern Waters, Cape Town, South Africa, and Hagler Bailly Pakistan.

Speirs-Bridge, A., Fidler, F., McBride, M. et al. (2010). Reducing overconfidence in the interval judgments of experts. *Risk Analysis* 30: 512–523.

Stalnaker, C., Lamb, B.L., Henriksen, J. et al. (1995). *The Instream Flow Incremental Methodology: a Primer for IFIM*, Biological Report 29. Washington, DC: U. S. Dept. of the Interior, National Biological Service.

Starfield, A.M. (1997). A pragmatic approach to modeling for wildlife management. *Journal of Wildlife Management* 61: 261–270.

Stearns, S.C. and Hendry, A.P. (2004). The salmonid contribution to key issues in evolution. In: *Evolution Illuminated: Salmon and their Relatives* (ed. I.A. Fleming and J.D. Reynolds), 3–19. New York: Oxford University Press.

Steen, P.J., Zorn, T.G., Seelbach, P.W., and Schaeffer, J.S. (2008). Classification tree models for predicting distributions of Michigan stream fish from landscape variables. *Transactions of the American Fisheries Society* 137: 976–996.

Steidl, R.J., Hayes, J.P., and Schauber, E. (1997). Statistical power analysis in wildlife research. *Journal of Wildlife Management* 61: 270–279.

Stenhouse, S.A., Bean, C.E., Chesney, W.R., and Pisano, M.S. (2012). Water temperature thresholds for coho salmon in a spring-fed river, Siskiyou County, California. *California Fish and Game* 98: 19–37.

Stevens, A. and Milne, R. (1997). The effectiveness revolution and public health. In: *Progress in Public Health*, (ed. G. Scally), 197–225. London: Royal Society of Medicine.

Stevens, D.L. Jr., Larsen, D.P., and Olsen, A.R. (2007). The role of sample surveys: why should practitioners consider using a statistical sampling design? In: *Salmonid Field Protocols Handbook* (ed. O. Pourret, P. Naim and B. Marcot), 11–23. Bethesda, MD: American Fisheries Society.

Stevens, D.L. and Olsen, A.R. (2004). Spatially balanced sampling of natural resources. *Journal of the American Statistical Association* 99: 262–278.

Steventon, J.D. (2008). Conservation of marbled murrelets in British Columbia. In: *Bayesian Networks: A Practical Guide to Applications*, (ed. O. Pourret, P. Naim and B. Marcot), 127–148. New York: John Wiley and Sons.

Stewardson, M.J. and Webb, J.A. (2010). Modelling ecological responses to flow alteration: making the most of existing data and knowledge. In: *Ecosystem Response Modeling in the Murray–Darling Basin* (ed. N. Saintilan and I. Overton), 37–49. Collingwood, Victoria: CSIRO Publishing.

Stewart-Koster, B., Olden, J.D., and Gido, K.B. (2014). Quantifying flow–ecology relationships with functional linear models. *Hydrological Science Journal* 59: 629–644.

Stewart-Koster, B.S., Bunn, S.E., MacKay, S.J. et al. (2010). The use of Bayesian networks to guide investments in flow and catchment restoration for impaired river ecosystems. *Freshwater Biology* 55: 243–260.

Stewart-Oaten, A. (1996). Goals in environmental monitoring. In: *Detecting Ecological Impacts: Concepts and Applications in Coastal Habitats* (ed. R.J. Schmitt and C.W. Osenberg), 17–27. New York: Academic Press.

Strange, E.M., Moyle, P.B., and Fein, T.C. (1993). Interactions between stochastic and deterministic processes in stream fish community assembly. *Environmental Biology of Fishes* 36: 1–15.

Sumi, T. 2008. Evaluation of efficiency of reservoir sediment flushing in Kurobe River. *In*: Proceedings of the 4th International Conference on Scour and Erosion, Tokyo, Japan, pp. 608–613.

Sutherland, W.J., Pullin, A.S., Doman, P.M., and Knight, T.M. (2004). The need for evidence-based conservation. *Trends in Ecology and Evolution* 19: 305–308.

Swirepik, J.L., Burns, I.C., Dyer, F.J. et al. (2015). Establishing environmental water requirements for the Murray–Darling Basin, Australia's largest developed river system. *River Research and Applications* 32: 1153–1165.

Tena, A., Batalla, R.J., and Vericat, D. (2012). Reach-scale suspended sediment balance downstream from dams in a large Mediterranean river. *Hydrological Sciences Journal* 57 (5): 831–849.

Tennant, D.L. (1976). Instream flow regimens for fish, wildlife, recreation and related environmental resources. *Fisheries* 1: 6–10.

Terrell, J. W., McMahon, T. E., Inskip, P. D., Raleigh, R. F. and Williamson, K. L. 1982. Habitat suitability index models: Appendix A. Guidelines for riverine and lacustrine applications of fish HSI models with the Habitat Evaluation Procedures. FWS/OBS-82/l0.A. r. USFWS.

Teoh, C. (ed.) (1989). *Proceedings of the Specialist Conference on Instream Needs and Water Uses.* Canberra: Australian Water Resources Council.

Tharme, R.E. (2003). A global perspective on environmental flow assessment: emerging trends in the development and application of environmental flow methodologies for rivers. *River Research and Applications* 19: 397–441.

Thielke, J. (1985). A logistic regression approach for developing suitability-of-use functions for fish habitat. In: *Proceedings of the Symposium on Small Hydropower and Fisheries*, (ed. F.W. Olson, R.G. White and R.H. Hamre), 32–38. Bethesda, MD: American Fisheries Society.

Thomas, J.A. and Bovee, K.D. (1993). Application and testing of a procedure to evaluate transferability of habitat suitability criteria. *Regulated Rivers: Research and Management* 8: 285–294.

Thompson, S.K. (2002). *Sampling.* New York: John Wiley and Sons.

Tompkins, M. R. 2006. *Floodplain connectivity and river corridor complexity: implications for restoration planning and floodplain management.* PhD dissertation, University of California Berkeley.

Tonina, D. and Jorde, K. (2013). Hydraulic modelling approaches for ecohydraulic studies: 3D, 2D, 1D and non-numerical models. In: *Ecohydraulics* (ed. I. Maddock, A. Harby, P. Kemp and P. Wood), 37–74. New York: John Wiley and Sons.

Tonkin, J.D. and Death, R.G. (2014). The combined effects of flow regulation and an artificial flow release on a regulated river. *River Research and Applications* 30: 329–337.

Trenberth, K.E. (1999). Conceptual framework for changes of extremes of the hydrologic cycle with climate change. *Climate Change* 42: 327–339.

Tullos, D., Walter, C., and Dunham, J. (2016). Does resolution of flow field observation influence apparent habitat use and energy expenditure in juvenile coho salmon. *Water Resources Research* 52: 5938–5950.

Turgeon, K. and Rodríguez, M.A. (2005). Predicting microhabitat selection in juvenile Atlantic salmon *Salmo salar* by the use of logistic regression and classification trees. *Freshwater Biology* 50: 539–551.

Tvedt, T. (2015). *Water and Society: Changing Perceptions of Societal and Historical Development.* London: IB Taurus.

USACE (US Army Corps of Engineers, Los Angeles District). 2017. *Malibu Creek Ecosystem Restoration Study Draft Integrated Feasibility Report with Environmental Impact Statement/Environmental Impact Report (EIS/EIR), Los Angeles and Ventura Counties, California.* US Army Corps of Engineers, Los Angeles District, January 2017.

United States Fish and Wildlife Service (USFWS) 1979. Cooperative Instream Flow Service Group: September 1977/December 1978. FWS/OBS 79/27. Fort Collins, CO: USFWS.

United States Fish and Wildlife Service (USFWS) (1981). *Standards for the development of habitat suitability index models. 103 ESM.* Sacramento, CA: USFWS, Division of Ecological Services.

United States Fish and Wildlife Service (USFWS) (2013). *Flow–habitat Relationships for Juvenile Spring-run Chinook Salmon and Steelhead/Rainbow Trout Rearing in Clear Creek Between Clear Creek Road and the Sacramento River.* Sacramento, CA: USFWS.

Urabe, H., Nakajima, M., and Tarao, M. (2010). Evaluation of habitat quality for stream salmonids based on a bioenergetics model. *Transactions of the American Fisheries Society* 139: 1665–1676.

Vadas, R.L. and Orth, D.J. (2001). Formulation of habitat suitability models for stream fish guilds: do the standard methods work? *Transactions of the American Fisheries Society* 130: 217–235.

Van Horne, B. (1983). Density as a misleading indicator of habitat quality. *Journal of Wildlife Management* 47:: 893–901.

Vannote, R.L., Minshall, G.W., Cummins, K., and Sedell, J.R. (1980). The River Continuum concept. *Canadian Journal of Fisheries and Aquatic Sciences* 37: 130–137.

Vaughan, I.P. and Ormerod, S.J. (2005). The continuing challenges of testing species distribution models. *Journal of Applied Ecology* 42: 720–730.

Vezza, P., Parasiewicz, P., Spairani, M., and Comoglio, C. (2014). Habitat modeling in high-gradient streams: the mesoscale approach and application. *Ecological Applications* 24: 844–861.

Vilizzi, L.G., Copp, G.H., and Roussel, J.M. (2004). Assessing variation in suitability curves and electivity profiles in temporal studies of fish habitat use. *River Research and Applications* 20: 605–618.

Vondracek, B. and Longanecker, D.R. (1993). Habitat selection by rainbow trout *Oncorhynchus mykiss* in a California stream: implications for the instream flow incremental methodology. *Ecology of Freshwater Fishes* 2: 137–186.

Voshell, J.R. (2002). *A Guide to Common Freshwater Invertebrates of North America*. Newark, OH: McDonald and Woodward.

Wall, C.E., Bouwes, N., Wheaton, J.M. et al. (2016). Net rate of energy intake predicts reach-level steelhead (*Oncorhynchus mykiss*) densities in diverse basins from a large monitoring program. *Canadian Journal of Fisheries and Aquatic Sciences* 73: 1081–1091.

Walters, C.J. (1986). *Adaptive Management of Renewable Resources*. New York: MacMillan.

Walters, C.J. and Hilborn, R. (1976). Adaptive control of fishing systems. *Journal of the Fisheries Research Board of Canada* 33: 5–159.

Walters, C.J. and Holling, C.S. (1990). Large-scale management experiments and learning by doing. *Ecology* 71: 2060–2068.

Wang, H.-W. and Kondolf, G.M. (2013). Upstream sediment-control dams: five decades of experience in the rapidly-eroding Dahan River Basin, Taiwan. *Journal of the American Water Resources Association* 50: 735–747.

Wang, Z.-Y. and Hu, C. (2009). Strategies for managing reservoir sedimentation. *International Journal of Sediment Research* 24: 369–384.

Ward, J.V. and Stanford, J.A. (1983). The intermediate disturbance hypothesis: an explanation for biotic diversity patterns in lotic ecosystems. In: *Dynamics of Lotic Ecosystems* (ed. T.D. Fontaine and S.M. Bartell). Ann Arbor, MI: Ann Arbor Science Publishers.

Ward, J.V., Tockner, K., and Schiemer, F. (1999). Biodiversity of floodplain river ecosystems: ecotones and connectivity. *Regulated Rivers: Research and Management* 15: 125–139.

Wasserstein, R.L. and Lazar, N.A. (2016). The ASA's statement on p-values: context, process, and purpose. *The American Statistician* 70: 129–131.

Waters, B.F. (1976). A methodology for evaluating the effects of different streamflows on salmonid habitat. In: *Proceedings of the Symposium and Specialty Conference on Instream Flow Needs*, 3–6 May 1976 1976 Boise, Idaho (ed. J.F. Osborn and C.H. Allman), 254–266. Bethesda, MD: American Fisheries Society.

Watts, R. J., Allan, C., Bowmer, K. H., Page, K. J. et al. 2009. *Pulsed flows: a review of environmental costs and benefits and best practice*. Waterlines Report Series No.16, May 2009. Canberra, Australian Government National Water Commission.

Waylen, P.R. and Caviedes, C.N. (1990). Annual and seasonal fluctuations of precipitation and streamflow in the Aconcagua River basin, Chile. *Journal of Hydrology* 120: 79–102.

Webb, A., Baker, B., Casanelia, S., Grace, M. et al. (2017a) *Commonwealth Environmental Water Office Long Term Intervention Monitoring Project: Goulburn River Selected Area Evaluation Report 2015–16*. Report Prepared for the Commonwealth Environmental Water Office. Melbourne: Commonwealth of Australia.

Webb, A., Casanelia, S., Earl, G. et al. (2016). *Commonwealth Environmental Water Office Long Term Intervention Monitoring Project: Goulburn River Selected Area evaluation report 2014–15*. Melbourne: University of Melbourne.

Webb, J.A. (2017). Rapid evidence synthesis in environmental causal assessments. *Freshwater Science* 36: 218–219.

Webb, J.A., Arthington, A.H., and Olden, J.D. (2017b). Models of ecological responses to flow regime change to inform environmental flow assessments. In: *Water for the Environment: From Policy and Science to Implementation and Management* (ed. A.C. Horne, J.A. Webb, M.J. Stewardson, et al.), 287–316. Cambridge, MA: Elsevier.

Webb, J.A., Bond, N.R., Wealands, S.R. et al. (2007). Bayesian clustering with AutoClass explicitly recognizes uncertainties in landscape classification. *Ecography* 30: 526–536.

Webb, J.A., de Little, S.C., Miller, K.A., and Stewardson, M.J. (2018). Quantifying and predicting the benefits of environmental flows: combining large-scale monitoring data and expert knowledge within hierarchical Bayesian models. *Freshwater Biology* 63: 831–843.

Webb, J.A., De Little, S.C., Miller, K.A. et al. (2015a). A general approach to predicting ecological responses to environmental flows: making the best use of the literature, expert knowledge, and monitoring data. *River Research and Applications* 31: 505–514.

Webb, J.A., Miller, K.A., De Little, S.C., and Stewardson, M.J. (2014). Overcoming the challenges of monitoring and evaluating environmental flows through science-management partnerships. *International Journal of River Basin Management* 12: 111–121.

Webb, J.A., Miller, K.A., de Little, S.C. et al. (2015b). An online database and desktop assessment software to simplify systematic reviews in environmental science. *Environmental Modelling and Software* 64: 72–79.

Webb, J.A., Miller, K.A., King, E. et al. (2013). Squeezing the most out of existing literature: a systematic re-analysis of published evidence on ecological responses to altered flows. *Freshwater Biology* 58: 2439–2451.

Webb, J.A., Schofield, K., Peat, M. et al. (2017c). Weaving the common threads in environmental causal assessment methods: towards an ideal method for rapid evidence synthesis. *Freshwater Science* 36: 250–256.

Webb, J.A., Stewardson, M.J., Chee, Y.E. et al. (2010a). Negotiating the turbulent boundary: the challenges of building a science-management collaboration for landscape-scale monitoring of environmental flows. *Marine and Freshwater Research* 61: 798–807.

Webb, J.A., Stewardson, M.J., and Koster, W.M. (2010b). Detecting ecological responses to flow variation using Bayesian hierarchical models. *Freshwater Biology* 55: 108–126.

Webb, J.A., Wallis, E.M., and Stewardson, M.J. (2012). A systematic review of published evidence linking wetland plants to water regime components. *Aquatic Botany* 103: 1–14.

Webb, J.A., Watts, R.J., Allan, C., and Warner, A.T. (2017d). Principles for monitoring, evaluation and adaptive management of environmental flows. In: *Water for the Environment: From Policy and Science to Implementation and Management*, (ed. A.C. Horne, J.A. Webb, M.J. Stewardson, et al.), 599–623. Cambridge MA: Elsevier.

Weinberg, A.M. (1972). Science and trans-science. *Minerva* 10: 209–222.

Welcomme, R.L., Winemiller, K.O., and Cowx, I.G. (2006). Fish environmental guilds as a tool for assessment of ecological conditions of rivers. *River Research and Applications* 23: 377–396.

Welsh, S.A. and Perry, S.A. (1998). Influence of spatial scale on estimates of substrate use by benthic darters. *North American Journal of Fisheries Management* 18: 354–359.

Wenger, S.J. and Olden, J.D. (2012). Assessing transferability of ecological models: an underappreciated aspect of statistical validation. *Methods in Ecology and Evolution* 3: 260–267.

Wesche, T.A. (1976). Development and application of a trout cover rating system for IFN determinations. In: *Proceedings of the Symposium and Specialty Conference on Instream Flow Needs* (ed. J.F. Orsborn and C.H. Allman). Bethesda, MD: American Fisheries Society.

Wesche, T.A., Goertler, C.M., and Hubert, W.A. (1987). Modified habitat suitability index model for brown trout in southeastern Wyoming. *North American Journal of Fisheries Management* 7: 232–237.

Whiting, P.J. and Dietrich, W.E. (1991). Convective accelerations and boundary shear stress over a channel bar. *Water Resources Research* 27: 783–796.

Whitledge, G.W., Bajer, P.G., and Hayward, R.S. (2010). Laboratory evaluation of two bioenergetics models for brown trout. *Transactions of the American Fisheries Society* 139: 929–936.

Wilcock, P.R., Kondolf, G.M., Matthews, W.V.G., and Barta, A.F. (1996). Specification of sediment maintenance flows for a large gravel-bed river. *Water Resources Research* 32 (9): 2911–2921.

Wilcock, P., Pitlick, J. and Cui, Y. 2009. *Sediment Transport Primer Estimating Bed-Material Transport in Gravel-bed Rivers*. USDA Rocky Mountain Research Station, General Technical Report RMRS-GTR-226, May 2009.

Wilcox, A.C. and Shafroth, P.B. (2013). Coupled hydrogeomorphic and woody-seedling responses to controlled flood releases in a dryland river. *Water Resources Research* 49 (5): 2843–2860.

Wild, T.B., Loucks, D.P., Annandale, G.W., and Kaini, P. (2016). Maintaining sediment flows through hydropower dams in the Mekong River Basin. *Journal of Water Resources Planning and Management* 142 (1).

Wilding, T.K., Sanderson, J., Merritt, D.M. et al. (2014). Riparian responses to reduced flood flows: comparing and contrasting narrowleaf and broadleaf cottonwoods. *Hydrological Science Journal* 59: 605–617.

Williams, G.P. (1978). Bank-full discharge of rivers. *Water Resources Research* 14: 1141–1154.

Williams, G. P. and Wolman, M. G. 1984. *Downstream Effects of Dams on Alluvial Rivers*. US Geological Survey Professional Paper 1286. Washington, DC: USGS.

Williams, J.G. (1991). Stock dynamics and adaptive management of habitat: an evaluation based on simulations. *North American Journal of Fisheries Management* 19: 329–341.

Williams, J. G. 1995. Report of the Special Master, Environmental Defense Fund v. East Bay Municipal Utility District, Alameda County Superior Court (California) Action 425955. Davis, CA.

Williams, J.G. (1996). Lost in space: minimum confidence intervals for idealized PHABSIM studies. *Transactions of the American Fisheries Society* 125: 458–465.

Williams, J. G. 2001. Chinook salmon in the lower American River, California's largest urban stream. *Fish Bulletin 179, v2*. Sacramento: California Department of Fish and Game.

Williams, J. G. 2006. Central Valley salmon: a perspective on Chinook and steelhead in the Central Valley of California. San Francisco Estuary and Watershed Science, 4, Article 2 (online).

Williams, J.G. (2010a). Sampling for environmental flow assessments. *Fisheries* 35: 434–443.

Williams, J.G. (2010b). Lost in space, the sequel; spatial sampling issues with 1-D PHABSIM. *River Research and Applications* 26: 341–352.

Williams, J.G. (2013). Bootstrap sampling is with replacement: a comment on Ayllón et al. (2011). *River Research and Applications* 29: 399–401.

Williams, J. G. 2017 Building hydrologic foundations for applications of ELOHA: how long a record should you have? River Research and Applications. online.

Williams, J.G. (2018). Comment on Sabo et al. (2017). *Science* 361 (6398): eaat1224.

Williams, J.G., Speed, T.P., and Forrest, W.F. (1999). Comment: transferability of habitat suitability criteria. *North American Journal of Fisheries Management* 19: 623–625.

Williams, L.R., Warren, M.L.J., Adams, S.B. et al. (2004). Basin visual estimation technique (CVET) and representative reach approaches to wadeable stream surveys: methodological limitations and future directions. *Fisheries* 19: 12–22.

Willmott, C.J., Robeson, S.M., Matsuura, K., and Ficklin, D.L. (2015). Assessment of three dimensionless measures of model performance. *Environmental Modelling and Software* 73: 167–174.

Wilson, K., Baker, S. and Kondolf, G. M. Idealized river meanders. (2019, in review), *Freshwater Science*

Wintle, B.C., Fidler, F., Vesk, P.A., and L Moore, J. (2013). Improving visual estimation through active feedback. *Methods in Ecology and Evolution* 4: 53–62.

Wohl, E., Bledsoe, B.P., Jacobson, R.B. et al. (2015). The natural sediment regime in rivers: broadening the foundation for ecosystem management. *Bioscience* 65 (4): 358–371.

Wolman, M.G. and Gerson, R. (1978). Relative scales of time and effectiveness of climate in watershed geomorphology. *Earth Surface Processes* 3: 189–208.

Wolman, M.G. and Miller, J.P. (1960). Magnitude and frequency of forces in geomorphic processes. *Journal of Geology* 68 (1): 54–74.

Wooton, J.T., Parker, M.S., and Power, M.E. (1996). Effects of disturbance on river food webs. *Science* 273: 1558–1561.

Wright, K.A., Goodman, D.H., Som, N.A. et al. (2017). Improving hydrodynamic modelling: an analytical framework for assessment of two-dimensional hydrodynamic models. *River Research and Applications* 33: 170–181.

Wu, F.C. and Chou, Y.J. (2004). Tradeoffs associated with sediment-maintenance flushing flows: a simulation approach to exploring non-inferior options. *River Research and Applications* 20 (5): 591–604.

Xu, C.L., Letcher, B.H., and Nislow, K.H. (2010). Size-dependent survival of brook trout *Salvelinus fontinalis* in summer: effects of water temperature and stream flow. *Journal of Fish Biology* 76: 2342–2369.

Yang, A. 2010. The integrated river system modeling framework: a report to the Murray–Darling Basin Authority. *CSIRO Water for a Healthy Country Flagship Report*. Canberra, Australia: CSIRO.

Yarnell, S.M., Petts, G.E., Schmidt, J.C. et al. (2015). Functional flows in modified riverscapes: hydrographs, habitats and opportunities. *BioScience* 65: 963–972.

Young, W. J., Scott, A. C., Cuddy, S. M. and Rennie, B. A. 2003. Murray Flow Assessment Tool – a technical description. *Client Report, 2003.* Canberra: CSIRO Land and Water.

Zampatti, B. and Leigh, S. (2013). Effects of flooding on recruitment and abundance of Golden Perch (*Macquaria ambigua ambigua*) in the lower River Murray. *Ecological Management and Restoration* 14: 135–143.

Zarfl, C., Lumsdon, A.E., Berlekamp, J. et al. (2015). A global boom in hydropower dam construction. *Aquatic Sciences* 77: 161–170.

Zorn, T.G. and Seelbach, P.W. (1995). The relation between habitat availability and the short-term carrying capacity of a stream for smallmouth bass. *North American Journal of Fisheries Management* 15: 773–783.

Zorn, T. G., Seelbach, P. W., Rutherford, E. S., Wills, T. C. et al. (2008) *A regional-scale habitat suitability model to assess the effects of flow reduction on fish assemblages in Michigan streams.* Fisheries Research Report 2089. Michigan Department of Natural Resources.

Zorn, T.G., Seelbach, P.W., and Rutherford, E.S. (2012). A regional-scale habitat suitability model to assess the effects of flow reduction on fish assemblages in Michigan streams. *Journal of the American Water Resources Association* 48: 871–895.

Zorn, T.G., Seelbach, P.W., and Wiley, M.J. (2002). Distribution of stream fishes and their relationship to stream size and hydrology in Michigan's Lower Peninsula. *Transactions of the American Fisheries Society* 131: 70–85.

Zorn, T.G., Seelbach, P.W., and Wiley, M.J. (2011). Developing user-friendly habitat suitability tools from regional stream fish survey data. *North American Journal of Fisheries Management* 31: 41–55.

Zuir, A.F., Leno, E.N., and Elphick, C.S. (2010). A protocol for data exploration to avoid common statistical problems. *Methods in Ecology and Evolution* 1: 3–14.

Zurwerra, A., Meile, T. and Käser, S. 2016. Künstliche Hochwasser. Massnahme zur Beseitigung ökologischer Beeinträchtigungen in Restwasserstrecken unterhalb von Speicherseen. Auslegeordnung Grundlagen und Handlungsbedarf (PRONAT and BG, 2016).

Index

Page references in *italics* refer to Figures; those in **bold** refer to Tables

abundance–environment relations (AERs) 64–5
accessibility of methods 176
acoustic Doppler current profiler (ADCP) 52
acoustic Doppler velocimetry (ADV) 52
adaptation to extreme flows 46–7
adaptive management 13, 78, 102–7, 177–8
adjusted R^2 132
adverse resource impact 101
Akaike Information Criterion (AIC) 62, **63**, 124, 132, 133
American River (California) 13, 71, 103–4
Apalachicola River (Florida) 28
Arase Dam (Japan) 150
area under the curve (AUC) 132
Arroyo Seco River(California), discharge in 4, *4*
assessments, recommendations for 181–2
Aswan High Dam 33

BAGS Excel based software 72
"bankfull" discharge 28
baseflow 22
bass (*Micropterus*) 42
Battle Creek (Michigan) 26

Bayes' rule 174
Bayesian Belief Network (BBN) structure, 164
  vs. Hierarchical Bayesian Models 167
Bayesian belief network, transforming into 166
Bayesian decision networks (BDNs) 69
Bayesian Information Criterion (BIC) 62, 124, 132, 133
Bayesian methodology 18, 20, 59, 105, 107, 164
Bayesian Mixture Modeling 30, 53
Bayesian Network (BN) models 66–9, *68*, *69*, 88, 105, 110, 176, 178
  improving use of 173–4
behavioral adaptations of stream organisms 45–6
behavioral carrying-capacity tests 128
Big Coldwater Creek 25
Bill Williams River (Arizona), flow release 156
bioenergetic equation 44
bioenergetic models 88–92
bivariate flow–response relationships 164
bootstrap method, 18, 60, 62, 79

bottom-up methods 77, 78
Bridge River (British Columbia) 70, 103, 127
Brisbane Declaration (2007) 1
Bristol Bay (Alaska), sockeye salmon fishery study 2
Brows Beck 59
Building Block Methodology 19

California Environmental Quality Act (CEQA) (1970) 12, 13, 15
California State Water Resources Control Board (SWRCB) 14
capability models 17, 66
Carmel River (California) 12, 13, 14, 151
CASiMiR 18, 79, 87
catfishes 41, 42
causal criteria 60, **60**
causal relationships
  quantification of 166
  updating using empirical data 166–8
chasing wave hydrographs 156
checklist for EFA 182–3
Chinook salmon (*Oncorhynchus tshawytscha*) 5, *37*, 91
chi-squared test 121
Cladophora glomerata 3

Environmental Flow Assessment: Methods and Applications, First Edition. John G. Williams, Peter B. Moyle, J. Angus Webb and G. Mathias Kondolf.
© 2019 John Wiley & Sons Ltd. Published 2019 by John Wiley & Sons Ltd.

Clean Water Act (1972) (US) 12
Clear Creek (California) 91, 146, 155
clients in disputes 7–8
climate change 4–5, 28
clustering algorithms 53
Cochrane Reviews 55
coefficient of determination (R^2) 89,
  125, 129, 130, 132
  adjusted 132
coho salmon, physiological
  adaptations of 44–5
Collaboration for Environmental
  Evidence (CEE) 55–6, 173
Colorado River 56, 145, 153
Colorado River Compact (1922) 4
combat biology 7
common carp (*Cyprinus carpio*)
  42, 49
computational fluid dynamics (CFD)
  models 52, 71, 110
Computer Aided Simulation Model
  for Instream Flow
  Requirements *see* CASiMiR
conceptual foundation 51
conditional probability tables
  (CPTs) 67, 68
Conference on Instream Flow Needs
  (1976) 15
confidence intervals 79
Convention on Biological
  Diversity 105
Copernican model 120
Cosumnes River (California) 36, *37*
Coweeta Creek (North Carolina) 89
crayfish 42
cross-validation 123

dace *42*
dams and channel morphology
  143–61
  flows for managing vegetation 159
  problem diagnosis and setting
  objectives 145–6
  sediment load management 146
  existing dams 146–7

obsolete dams 150–2
  proposed dams 147–50
darters 41
  microhabitat 45
data
  availability 111
  quality 111, 123
Delta Fish and Wildlife Protection
  Study, California 11–12
Demonstration Flow Assessment
  59, 130
descriptive tools 52–5
  graphical tools and images 52–3
  habitat classifications 54–5
  methods classifications 55
  species classifications 55
  stream classifications 53–4
direct calibration methods 152
dispute resolution 7
disturbance regime 17
Downstream Response To Imposed
  Flow Transformation *see*
  DRIFT methodology
DRIFT methodology 19, 72, 78,
  95–7
drift-foraging models 18, 65–6,
  88–92, 179
dynamic energy budget models
  71, 178
dynamic occupancy models 70, 178

Ebro River (Spain) 26, 145,
  156, 158
Eco Evidence 56, 173, 175, 176
Ecological Limits of Hydrologic
  Alteration (ELOHA)
ecological simulation models 110
Ecosystem Diagnosis and Treatment
  (EDT) model 73, 91
ecosystem restoration, effects on
  fish 110
Eel River (California) 98
eels 42
El Niño Southern Oscillation
  (ENSO) 26

electrofishing 61, 100, 119, 126, 133
ELOHA 20, 30, 53, 65, 70, 78, 81,
  97–102, 164, 176, 178
  in Michigan 99–102
  in Southeast Queensland 98–9
Elwha River (Washington) 150
Elwha River Ecosystem and Fisheries
  Restoration Act 151
Endangered Species Act (1973)
  (US) 12, 73
Environmental Defense Fund et al. v.
  East Bay Municipal Utility
  District (1989) 13
environmental flow methods
  (EFMs) 9, 77–107, 122
  classification 77–80
  frameworks 92–107
  habitat simulation methods 83–92
  hydraulic rating methods 82
  hydrological methods 80–2
  testing 130
environmental flows
  definition 1
  dynamic and open 2–3
  limitation of models and objective
  methods 8–9
  literature problems 8
  scientific issues 2
  social issues 608
Escoumins River (Quebec) 137, 138
estimation models 110
Eurasian watermilfoil (*Myriophyllum*
  spicatum) 49
European Union Water Framework
  Directive 13
evidence, existing,
  improving 163–76
evidence-based conceptual model of
  ecological responses 164
evidence-based EFA 104–7
evidence-based practice 173, 178
evolution of fish 3–4
expectation maximization (EM)
  algorithm 168, 172
experimental tests 126–8

experiments 56–8
  flow experiments 56
  laboratory experiments 56
  thought experiments 56–8
expert fatigue 170
expert opinion, improving, 163–76
exploratory data analysis 115

fathead minnows 41
Feather River (California) 119
field-of-dreams hypothesis 163
fish assemblages 47–9
  filters 47, *48*
Flint River (Georgia) 59, 70
flood frequency curves 22
floodplain zone 40
flow experiments 56, 126–8
flow–duration curves 22, 53
flow–response model 164
flow regimes 22–30
  hydrologic classifications 29–30
  describing or depicting 22–4
  variation across climates and
    regions 25–8
  anthropogenic changes in 28–9
FLOWS methodology19–20, 92, 104,
    105, 163
Frameworks for EFA 92
Fraser River (British Columbia) 127
Froude Number 74
functional flow approach 78
fuzzy logic 87–8

generalized random tessellation
    stratified (GRTS) design
    61, 119
Giger, Richard 14–15
Glen Canyon Dam (Colorado) 150,
    153, 157–8, 159
golden perch (*Macquaria
    ambigua*) 164, 168–74
Goulburn River, SE Australia (case
    study) 168–74
  Bayesian belief network 168
    168–9

evidence-based conceptual model
    of responses to flow
    variation 168
expert-based quantification of
    effects of flow and non-flow
    drivers 169–70
  inclusion of monitoring data to
    update BBN 171–2
Grande Eau River(Switzerland) 158
graphical tools and images 52–3
Great Lakes Compact 13
green sunfish 41
Green–Colorado River system 146
guild concept 55

habitat association models (HAMS)
    65, 83, 84–8, 179
habitat classifications 54–5
Habitat Probabilistic Index
    (HPI) 134–41, *137, 138*
habitat rating methods 78
habitat selection 5
habitat simulation methods 78,
    83–92
habitat suitability criteria (HSC) 85
Habitat Suitability Index (HSI) 17,
    134, 137
Habitat Suitability Index model 17
HARPA 87
HECRAS modeling suite 163
hierarchical Bayesian models 69–70,
    164, 175, 178, 179
  as best practice 174–5
hierarchical modeling 66
holistic methods 18–19, 78
Hunt Creek (Michigan) 126
hydraulic drag 74
hydraulic habitat indices, 73–5
hydraulic models 15, 16, 38, 52, 65,
    71–2, 78, 83, 89, 109, 112,
    119, 163
  1-D 84, 85, 91
  2-D 18, 84, 111, 132–4
  testing 129–30
hydraulic rating methods 82–3, 179

hydrographs 22, *24–5*
hydrological indices 75
hydrological methods 78, 80–3
hydrological models 72
hydromorphologic units 18
hyetographs 22, *24–5*
hyporheic zone 36, 40, 41
hyporheos 36

Ideal Despotic Distribution
    (IDD) 56–8
Ideal Free Distribution (IFD) 56–8,
    64, 134
incremental method 77, 79
Indicators of Hydrologic Alteration
    (IHA) 75, 81
Individual-based models (IBM) 66,
    88, 90–2, 128, 178
InSALMO 91, 179
InSALMO-FA 91
InSTREAM 72, 90–1
Instream Flow Council 15
Instream Flow Incremental
    Methodology (IFIM) 7, 16,
    19, 78, 92–5, *93*
instream flows 1, 2, 31, 161, 177
intermediate disturbance
    hypothesis 17

Japanese clams (*Corbicula
    fluminea*) 49
Jogne River (Switzerland) 158

Klamath River (Oregon and
    California) 150
Kuma River (Japan) 150

large armored caddisflies
    (*Dicosmoecus gilvipes*) 3
Latrobe River (Victoria,
    Australia) 66
legal basis for environmental
    flows 12–13
Lewiston Dam (California) 156
LIDAR 53

literature
  as basis of evidence-based
    conceptual model 165
  improved use of knowledge
    from 172–3
  reviews 55–6, 179
logistic regression 114, 115, 124–6
Los Padres Dam (California) 14

Malibu Creek (California) 151
management of environmental
  flows 37–8
Manning's *n* 62
Markov chain Monte Carlo
  algorithm 69, 70
Markov chain Monte Carlo Modeling
  (MCMC) 62
mark–recapture estimates of
  populations 133
Martis Creek 48, 68
Matilija Dam 151
Mediterranean-climate precipitation
  patterns 26
Mekong River 37, 117, 148
Merced River (California) 26, 146
Mercer Creek (Washington) 28,
  29, 30
MesHABSIM 18
mesohabitat models 18, 77, 86–8
Mesohabitat Simulation Model *see*
  MesoHABSIM
MesoHABSIM 55, 79, 86–7, 131, 163
meso-scale methods
metaspecies 55
methods
  classifications 55
  for EFA 178
Michigan Creek (Colorado Rocky
  Mountains) 25
microhabitat 77, 78
microhabitat models 84–6
microhabitat selection 45
micro-scale methods 79
Millennium Drought (Australia)
  7, 105

minnow 41
  microhabitat 45
model testing 120–6
  against other models 123–6
  consistency with prior
    knowledge 129
  data quality and 123
  dependence on method being
    applied 122
  difficulties in 120
  hydraulic 129–30
  limited utility of significance
    tests, 121–2
  multiple 122
  plausibility 123
  purpose of 120
  replication 123
  validation 120–1
  with independent data 123
model-based methods 77, 79–80
modeling 8–9, 63–75, 110–14
  assessment of natural system 111
  behavioral issues in 114–15
  budget distribution 112
  data availability and quality 111
  data-dependent activities in
    development 115
  documention of development 113
  need for 113
  purposes of 110–11
  results uncertainty 112–13
  results, use of 113
  testing 113
  usefulness of 113
  user of 113
Mokelumne river 71
monitoring
  long-term 58–9
  recommendations on 180–1
Montana method 80–1
Monte Carlo analysis 62
morphogenic flows,
    constraints 159–61
  cost of power production and
    water use 159–60

spawning gravels 160–1
  flooding and bank erosion
    prevention 161
morphogenic flows, potential
    objectives **145**
morphogenic flows,
    specifying 152–9
  duration 155
  hydrograph 155–6, *156, 158*
  magnitude 153–5
  objectives 153
  recurrence 158–9
  seasonality 156–8
  three common approaches
    to 152–3
morphological adaptations of stream
    organisms 43–4
Murou Dam 146
Murray Flow Assessment Tool
    (MFAT) 19
Murray–Darling Basin (Australia) 7,
    3, 19, 92, 105
Murray–Darling Basin Authority
    (MBDA) 105
mussels 42, 43

narrative review 56
*National Audubon Society v. Superior
    Court* 12
National Environmental Policy Act
    (NEPA) (1970) (US) 12,
    13, 15
Natural Flow Paradigm 78
net energy intake 44
net energy intake (NEI) models 88,
    89–90, 179
Netica software 68
Newaukum Creek (Washington) 29
Nile River 33, 36

objective methods 8
openBUGS 70
opinion-based methods 77, 79–80
optimal habitat ratios 58
Orestima Creek 26

Pacific Northwest Aquatic Monitoring Partnership 53
paddlefish 42
Paria River 158
Pearson product moment correlation 129
peer review 8
percent habitat saturation (PHS) 58
permanent channel zone 40
PHABSIM 7, 14, 16–17, 19, 52, 55, 62, 64, 65, 71, 79, 83–7, 92–3, 115, 117, 122, 124, 131, 179
Physical Habitat Simulation system see PHABSIM
physiological adaptations of stream organisms 44–5
piggy-backing on existing knowledge 175
population thinking in salmon management 51, **52**
precipitation 22
prediction maps 132
preference curves 84, 135
principal components analysis 53, 75
probability distribution 79
probability sampling 61
production thinking in salmon management 51, **52**
professional opinion 59
Ptolemaic model of the universe 120
Putah Creek (California) 131
p-values 61, 116, 117, 122

quantifying environmental flows 14–20
quantile regression 64–5, 66

r² values 134–5, 137, 138 see also coefficient of determination
rainbow trout 48
random sampling 61
Range of Variation method 81
receiver operator characteristic curve 132

regression modeling 75
regression tree analysis 9, 100
replication crisis 62
representative reach approach 61
reproduction 46
resourcing improved practice 175–6
Reynolds Number 74
RHABSIM 84
RHYHABSIM 19, 84
riffle sculpin (*Cottus gulosus*) 45–6 microhabitat 45 46
riffle-pool ratio 14
Ringe Dam (California) 151
riparian areas 40–1
River Hydraulics and Habitat Simulation system (RHYHABSIM) 84
River2D 84
Riverine Habitat Simulation system (RHABSIM) 84
river-scale methods 79
roach (*Lavinia (Hesperoleucas) symmetricus*) 3
runoff 22
Russian River (California) 70

'S' curves comparison (sinuosity and symmetry) *34*
Sacramento River (California) 5, 26, *27*, 33,40, 91, 146
Sacramento–San Joaquin River basin (California) 26
Sainte-Marguerite River model 138, 139, 141
sample-based method 77, 78–9
sampling 118–20
cleaning data sets 119–20
general considerations 118
spatial scale 119–20
sampling error 61
San Clemente Dam (California) 12, 14, 151
San Joaquin River (California) 11, 12, 26, *27*, 56, 104

scope of environmental flow assessments 13
sculpins 41, *42*, 43 microhabitat 45
Searsville Dam on San Francisquito Creek 151
sediment entrainment methods 152, 153
sediment transport models 72
self-adjustedchannel methods 152
self-formed channels 33
shad 42
Shapovalov, Leo 14
Shasta Dam (California) 5
Shasta River (California) 45
shear velocity 74
Shields Number 62
simulation models 110
Smith River (Virginia) 71
Snake River 53
snowmelt runoff 22
social issues 6–8
social objectives, evolution of 6
Solfatara Creek, Wyoming 31, *32*, 33
South African Building Block Methodology 78
South Fork Eel River 2, 3, 59
spatial patterns and variability within streams 30–7
bed topography and hyporheic exchange 36–7
lateral connectivity with floodplain and off-channel water bodies 33–6, *36*
spatial complexity of flows 30–1
variety of channel forms 31–3
spatial scales 5–6
species classifications 55
species distribution models (SDMs) 83, 120
goodness of fit 132
imperfect detection 133
prevalence 132–3
spatial scale 133–41
testing 131–2

speckled dace (*Rhinichthys osculus*),
microhabitat 45 46
Spöl (Switzerland), 145
Spöl, River 158, *158*, 159
standard errors 79
standard setting methods 77, 79
state-dependent life history
models 71, 178
statistical significance 62
statistics 60–3
hypothesis testing 61–2
model selection and averaging 62
resampling algorithms 62–3
sampling methods 61
sampling 61
steelhead (*Oncorhynchus mykiss*) 3, 91
storm runoff 22, *23*
stormflow 22
stratified random sampling 61
stream classifications 53–4
stream continuum conceptual
model 40
stream ecosystems, structure of 40–3
cross-channel gradients, 40–1
upstream–downstream
gradient 41–3
stream gradient system
depositional zone 42–3

erosional zone 41
intermediate zone 41–2
stream organisms, adaptations
of, 43–4
behavioral 45–6
morphological 43–4
physiological 44–5
Stung Treng, Mekong River 37
subjective methods 9
suckers 41, *42*, 42
suitability curves 84
systematic review 56, 175

temperature models 72
temporal scales 5–6
Tennant Method 55, 79, 80–1, 161
thought experiments 56–8
3-D laser scanning 53
Tonle Sap River 37
top-down methods 77, 78
Trinity Dam (California) 146
Trinity River (California) 61, 146,
150, 154, 155, 156, *157*, 161
Lowden Ranch rehabilitation
site 156, *157*
trout (*Salmo trutta*) 41, *42*, 43, *48*
Tuolumne River (California) 11
turbulent kinetic energy 74

Uda River 146
uncertainty, confronting and
managing 177

vegetation, encroachment
downstream of dam 159, *160*
velocimeters 74
Ventura River 151
Victorian eFlows and Monitoring
Assessment Program 67
virtual ecosystem experiments 128

Water Act (2007) (Australia) 13, 105
Water Footprint Network 81
Water Resources Act (1963) (UK) 13
weighted usable area (WUA) 16, 85
West Brook 59
Whiskeytown Reservoir 155
whole river methods 77
whole-stream methods 77
whole-system methods 78–9
wildlife habitat suitability
modelling 110
winBUGS 70

Yalakon River 127
Yolo Bypass (Sacramento River
system) 34, 40